Preparing Agriculture and Agriscience Educators for the Classroom

Andrew C. Thoron
Abraham Baldwin Agricultural College, USA

R. Kirby Barrick
University of Florida, USA (retired)

IGI Global
PUBLISHER of TIMELY KNOWLEDGE

A volume in the Advances in Higher Education and Professional Development (AHEPD) Book Series

Published in the United States of America by
IGI Global
Information Science Reference (an imprint of IGI Global)
701 E. Chocolate Avenue
Hershey PA, USA 17033
Tel: 717-533-8845
Fax: 717-533-8661
E-mail: cust@igi-global.com
Web site: http://www.igi-global.com

Library of Congress Cataloging-in-Publication Data

Names: Thoron, Andrew, 1980- editor. | Barrick, R. Kirby, editor.
Title: Preparing agriculture and agriscience educators for the classroom / Andrew Thoron, R. Kirby Barrick.
Description: Hershey, PA : Information Science Reference, [2022] | Includes bibliographical references and index. | Summary: "This text will focus on highlighting concepts and showcasing applications for agricultural teachers in planning their curriculum"-- Provided by publisher.
Identifiers: LCCN 2022970044 (print) | LCCN 2022970045 (ebook) | ISBN 9781668434208 (hardcover) | ISBN 9781668434215 (library binding) | ISBN 9781668434222 (ebook)
Subjects: LCSH: Agriculture--Study and teaching. | Teachers--Training of.
Classification: LCC S531.3 .P74 2022 (print) | LCC S531.3 (ebook) | DDC 630.76--dc23/eng/20220222
LC record available at https://lccn.loc.gov/2022970044
LC ebook record available at https://lccn.loc.gov/2022970045

This book is published in the IGI Global book series Advances in Higher Education and Professional Development (AHEPD) (ISSN: 2327-6983; eISSN: 2327-6991)

British Cataloguing in Publication Data
A Cataloguing in Publication record for this book is available from the British Library.

Advances in Higher Education and Professional Development (AHEPD) Book Series

Jared Keengwe
University of North Dakota, USA

ISSN:2327-6983
EISSN:2327-6991

Mission

As world economies continue to shift and change in response to global financial situations, job markets have begun to demand a more highly-skilled workforce. In many industries a college degree is the minimum requirement and further educational development is expected to advance. With these current trends in mind, the **Advances in Higher Education & Professional Development (AHEPD) Book Series** provides an outlet for researchers and academics to publish their research in these areas and to distribute these works to practitioners and other researchers.

AHEPD encompasses all research dealing with higher education pedagogy, development, and curriculum design, as well as all areas of professional development, regardless of focus.

Coverage

- Adult Education
- Assessment in Higher Education
- Career Training
- Coaching and Mentoring
- Continuing Professional Development
- Governance in Higher Education
- Higher Education Policy
- Pedagogy of Teaching Higher Education
- Vocational Education

The Advances in Higher Education and Professional Development (AHEPD) Book Series (ISSN 2327-6983) is published by IGI Global, 701 E. Chocolate Avenue, Hershey, PA 17033-1240, USA, www.igi-global.com. This series is composed of titles available for purchase individually; each title is edited to be contextually exclusive from any other title within the series. For pricing and ordering information please visit http://www.igi-global.com/book-series/advances-higher-education-professional-development/73681. Postmaster: Send all address changes to above address.

Titles in this Series

For a list of additional titles in this series, please visit: www.igi-global.com/book-series/advances-higher-education-professional-development/73681

Essential Practices for Building STEM Enrichment and College Bridge Programs
Calvin Briggs (Southern Center for Broadening Participation in STEM, USA) and Christopher K. Bass (Clark Atlanta University, USA)
Information Science Reference • © 2022 • 300pp • H/C (ISBN: 9781799871224) • US $195.00

Leadership Wellness and Mental Health Concerns in Higher Education
Cynthia J. Alexander (Acadia University, Canada) and Amy Tureen (University of Nevada, Las Vegas, USA)
Information Science Reference • © 2022 • 365pp • H/C (ISBN: 9781799876939) • US $195.00

Voicing Diverse Teaching Experiences, Approaches, and Perspectives in Higher Education
Wilfredo Alvarez (Utica University, USA) and Patrick S. De Walt (California State University, Fresno, USA)
Information Science Reference • © 2022 • 273pp • H/C (ISBN: 9781799890003) • US $195.00

Preparing Globally Competent Professionals and Leaders for Innovation and Sustainability
Michael J. Guo-Brennan (Troy University, USA) and Linyuan Guo-Brennan (University of Prince Edward Island, Canada)
Information Science Reference • © 2022 • 312pp • H/C (ISBN: 9781668445280) • US $195.00

Achieving Equity in Higher Education Using Empathy as a Guiding Principle
Catherine Ward (California State University, Fullerton, USA)
Information Science Reference • © 2022 • 335pp • H/C (ISBN: 9781799897460) • US $195.00

Teacher Reflections on Transitioning From K-12 to Higher Education Classrooms
Amy D. Broemmel (The University of Tennessee, Knoxville, USA) and Nora Vines (The University of Tennessee, Knoxville, USA)
Information Science Reference • © 2022 • 448pp • H/C (ISBN: 9781668434604) • US $215.00

Promoting Entrepreneurship to Reduce Graduate Unemployment
Isaac Wasswa Katono (Uganda Christian University, Uganda)
Business Science Reference • © 2022 • 363pp • H/C (ISBN: 9781799895817) • US $195.00

Handbook of Research on Improving Engineering Education With the European Project Semester
Benedita Malheiro (Instituto Superior de Engenharia do Porto, Instituto Politécnico do Porto, Portugal) and Pedro Fuentes-Durá (Universitat Politècnica de València, Spain)
Engineering Science Reference • © 2022 • 390pp • H/C (ISBN: 9781668423097) • US $265.00

IGI Global
PUBLISHER of TIMELY KNOWLEDGE

701 East Chocolate Avenue, Hershey, PA 17033, USA
Tel: 717-533-8845 x100 • Fax: 717-533-8661
E-Mail: cust@igi-global.com • www.igi-global.com

Table of Contents

Detailed Table of Contents

This chapter focuses on the phenomenal opportunity that is becoming and being an agriculture teacher. This chapter examines what it means to be a profession and evaluates agricultural education using seven criteria. This chapter also explains what it means to be a professionally prepared teacher. The traditional approach to teacher education is described, alternative pathways to certification are acknowledged, and the advantages and challenges posed by each approach are considered. The most important roles of a school-based agricultural education teacher are examined within the three-component model of agricultural education. The chapter also examines the characteristics of a good agriculture teacher, while emphasizing that it is important for agricultural education to be committed to becoming a diverse, equitable, and inclusive profession. Finally, the chapter summarizes key principles from codes of ethics and the Ag Teacher's Creed. These principles can guide teachers in making good behavioral choices.

This chapter investigates how school-based agricultural education systems and FFA programs have been structured, organized, and administered in the United States. Specifically, the roles and responsibilities of federal, state, and local education authorities are described and how school-based agricultural education (SBAE) and FFA programs operate within those sectors. The chapter also chronicles how supervision models have evolved and changed after the passage of the Smith-Hughes Act of 1917. Agricultural education stakeholder groups driving policy at the state level were also explored. Finally, an emphasis is placed on the diversity of each state and local entity's structure and organization throughout the chapter.

Preface

Agriculture has been taught in schools in the United States for more than a century. In the late 1800s, students, primarily in elementary schools, were offered instruction in the natural resources and environment as well as in farming practices. As more high schools were established in rural areas, farming became a more prevalent subject in the school curricula.

With the passage of the National Vocational Education Act of 1917, commonly called the Smith-Hughes Act, vocational agriculture programs began to flourish. The Act provided for oversight of programs from the Federal Board for Vocational Education as well as similar state boards. Standards were set, and supervision was provided by the Federal and state agencies. The preparation of vocational agriculture teachers for secondary schools was initiated with funding support from the states and the Agricultural Experiment Station system, primarily through land-grant colleges of agriculture. Students enrolled in the vocational agriculture programs were taught farming principles and practices as well as agricultural mechanization (farm shop). A unique requirement of vocational agriculture was that the students be engaged in a planned and supervised farming program, since nearly all were farm boys at that time. Then in 1928, the Future Farmers of America (now the National FFA Organization) was formed, providing activities and incentives for vocational agriculture students to excel in leadership and vocational skills development.

Programs of vocational agriculture greatly expanded in the second half of the 20th century. "Off-farm" agriculture topics were approved for the curricula, including areas such as agricultural business, horticulture, agricultural mechanics, natural resources, and food science. The term "vocational agriculture" was changed to "agricultural education" to reflect the national trend in career and technical education. Today the program is generally referred to as school-based agricultural education, differentiating it from college-level programs, with an expanded focus on teaching the science in agriculture and its application in the real world.

Teacher education in agriculture grew considerably over the years, with a pressing need to provide adequate numbers of teachers for the expanding school-based programs. A complete teacher education program typically consists of teacher preparation leading to state licensing or certification of teachers, graduate education, professional development, and research. This text is designed to be utilized in undergraduate and similar teacher preparation courses and can be part of advanced degree programs in teacher education in agriculture.

We dedicate this book to the school-based agricultural education teachers – past, present, and future—who make a positive difference in the lives of thousands of students everyday.

ORGANIZATION OF THE BOOK

Chapter 1. Becoming a School-Based Agriculture Teacher

This chapter focuses on the phenomenal opportunity that is becoming and being an agriculture teacher. This chapter examines what it means to be a profession and evaluates Agricultural Education using seven criteria. This chapter also explains what it means to be a professionally prepared teacher. The traditional approach to teacher education is described, alternative pathways to certification are acknowledged, and the advantages and challenges posed by each approach are considered. The most important roles of a school based agricultural education teacher are examined within the three-component model of agricultural education. The chapter also examines the characteristics of a good agriculture teacher, while emphasizing that it is important for Agricultural Education to be committed to becoming a diverse, equitable, and inclusive profession. Finally, the chapter summarizes key principles from codes of ethics and the Ag Teacher's Creed. These principles can guide teachers in making good behavioral choices.

Chapter 2. Organizing and Administering School-Based Agricultural Education Systems and the FFA

This chapter investigates how school-based agricultural education systems and FFA programs have been structured, organized, and administered in the United States. Specifically, the roles and responsibilities of federal, state, and local education authorities are described and how school-based agricultural education (SBAE) and FFA programs operate within those sectors. The chapter also chronicles how supervision models have evolved and changed after the passage of the Smith-Hughes Act of 1917. Agricultural education stakeholder groups driving policy at the state level were also explored. Finally, an emphasis is placed on the diversity of each state and local entity's structure and organization throughout the chapter.

Chapter 3. Promoting the Relationship Between School-Based Agricultural Education and the Community

This chapter explores the past, current, and future relationship between agricultural education programs and the communities in which programs operate. Throughout the chapter, opportunities to leverage communities to optimize the design and operation of an agricultural program are emphasized. Specifically, this chapter explores the development and utilization of a community profile, methods for identifying community needs and structuring a program to meet identified needs, state and national curriculum and program standards, and building and utilizing stakeholder groups. The chapter concludes with an opportunity to utilize the knowledge gained in the development of a program philosophy statement designed to guide the facilitation of an agricultural education program.

Chapter 4. Utilizing What We Know About How People Learn

Teaching and learning are skills that can be improved that begin at birth. Formal education in schools has employed strategies for teachers to help students acquire new information, process that information, and transfer it to new situations. The prior experiences of students help to provide a foundation for all future learning. Problem-solving and inquiry-based instruction can be strategies to help students engage in their learning. The principles of teaching and learning are outlined in this chapter that guide teachers in their understanding of how students best learn. Helping all students succeed is critical, and teachers can engage with the community to provide resources to students, including formal programming and individualized education plans.

Chapter 5. Designing the Curriculum and Developing a Course

The purpose of this chapter is to describe various types of school-based agricultural education (SBAE) programs and explain the process to develop a comprehensive agriscience course in a career pathway of a SBAE program. Current issues and challenges associated with curricular decision making are posed and curriculum is clearly defined. Strategies to develop a school instructional program plan, facility use calendar, and a curriculum map are presented as steps to build the scope and sequence and align content with the course objectives. The principles of the Understanding by Design curricular model are demonstrated in three stages (identify the desired results, determine acceptable evidence, plan learning experiences and instruction). The chapter concludes by identifying the components of a quality course syllabus to arrange relevant information to design a document that can be utilized to describe the course, communicate expectations, and outline guidelines for learner success.

Chapter 6. Planning for Teaching

This chapter consists of a clear and direct path to assist educators in creating specific unit and lesson plans. The chapter clearly outlines the differences among unit plans, lesson plans, the use of the three-circle model in agricultural education, and how to evaluate classroom resources. There are templates and examples available for the educator to utilize in planning for their curriculum, program, and resources. If an educator utilizes this chapter effectively, many problems with student engagement, pedagogical understanding, and planning will be eliminated.

Chapter 7. Planning Daily Lessons

The planning of daily lessons involves a variety of components to deliver a successful lesson. This chapter will look at the integration of a daily lesson plan with a focus on how to write the objectives, essential questions, the integration of Bloom's taxonomy in lesson objectives and connecting objectives and essential questions to an assessment. An additional focus is placed on the components of the daily lesson plan looking specifically at the alignment of standards, accommodations for students' IEP and 504 plans, integrating interest approaches and lesson activities.

Chapter 8. Developing Instructional Methods

Determining which methods of teaching to use is an important step in developing content for your classroom. Teacher-focused methods involve more direct instruction to students and include methods such as lecture, questioning, demonstration, resource people, field trips, and checking for understanding. Learner-focused methods of teaching include more student input and collaboration and include discussion, debate, problem solving approach, contextualized teaching and learning, flipped classes/team-based learning, case study, inquiry-based learning, experiential learning, problem-based learning, project-based learning, cooperative learning, and role playing. Each method of teaching has different strengths and weaknesses and allows teachers to achieve their learning objectives in different ways and increase their variability.

Chapter 9. Managing the Classroom and Laboratory

This chapter focuses on management strategies for classroom and laboratory instruction for school-based agricultural education teacher preparation. Specifically, the focus is on teaching methods, techniques, and strategies to proactively sustain positive student engagement in activities to prevent behavioral issues that disrupt the learning process. Characteristics and guidelines for instructional technology are discussed for effective management of student learning. Management practices for both instructional spaces and student behavior are incorporated. Highlights of underlying causes of behavior such as issues within social and psychological disruptions provide context for student engagement efforts. A portion of the chapter differentiates appropriate professional interactions of teachers with students. Behavior scenarios are utilized to provide application of theoretical concepts.

Chapter 10. Assessing Student Learning

The purpose of this chapter is to present assessment concepts and provide guidelines for effective assessment practices within school-based agricultural education. The chapter explains common terms and different types of assessments. It also provides guiding principles for how to effectively construct various types of assessments including both traditional and alternative types of assessments. In addition, this chapter provides guidelines and principles for managing grades in school-based agricultural education programs. This chapter provides practical recommendations and examples within a school-based agricultural education context.

Chapter 11. Evaluating a School-Based Agricultural Education Program

Systems of education intentionally evaluate students and teachers in the teaching and learning process. For many of the same reasons, the case can be made that programs should undergo systematic evaluation to discover where and how growth can occur. The purpose of this chapter is to provide a practical overview of the SBAE program evaluation process. The authors are careful not to be overly prescriptive in the recommendations for how these should be conducted, in part because effective program evaluation is tailored to the local program. Still, the case is made for why program evaluations should be conducted, what the research indicates good program evaluations look like, and how one should go about conducting one at the local level.

Chapter 12. Designing and Implementing Clinical and Field Experiences

The knowledge, skills, and dispositions needed to lead and deliver School-based Agricultural Education (SBAE) is introduced to teacher candidates through clinical and field experiences. These experiences are designed to introduce teacher candidates to the roles and responsibilities of the agricultural education teacher and FFA advisor. Experiences are intended to expose candidates to the three-component model of agricultural education and introduce them to the agricultural education career. Nationally accredited teacher preparation programs integrate clinical and field experiences into plans of study and degree plans to meet teacher certification and program accreditation requirements. This chapter focuses on the early field experience and student teaching internship components utilized by teacher preparation programs.

Chapter 13. Assessing Teacher Effectiveness in Agricultural Education

Effective teaching is a multidimensional concept that is often considered to be elusive. Those who are deemed effective teachers reflect frequently and often on their practice, continuously develop their human capital through professional development and conversations with others, and regularly solicit feedback and advice on their teaching from experienced and trusted mentors. When considering the development, support, and evaluation of school-based agricultural education (SBAE) teachers, it is essential to consider the sources of feedback along with formative and summative assessments to further develop career specific human capital to prepare for and sustain a career as an effective SBAE teacher. Therefore, this chapter will identify sources of feedback, describe methods of effective teaching evaluation, and explore opportunities to use evaluations and feedback for continual improvement for pre-service and in-service SBAE teachers.

Chapter 14. Providing Teacher Certification and Professional Development Programs

Agriculture teacher preparation programs, whether traditional or lateral entry, should prepare school-based agriculture teachers with the technical content, pedagogical knowledge, and professional skills to be successful educators in the classroom. This chapter explores the preparation of agriculture and agriscience educators for the classroom through an overview of the types of teacher certification options, describing the components of an agriculture teacher preparation program, including the Standards for School-Based Agricultural Teacher Preparation, exploring opportunities for agriculture teacher professional development, and investigating opportunities to enhance teaching excellence and technical skill enhancement through systematic needs assessment.

Chapter 15. Conducting Graduate Education Programs in Agricultural Education

Graduate education fulfills a specific need within the professional growth of students but carries different considerations when pursuing a degree or designing a degree program as compared to undergraduate programs. Agricultural Education graduate programs possess a nuance in balancing the need to serve a wide audience, use limited departmental resources, and encapsulate the breadth of school-based agricultural education to develop effective future teacher educators and scholars. This chapter serves to provide potential graduate students with an understanding of what to reflect on when pursuing a graduate degree, as well as the various curriculum components and experiences which they might engage in. Additionally, the authors have outlined programmatic decisions for faculty to consider when creating or revising agricultural education graduate programs.

Chapter 16. Conducting Programmatic Research in Agriculture Teacher Education

This chapter focuses on programmatic research in agriculture teacher education programs. Included in this chapter are the definition of research, why agriculture teacher education programs should conduct research, approaches to programmatic research, and practical applications for using research in teacher educations programs. The authors break down various methodologies for conducting programmatic research and offer many practical examples of how these can be easily implemented into any agriculture teacher education program. This chapter walks the reader through developing researchable questions, using the scientific method, choosing appropriate research design and methodology, and simple steps to disseminate and communicate research findings. The chapter also touches on the relationship between research and evaluation and how evaluation is used to assess programmatic effectiveness.

Chapter 17. Comparing Agricultural Education Programs Around the World

Agricultural education programs are offered in many countries around the world. Students are taught hands-on vocational agricultural skills that allow them to enter the agricultural workforce or advance to higher education. Interpersonal communication and leadership skills are offered in some programs but are not a common component of secondary agricultural education curriculum in other countries. Agricultural teachers come to the profession after working in agricultural industry, completing a college degree in technical agriculture or general teacher education, or by completing a formal teacher education program in agriculture. A wide variety of different youth organizations is offered in other countries for students studying agricultural education.

Andrew C. Thoron
Abraham Baldwin Agricultural College, USA

R. Kirby Barrick
University of Florida, USA (retired)

Chapter 1
Becoming a School-Based Agriculture Teacher

Greg Miller
Iowa State University, USA

ABSTRACT

This chapter focuses on the phenomenal opportunity that is becoming and being an agriculture teacher. This chapter examines what it means to be a profession and evaluates agricultural education using seven criteria. This chapter also explains what it means to be a professionally prepared teacher. The traditional approach to teacher education is described, alternative pathways to certification are acknowledged, and the advantages and challenges posed by each approach are considered. The most important roles of a school-based agricultural education teacher are examined within the three-component model of agricultural education. The chapter also examines the characteristics of a good agriculture teacher, while emphasizing that it is important for agricultural education to be committed to becoming a diverse, equitable, and inclusive profession. Finally, the chapter summarizes key principles from codes of ethics and the Ag Teacher's Creed. These principles can guide teachers in making good behavioral choices.

INTRODUCTION

This section will introduce the meaning of school-based agricultural education, indicate its scope and size, suggest who might be drawn to the profession and why, identify career opportunities beyond the classroom, and provide a list of objectives for the chapter. People who teach agriculture to students in grades K-12 in public, private, and parochial schools are commonly known as school-based agriculture teachers. Formal programs of instruction in agriculture at this level are referred to as school-based agricultural education programs. Most of the full-time school-based agriculture teachers and their programs are at the high school and middle school levels. The most recently available data (Foster et al., 2021) shows that more than 13,000 people in the United States are employed as school-based agricultural educators, and there are approximately 1500 jobs available each year. Historically the demand for school-based agriculture teachers has been equal to or greater than the number of qualified candidates. In addition, about 25% of newly qualified teachers do not immediately pursue employment in school-based agricul-

DOI: 10.4018/978-1-6684-3420-8.ch001

tural education. This generally creates a shortage of teachers and makes it difficult to fill all openings. On a positive note, this demonstrates that people with a degree in Agricultural Education are in demand, and it also presents significant opportunity for those aspiring to be school-based agricultural educators.

Teaching school-based agricultural education is a fantastic occupation. People who love agriculture-related subjects and have a passion for sharing what they know with others may find that teaching is an illustrious career path. School-based agriculture teachers are primarily motivated to become teachers by intrinsic and altruistic factors. They likely have a strong interest in agriculture, enjoy learning, view teaching as a challenging profession, believe that they will find fulfillment in doing something that they enjoy, believe they have an opportunity to make a difference, and want to serve others. Agriculture teachers are also motivated by extrinsic factors such as pay and benefits, availability of resources, rewards and recognition, and job security. Extrinsic factors, however, are not as influential as intrinsic and altruistic factors. School-based agriculture teachers continue teaching because they are satisfied and fulfilled, they continue to grow in their knowledge and skills, they have an appropriate level of autonomy and decision-making authority, they have a supportive administration and community, and they have developed a satisfactory work-life balance (Ismail, 2018).

Many people start out teaching agriculture in elementary, middle, and high schools. Some remain there for their entire career and find it to be extremely rewarding. Teaching at this level is anything but boring. There is always something new to learn about agriculture and education. Technology continues to change how things are done. Teachers get a new class of students every year, and they get to see the successes of those who came through their program. The bottom line is that teachers make a positive difference for their students and community.

Some teachers move on to different roles within and outside of school-based agricultural education. They may become community college instructors or university professors. Others might pursue a position on their state staff representing agricultural education or career and technical education. At the state level, titles might include director, state supervisor, district supervisor, agricultural education consultant, FFA advisor, or FFA executive secretary. At the national level, many opportunities exist through the National FFA Organization, the National Association of Agricultural Educators, and the Association for Career and Technical Education. Roles at this level could include organizational management, curriculum development, professional development for teachers, and conference planning. Some agriculture teachers move into different roles within education such as guidance counselor, principal, or superintendent. Some pursue a path aligned with their agricultural expertise. They might work for agribusinesses in a plethora of diverse ways, they could work for state or federal government agencies, or they might become entrepreneurs. The point is teaching K-12 agricultural education is a wonderful place begin a professional career. It also opens many other doors for those who are interested in outside opportunities.

This chapter will focus on the school-based agricultural education profession, particularly as it relates to the K-12 environment. The objectives of this chapter are to: 1.) evaluate the professional status of school-based agricultural education, 2.) describe the requirements to become a professionally prepared teacher, 3.) list, explain, and prioritize the roles of a school-based agricultural education teacher, 4.) identify the characteristics of a good teacher, and 5.) formulate a code of ethics of teaching agriculture.

TEACHING AS A PROFESSION

This section will address the meaning of the word profession and evaluate school-based agricultural education against seven criteria for professions. What do people think of when they hear the word profession? Many times, people refer to someone's job as their profession. In some cases, the distinguishing characteristic of a profession is the fact that someone is paid for their employment. The word profession may be used to refer to all persons engaged in a particular occupation. Applying common usage of the term, teaching is easily a profession. However, educators are often interested in being recognized for something more. They want to know how well they measure up against important criteria in comparison to professions recognized as having relatively high status. There is a professional model and seven criteria that can be used to assess an occupation (Ingersoll & Collins, 2018). Where does school-based agricultural education stand relative to these criteria?

1. Credentials
   Agriculture teachers are licensed by their state after having demonstrated that they possess an appropriate level of knowledge and skill related to teaching and to their subject matter content area. They must normally complete an accredited teacher education program and achieve passing scores on examinations related to pedagogy and content knowledge.
2. Induction
   New agriculture teachers are typically assigned a mentor within their school district. They may also receive additional supervision and programming aimed at assisting them as they transition into a full-time teaching role. Agriculture teachers have opportunities beyond what is provided by the local school system. In many states there is a beginning teacher program facilitated by a college or university or the state department of education. The state association of agricultural educators may also assign a mentor to new teachers and offer programming specifically aimed at new teachers throughout their first two or three years on the job.
3. Professional Development
   Agriculture teachers must engage in ongoing professional development to renew their teaching license. Some teachers do this by pursuing graduate degrees while others prefer to earn continuing education credits (CEUs). Opportunities for professional development are often facilitated by the National Association of Agricultural Educators and state associations of agricultural educators.
4. Specialization
   While there are exceptions, school-based agricultural educators are best described as generalists in agriculture. In a sense this is their specialization. Agriculture broadly defined is the teaching focus, and it is consistent with agriculture teacher preparation and licensing. Post-secondary instructors are more likely to have a more focused specialization such as agronomy or animal science.
5. Authority
   Agriculture teachers have significant authority in deciding what they teach and how they teach it. They have a great degree of autonomy in running the agriculture program. However, teachers at the secondary level have less of a role in making decisions that impact the broader school enterprise particularly when it comes to hiring and evaluating their peers. Post-secondary instructors typically have more power in these areas.

6. Compensation
   Teachers are typically not compensated as well as people who pursue many other professional careers. Agriculture teachers, however, do have opportunities to enhance their compensation through extended contracts. Sometimes agriculture teacher compensation is based on working a longer day and working more days per year. This is in recognition of their duties related to FFA and Supervised Agricultural Experience (SAE).
7. Prestige
   When compared to other occupations, teaching is somewhere in the middle relative to prestige (Ingersoll & Collins, 2018). Agriculture teachers rate higher than most other teachers on the criterion of prestige. In many communities across the country, very few people have higher status than the agriculture teacher.

Because education in general falls short of meeting all the criteria, scholars have concluded that teaching is a semi-profession (Ingersol & Collins, 2018; Heller, 2020). School-based agricultural education comes nearer to meeting the criteria than the broader education community. It is worth knowing that there is some debate about the status of education as a profession, however, this does not change the fact that education is critically important. Teaching is the profession that makes all other professions possible. In addition, it is well known that the Agricultural Education profession makes a positive difference in students' lives every day.

REQUIREMENTS TO BECOME A PROFESSIONALLY PREPARED TEACHER

This section will describe two approaches to becoming a professionally prepared teacher, compare the advantages and challenges of each approach, and explain the need for ongoing professional development. The most common approach to becoming a professionally prepared teacher is to complete an approved teacher education program in agriculture at an accredited college or university. This is referred to as the traditional approach. In some states people complete the program and obtain a teaching license while earning a bachelor's degree. In other states, people are required to earn a bachelor's degree in agriculture or a related area and then either pursue the teaching license through a fifth-year program or by completing a master's degree. Accredited agricultural teacher education programs have some fundamental components in common. These include:

1. General education
   All teachers need to be generally well educated. First-year and sophomore level courses in subject areas such as English, math, biology, and chemistry are the types of courses that are typically found in the general education component.
2. Agriculture content
   Teachers are expected to have expertise in their content area. Agriculture content expectations vary by program and often reflect subject matter that would be relevant to a particular state or region of the country. Agriculture courses that could be included in this part of the program often come from animal science, agronomy, horticulture, economics, agricultural mechanics, entomology, natural resources, and leadership.

3. Professional education
   Teachers need to have expertise in planning and facilitating educational experiences. Courses that support teachers in doing this effectively may focus on educational psychology, special education, measurement, educational policy and law, foundations of career and technical education, program planning, and teaching methods. In addition to traditional courses, preservice teachers will be expected to engage in field experiences over time. Some or all of these experiences may be connected to a specific traditional course. The culminating field experience is student teaching.

At the undergraduate level, the amount of time devoted to the three main components varies by institution. However, a ballpark estimate would be about 35% of the program focused on general education outcomes, 35% focused on developing agricultural content knowledge and 30% on professional education outcomes. For students pursuing licensure through an approved post baccalaureate program the entire program would be dedicated to professional education outcomes.

Traditional teacher education programs in agriculture are in an academic unit (e.g., college, school, department) focused on agriculture or education. Regardless of the academic home, there will be collaboration with the broader teacher education program which is located within the institution's education unit. The teacher education program will articulate a conceptual framework including a mission and vision for all teacher education programs. These statements reveal what makes a particular program unique. In addition, standards for aspiring teachers will be articulated. While education is a function of government that is left to the individual states, most states will have similar standards. They are often based on the *InTASC Model Core Teaching Standards and Learning Progressions for Teachers 1.0* (Council of Chief State School Officers, 2013). These standards impact decisions about all aspects of an agriculture teacher education program. Standards specific to agricultural education are detailed in the *Standards for School-Based Agricultural Education Teacher Preparation Programs* (AAAE, 2021).

There are alternative pathways to becoming a school-based agricultural educator. In fact, research shows that about 23% of newly hired agriculture teachers came through an alternative route to licensure (Smith et al., 2019). Some people within the profession have questioned the quality of alternative certification programs, but research on alternative certification programs has not demonstrated a negative impact of these programs on student learning (Bowling & Ball, 2018).

There are three common pathways that alternatively licensed school-based agriculture teachers have followed. These include:

1. Career and technical
   The career and technical approach can license a person to teach based solely on related-industry work experience. There is generally no expectation that the teacher holds a bachelor's degree, and they may or may not be required to go through a program designed to develop their pedagogical skills.
2. Bachelor's degree in agriculture
   Persons who hold a bachelor's degree in agriculture could be hired to teach agriculture with the expectation that they complete coursework in pedagogy within a specified period.
3. Adding agriculture to an existing license
   Teachers in other subject areas can become agriculture teachers by passing a state approved subject matter test or by taking a minimum number of credits of agriculture-related college coursework.

Traditional and alternative routes to certification are both legitimate ways to become a school-based agricultural education teacher. The two pathways offer their own advantages and challenges – especially early in the career. Advantages for traditional teacher preparation include: (1) an optimal balance of general education, content knowledge, and professional education, (2) a network of professional colleagues who completed their programs together, and (3) a gradual transition from student to teacher through carefully planned field experiences. Advantages for alternative certification may include: (1) having greater knowledge and skill in the content area of agriculture. (2) having a greater network of connections to industry and the local community, and (3) a level of maturity that may be beneficial from a classroom management perspective. Challenges faced by traditionally prepared teachers may include: (1) being close in age to their high school students which may pose challenges related to classroom management and (2) having a lower level of knowledge and skill in technical agriculture. Challenges faced by teachers coming from an alternative certification route may include (1) not appreciating all three components of the program model to include classroom and laboratory instruction, SAE, and FFA, and (2) may struggle with lesson planning and execution as a result of not having completed a professional education core.

Regardless of the path taken to become an agriculture teacher, teachers will need a license issued by their state to practice. In addition to completing an approved traditional or alternative licensure program, teachers may have to achieve an acceptable score on a standardized test focused on pedagogy and content knowledge. Because agriculture is considered part of career and technical education, some states may require a minimum number of hours of work experience in agriculture for both traditionally and alternatively certified teachers. A background check will also be completed to make sure that there have been no disqualifying criminal charges or convictions. When the license is issued, it will indicate which subject areas the teacher is qualified to teach, what grade levels they may teach, and the expiration date for the license. Additional information about traditional and alternative teacher certification is included in Chapter 14.

Going into the profession teachers should realize that their education will never be complete. Teachers are expected to engage in ongoing professional development to remain up to date with advancements in education and agriculture. To renew their license, teachers will be expected to earn a minimum number of CEUs. These credits could be earned by participating in approved programs that award CEUs such as workshops offered at state and national agriculture teacher conferences, professional development offered through the local school district, or programs offered through a regional education service agency. Credit earned toward graduate degrees may be used instead of CEUs. The advantage to taking graduate courses is that teachers have the potential to advance along their school district's salary schedule. The value of a master's degree varies by district, but it would be fair to say that, on average, the master's degree will lead to a salary increase of several thousand dollars.

THE ROLES OF A SCHOOL-BASED AGRICULTURAL EDUCATION TEACHER

This section will explain seven key roles of a school-based agricultural education teacher and propose priorities across the three most predominant roles. The three main components of a school-based agricultural education program are classroom and laboratory instruction, FFA, and SAE. Students are taught about the three components from the moment they enroll in a school-based agriculture course. Representations of the three components have ranged from the three-circle model (Croom, 2008) to a tricycle (Retallick, 2003). Given the general consensus that these are the three program components, it is fair to emphasize and prioritize the roles of a school-based agricultural educator within the three-component model. This chapter introduces important roles while later chapters examine them in greater detail.

Classroom/Laboratory Teacher

Roles related to classroom and laboratory instruction are the most important relative to the three-component model of Agricultural Education. This is where teachers spend most of their time. In fact, researchers found that agriculture teachers across three levels of experience, spent the majority of their time on classroom and laboratory instruction (Torres et al., 2008). This role involves considerable planning. Teachers may use professionally developed instructional materials such as the *Curriculum for Agricultural Science Education* (CASE, 2022), they may adapt professionally developed materials to fit their own situation, or they can develop instructional materials from scratch. Most teachers do all three of these things. Regardless of the approach, it takes time to think through lessons and organize the resources that are needed to successfully execute the lesson. Agriculture teachers develop their curriculum to achieve specific content standards. A great resource to use in identifying appropriate standards is the *National AFNR Career Cluster Content Standards* (The Council, 2021).

Of course, the biggest chunk of time is spent working directly with students. During the school year, teachers are in classrooms and laboratories teaching students for six hours a day five days a week. Embedded within teaching is the need to evaluate student performance. Much of this can be accomplished during class time, but many teachers spend additional time beyond the school day grading assignments and providing feedback. Grading systems vary across schools. Some schools determine grades using percentages while others have adopted standards-based grading. The grading system can impact many distinct aspects of teaching and evaluation.

Chapters 5 through 10 will address specific roles that teachers must perform when functioning as a classroom or laboratory instructor. These chapters will address curriculum development, lesson planning, instructional materials, classroom/laboratory management, and assessment of learning.

FFA Advisor

Serving as an FFA advisor is the second most significant role that a school-based agricultural education teacher must perform. The amount of time spent on this role is second only to the amount of time spent on classroom and laboratory instruction. Research showed that first year teachers spent about 16% of their time on FFA while experienced teachers devoted 24% of their time to FFA (Torres et al., 2008). In well-established programs, FFA is typically particularly important to students, parents, school administration, and the community. FFA gives students an opportunity to develop their leadership skills in a variety of ways and at different levels. FFA also provides a place for students to apply what they are learning through career development events. These events let students see how well they have mastered key concepts and skills against industry standards and in friendly competition with students from other FFA chapters. In addition, FFA does a wonderful job of recognizing students for their accomplishments as individuals, through their contribution to teams, and through their contributions to their chapters and beyond. FFA is the most public aspect of school-based agricultural education and has been instrumental in showcasing the overall program in a positive light. Chapters build a program of activities around three divisions that include growing leaders, building communities, and strengthening agriculture (FFA, 2021).

The program of activities guides the chapter in meeting its goals for the year. This document should be shared with all members and stakeholders of the FFA chapter.

FFA is the one aspect of an agricultural education program that can sometimes be all consuming. Teachers should not aspire to participate in every event or program that FFA offers. Teachers should develop a guiding philosophy on how to decide what to participate in. One way teachers can determine whether to participate is to ask themselves if the event or activity is related to the curriculum in the agriculture program. If someone teaches floriculture, it makes perfect sense to have students participate in the floriculture CDE. Another factor to consider is student interest. Teachers might not teach anything about horses, but they may have a group of students who are interested in horses. Many agriculture teachers would facilitate their students' participation in the horse evaluation CDE based on student interest. Furthermore, they might have a community member volunteer to coach and supervise the team.

Chapters 2, 6, and 11 will address additional areas related to FFA. Furthermore, an extensive amount of information is available on the National FFA Organization website. State FFA associations will also have a website with state specific information and resources.

SAE Supervisor

When it comes to the three-component model, Supervised Agricultural Experience (SAE) gets the least amount of attention. Research has shown that experienced teachers dedicated only about four percent of their time to SAE. First year teachers spent about two percent of their time on SAE (Torres et al., 2008). This does not mean that SAE is unimportant. In fact, SAE is a fantastic, individualized learning opportunity for students. Accomplishments in SAE are required for advancement into higher degrees of membership in FFA. In addition, many teachers are good at integrating students' SAEs into classroom and laboratory instruction.

Teachers have a solid grasp on the espoused theory of SAE, but they do not implement it according to the theory (Retallick, 2010). Teachers have faced real challenges in executing the SAE component of their program. Some of these challenges include inflexible viewpoints on what qualifies as SAE, lack of adequate extended contracts, inadequate budget to support on-site SAE visits, and limited facilities to support individual SAE projects including school-based enterprises. Research has noted additional challenges such as student demographics, school priorities and structures, and student reactions to SAE (Retallick, 2010).

SAE for All (The Council, 2021) is the most recent effort to reconceptualize and promote the SAE component of school-based agricultural education. It is an excellent resource for school-based agriculture teachers and students. The most notable change coming out of this initiative is the new types of SAEs. The most intriguing type of SAE is foundational which is intended for all students enrolled in agricultural education. *SAE for All* resources include a guide for teachers and students along with supporting materials. To the extent possible and within the constraints of their specific situation, teachers should strive to develop the SAE component of their program. Innovative and successful school-based agriculture teachers will adapt and utilize the *SAE for All* materials in a manner best suited to their own specific program.

Chapters 5, 6, and 11 will address additional areas related to SAE. Furthermore, an extensive amount of information is available on the National Council for Agricultural Education website.

Mentor

There is a saying in education that goes something like this: "elementary teachers love their students, high school teachers love their subjects, and college professors love themselves." Are there some grains of truth in this saying? Many school-based agricultural educators come to the profession because of their love of agriculture. While the love of agriculture is stronger in the beginning, agriculture teachers quickly develop a love for students. They become mentors to students and pass on a love for agriculture and the many agriculture-related opportunities that students can pursue. Agriculture teachers provide mentoring and guidance through classroom and laboratory instruction, by serving as an FFA advisor, and through SAE in a uniquely individualized and differentiated way. Some agriculture teachers may need to work on prioritizing work-life balance (love for themselves). This is especially true for early career teachers. Having a good work-life balance is good for the teacher and their students. Taking care of oneself is not selfish. When teachers are at their best, they can be more effective and efficient. Taking care of oneself may lead to a more sustainable, rewarding, and impactful career.

Serve the School

Even though it is common to find agriculture classroom and laboratory facilities that are physically separated from the rest of the school building, agriculture teachers must not view themselves or their program as an island. Agriculture teachers must advocate for their own program and educate their peers and administrators on what their program is all about. However, it is equally important that the agriculture teacher get to know about the rest of the school and understand their role in achieving outcomes that cut across disciplinary boundaries. Agriculture teachers like other teachers should expect to have their share of responsibilities like hall, lunchroom, and bus duty. When possible, agriculture teachers should attend school functions such as assemblies, athletic events, and award ceremonies. Doing so can help in forging relationships with students, parents, teachers, and administration.

Engage the Community

Agriculture teachers need to engage with their local community. Establishing and utilizing an advisory council is the most important thing that a teacher can do to engage the community. The advisory council can assist the agriculture teacher in developing and maintaining a curriculum that is suitable to the local community. They can also assist the teacher in evaluating all aspects of the program and turning the evaluation results into specific plans for improvement. Advisory committees are often very influential in garnering the political and financial support needed to implement improvement plans. Many agriculture programs have an FFA Alumni affiliate. The Alumni group often supports agriculture programs in raising and managing funds for the benefit of students. Alumni members are also a useful source of volunteers. If volunteerism is cultivated in support of the local agriculture program, the impact of the program on students can be magnified significantly.

Actively Engaged in Professional Organizations

Agriculture teachers should strongly consider joining the National Association of Agricultural Educators (NAAE) and the Association for Career and Technical Education (ACTE). The NAAE membership includes agricultural educators at all levels as well as the state and national leaders who support school-based agricultural education. NAAE has more than nine thousand members (NAAE, 2021). ACTE is the largest educational organization in the United States focused on workforce development. NAAE is affiliated with ACTE as one of ACTE's 11 divisions (ACTE, 2021). These organizations provide many benefits and services including workshops designed to enhance knowledge and skill, connections with other professionals, curriculum materials such as CASE, information about issues and developments impacting the profession, recognition of member accomplishments, scholarships for preservice teachers, professional liability insurance, and advocacy with state and federal government.

Depending on the state, some teachers can join a teacher's union. At the local level, teachers engage in collective bargaining to address issues such as salary, benefits, and other terms of employment. Teacher unions may offer professional development opportunities that count toward license renewal. Unions may also offer access to legal services. Memberships may be at the local, state, and national levels. Two of the largest and most widely known teacher unions are the National Education Association (NEA, 2021) and the American Federation of Teachers (AFT, 2021).

CHARACTERISTICS OF A GOOD TEACHER

This section emphasizes the importance of diversity, equity, and inclusion within Agricultural Education, identifies three characteristics that effective teachers possess regardless of grade level or subject matter, and provides a more extensive research-based list of characteristics of effective school-based agricultural education teachers. Before listing characteristics of good teachers, it is important to note that anyone who really wants to be an agriculture teacher can achieve this goal. There is no prototype that must be followed. Good agriculture teachers come from a variety of backgrounds and all walks of life. They could come from rural, suburban, or urban homes. They may or may not have taken agriculture courses while in high school. They could be extraverts or introverts. It does not matter what level of education their parents or grandparents achieved. Personal characteristics such as age, sex, gender identity, sexual orientation, race, national origin, and religion ought to have no impact on determining who is or who would be a good agriculture teacher. The Agricultural Education profession is strong and will become even stronger as it continues to demonstrate a commitment to a diversity, equity, and inclusion.

People who love agriculture and are enthusiastic about sharing their love of agriculture with others, have the potential to be good agriculture teachers. While the previous statement is a good summary, more detail on the characteristics of effective teachers is warranted. Research has shown that regardless of subject matter or grade level effective teachers possess three characteristics. These include expecting appropriately high levels of student achievement, being an excellent manager of the classroom and laboratory, and being able to design lessons that promote student learning (Wong & Wong, 2018). Specific to agriculture, a recent study generated fifty-eight characteristics of effective school-based agricultural education teachers in eight categories as shown in Table 1 (Eck et al., 2019).

It is important to understand where characteristics in Table 1 came from. The researchers started by asking thirty-five award winning agricultural educators to answer a single question. "What are the characteristics of an effective agricultural education teacher" (Eck et al., 2019, p. 5). From these answers the researchers led the award-winning agricultural educators through a process to reach consensus. So, these are the opinions of a distinguished group of agricultural educators concerning the characteristics of effective school-based agricultural educators.

Table 1. Categorized characteristics of effective SBAE teachers

Category	Identified Characteristic
Instruction	Is passionate about education. Provides a variety of learning opportunities to meet the needs of all students. Guides students to grow personally. Is a leader for students. Demonstrates pedagogical knowledge. Is a good communicator. Demonstrates sound educational practices. Is prepared for every class. Demonstrates classroom management. Understands experiential learning theory. Is motivated for student success. Is knowledgeable about agriculture. Is first and foremost a classroom teacher. Is innovative.
FFA	Advises the FFA chapter. Is not just a facilitator of record keeping for degrees and awards. Instructs students through FFA. Is passionate about FFA. Advises the FFA officers. Prepares students to be leaders.
Supervised Agricultural Experiences	Instructs students through supervised agricultural experiences.
Program Planning	Uses the complete agricultural education model as a guide to programmatic decisions and practices. Is resourceful as an administrator of the program.
Balance	Leads a balanced life. Has the ability to say no. Is never afraid to ask for help. Demonstrates a willingness to put in extra hours.
Diversity and Inclusion	Understands student needs. Is an advocate for all students. Values students regardless of gender. Values students regardless of economic status. Values students from all ethnic/racial groups. Understands diversity. Is culturally relevant. Cares about all students. Understands there is not an award for all students, but that does not mean they are not valuable.
Professionalism	Is a purposeful lifelong learner. Demonstrates adaptability. Is a dedicated professional. Is an advocate for public education. Is engaged in an appropriate professional organization.
Personal Dispositions	Is fair. Is student focused. Is trustworthy. Is honest. Is passionate about agriculture. Is respectful. Shows empathy. Is dependable. Is responsible. Is relatable. Is genuine. Is a hard worker. Is organized. Is helpful. Has patience. Is engaging. Shows integrity.

Note. The title and content of this table are quoted directly from "Identifying the characteristics of an effective agricultural education teacher: A national study," by C. J. Eck, S. Robinson, J.W. Ramsey, and K.L. Cole, 2019, *Journal of Agricultural Education*, *60*(4), 10-11.

These characteristics are helpful in gaining a general understanding of what effective agriculture teachers are like. However, some questions should be answered before applying this information. How can each of the characteristics be defined and measured? Effective towards what end? Can the same set of characteristics be applied to different outcomes with good result? Can the same characteristics be applied in different ways? Can it be assumed that the characteristics alone or in combination will result in higher levels of student achievement relative to important outcomes? Do effective teachers have to apply all the characteristics all the time? What other questions should be answered before applying the information about effective teacher characteristics?

A CODE OF ETHICS OF TEACHING AGRICULTURE

This section identifies ethical principles articulated by national organizations, draws attention to state and local information relevant to ethics, presents the Ag Teacher's Creed, and encourages school-based agricultural education teachers to formulate a personal code of ethics for teaching. According to Merriam-Webster's (2021) online dictionary, a code of ethics is "a set of rules about good and bad behavior." A quick Google search will locate codes of ethics for a variety of organizations, occupations, and professions including education. A code of ethics is an invaluable guide to teachers in making good behavioral choices.

Codes of ethics are promulgated by different organizations at the national level. One of these organizations is the NEA. NEA is the largest labor union in the United States representing three million educators (NEA, 2021). The NEA published their *Code of Ethics for Educators* on September 14, 2020. This code is guided by two principles: (1) "commitment to the student" and (2) "commitment to the profession" (NEA, 2021). Eight standards of conduct focus on the student. To summarize, educators support students in their effort to learn and pursue truth, protect students from harm and any form of illegal discrimination, and do not take advantage of students for personal gain. There are also eight standards of conduct related to the profession. In summary, educators work to advance their profession, they tell the truth about their credentials and qualifications, they demand the truth from others and do not allow unqualified persons to act as educators, and they treat other educators fairly and with respect.

The Association of American Educators (AAE, 2021) has also published a code of ethics. AAE is not a union. Their website indicates that they are "the largest national organization of its kind." Their *Code of Ethics for Educators* is very consistent with the NEA code, but there are some noticeable differences. Teacher strikes and related tactics would be inconsistent with the AAE code of ethics. The AAE code also places more emphasis on the educator's role in moral education, adherence to policy and law, and includes a principle focused on parents and the community.

Another alternative is the *Model Code of Ethics for Educators* (MCEE). The MCEE is published by the National Association of State Directors of Teacher Education and Certification (NASDTEC, 2021). The MCEE includes five principles and eighty-six specific statements describing ethical conduct. A particularly unique contribution of the MCEE is a principle focused on using technology in an appropriate manner. Given the pervasiveness of technology at this time, the principle is on point. It addresses among other things, social media presence, protecting students from illegal content, protecting students from cyber bullying, and maintaining confidentiality of student records.

Familiarity with national codes of ethics can be useful to teachers as a resource. Teachers should read them to develop a level of awareness and sense of right conduct. Teachers should refer to them when they experience a dilemma and need a frame of reference for determining the appropriate attitude and conduct. Even more important than being aware of the national codes is determining what code applies to a specific location. Each state will have a code of ethics for teachers. It may be articulated through state law, administrative code, or through a professional standards board. The state code of ethics will likely incorporate principles and standards found in national codes related to the profession, students, and the community.

Many school districts provide teachers with a handbook. The handbook may or may not specifically address a code of ethics, but it will address rules, policies, and procedures that are important to follow. Agricultural educators should obtain any such handbook and utilize it in making decisions that are governed by the handbook.

Agricultural educators have something that is unique to their discipline - the *Ag Teacher's Creed*. This creed is framed and on display in many school-based agricultural education programs across the United States. While the Creed is not a code of ethics per se, it describes behaviors that are consistent with the standards of behavior commonly included in codes of ethics for educators. Agricultural educators who follow this creed to the best of their ability reflect favorably on the agricultural education profession. The creed can help teachers to make good decisions about their behavior and increase the probability that they will choose right over wrong. The *Ag Teacher's Creed* is copied below with permission from the National Association of Agricultural Educators (NAAE, 2021).

Ag Teacher's Creed

I am an agricultural educator by choice and not by chance.

I believe in American agriculture; I dedicate my life to its development and the advancement of its people.

I will strive to set before my students by my deeds and actions the highest standards of citizenship for the community, state, and nation.

I will endeavor to develop professionally through study, travel, and exploration.

I will not knowingly wrong my fellow teachers. I will defend them as far as honesty will permit.

I will work for the advancement of agricultural education and I will defend it in my community, state, and nation.

I realize that I am a part of the school system. I will work in harmony with school authorities and other teachers of the school.

My love for youth will spur me on to impart something from my life that will help make for each of my students a full and happy future.

School-based agricultural education teachers should consider formulating a personal code of ethics for teaching agriculture. At a minimum, their personal code should incorporate principles from their state and local codes. Building in additional principles from the Ag Teacher's Creed and those articulated by national organizations such as the NEA, AAE, and MCEE can take ethics beyond the minimum and promote a higher level of professionalism.

SUMMARY

Teaching school-based agriculture is a wonderful place to begin a career. It is also a great place to finish, but depending on someone's interests, they will have many other options. Agricultural Education is a profession and part of the larger teaching profession. Agricultural Education is particularly strong on elements of the professional model including credentials, induction, professional development, and prestige.

There is more than one way to become an agriculture teacher, and there will be advantages and challenges associated with whatever approach is taken. Even so, anyone can become a successful teacher regardless of how they entered the profession. In addition, teachers should expect to engage in professional development to continue learning and growing throughout their career.

School-based agricultural education teachers conduct many specific roles. Most of these can be aligned with classroom and laboratory instruction, FFA, and SAE. Teachers also serve as mentors, contribute to the school and community, and are actively engaged in their professional organizations.

Research has identified characteristics of effective school-based agricultural education teachers in eight areas. Agriculture teachers need to make good behavioral choices. Their efforts in this regard should be guided by an appropriate code of ethics.

REVIEW QUESTIONS

1. How would you describe the professional status of school-based agricultural education teachers?
2. If you were designing a traditional agriculture teacher education program, what proportion of the curriculum would you devote to general education, agriculture content, and professional education? Why?
3. Which alternative pathway to becoming an agriculture teacher would work best in your state? Why?
4. From your perspective, what are the most important advantages to pursuing either a traditional or alternative pathway to becoming a teacher?
5. Why are teachers expected to engage in professional development throughout their career?
6. How would you explain the relative importance of the distinct roles performed by a school-based agricultural education teacher?
7. Why should agriculture teachers cultivate relationships within their school and community? What are some ways to do this?
8. Which professional organizations would you recommend to a fellow school-based agricultural educator? Why?
9. How would you characterize an effective agriculture teacher? On which characteristics are you strongest? Which ones do you aspire to improve?
10. If you were writing a code of ethics for school-based agricultural educators, what principles and standards of conduct would you include?

REFERENCES

Bowling, A. M., & Ball, A. L. (2018). Alternative certification: A solution or an alternative problem? *Journal of Agricultural Education*, *59*(2), 109–122. doi:10.5032/jae.2018.02109

Council of Chief State School Officers. (2013, April). *Interstate Teacher Assessment and Support Consortium InTASC Model Core Teaching Standards and Learning Progressions for Teachers 1.0: A Resource for Ongoing Teacher Development*. Washington, DC: Author.

Croom, D. B. (2008). The development of the integrated three-component model of agricultural education. *Journal of Agricultural Education*, *49*(1), 110–120. doi:10.5032/jae.2008.01110

Eck, C. J., Robinson, S., Ramsey, J. W., & Cole, K. L. (2019). Identifying the characteristics of an effective agricultural education teacher: A national study. *Journal of Agricultural Education*, *60*(4), 1–18. doi:10.5032/jae.2019.04001

Foster, D. D., Lawver, R. G., & Smith, A. R. (2021). *National Agricultural Education Supply and Demand Study, 2020 Executive Summary*. Retrieved from: http://aaaeonline.org/Teacher-Supply-and-Demand

Heller, R. (2020, March). A profession in process. *Phi Delta Kappan*, *101*(6), 4. doi:10.1177/0031721720909575

Ingersoll, R. M., & Collins, G. J. (2018). The Status of Teaching as a Profession. In J. Ballantine, J. Spade, & J. Stuber (Eds.), *Schools and Society: A Sociological Approach to Education* (6th ed., pp. 199–213). Pine Forge Press/Sage Publications.

Ismail, N. (2018). *Iowa high school agriculture teachers' motivations to teach, personal and professional needs, and career satisfaction in the teaching profession* [Doctoral dissertation, Iowa State University]. Iowa State University Digital Repository. https://dr.lib.iastate.edu/handle/20.500.12876/31003

Retallick, M. S. (2003). Reframing SAE: The tricycle principle. *The Agricultural Education Magazine*, *75*(6), 8–9.

Retallick, M. S. (2010). Implementation of supervised agricultural experience programs: The agriculture teachers' perspective. *Journal of Agricultural Education*, *51*(4), 59–70. doi:10.5032/jae.2010.04059

Smith, A. R., Lawver, R. G., & Foster, D. D. (2019). *National Agricultural Education Supply and Demand Study, 2018 Executive Summary*. Retrieved from: http://aaaeonline.org/Teacher-Supply-and-Demand

Torres, R. M., Ulmer, J. D., & Aschenbrener, M. S. (2008). Workload distribution among agriculture teachers. *Journal of Agricultural Education*, *49*(2), 75–87. doi:10.5032/jae.2008.02075

Wong, H. K., & Wong, R. T. (2018). The first days of school: How to be an effective teacher (5th ed.). Mountain View, CA: Harry K. Wong Publications, Inc.

KEY TERMS AND DEFINITIONS

Alternative Teacher Education: Any pathway to becoming a teacher other than the traditional approach. Alternative pathways may allow entry into teaching based solely on work experience. Teachers who take this approach are required to complete a minimum number of CEUs or college credits related to pedagogy within a specified period.

Code of Ethics: Rules for appropriate professional conduct.

Profession: All persons engaged in a particular occupation in exchange for pay.

School-Based Agricultural Education: Formal programs of instruction in agriculture for students in grades K-12 in public, private, and parochial schools.

Three Component Model: Regardless of how the model is depicted (e.g., A Venn Diagram, a Tricycle), it is widely believed that a school-based agricultural education program should include three components – classroom and laboratory instruction, supervised agricultural experience, and FFA.

Traditional Teacher Education: The pathway to becoming a teacher that involves completing an approved teacher education program at an accredited college or university. Approved programs require coursework and experiences that contribute to general education, agriculture content knowledge, and professional education competency.

Chapter 2
Organizing and Administering School-Based Agricultural Education Systems and the FFA

Jay K. Solomonson
Illinois State University, USA

Richie Roberts
Louisiana State University, USA

ABSTRACT

This chapter investigates how school-based agricultural education systems and FFA programs have been structured, organized, and administered in the United States. Specifically, the roles and responsibilities of federal, state, and local education authorities are described and how school-based agricultural education (SBAE) and FFA programs operate within those sectors. The chapter also chronicles how supervision models have evolved and changed after the passage of the Smith-Hughes Act of 1917. Agricultural education stakeholder groups driving policy at the state level were also explored. Finally, an emphasis is placed on the diversity of each state and local entity's structure and organization throughout the chapter.

INTRODUCTION

When FFA members have recited opening ceremonies during official functions, the public has become aware that U.S. agricultural education and the National FFA Organization "*reaches from the state of Alaska to the Virgin Islands and from the state of Maine to Hawaii*" (National FFA Organization, 2021c, p. 36). Although the National FFA Organization and other agricultural education entities has provided an effective, workable structure, the ways school-based agricultural education (SBAE) and the FFA have been organized and administered in states and local spaces has varied considerably. Since education in the U.S. is a state and local responsibility, each state and ***local education agency*** (LEA) has made decisions on the administration of their various legislative initiatives and the organization of their educational

DOI: 10.4018/978-1-6684-3420-8.ch002

agencies and programs to best meet the needs of their constituents, including those that have involved agricultural education and FFA.

This chapter explores how SBAE and the FFA are organized and administered in the U.S. Upon completion of this chapter, the reader should be able to (1) describe the roles and responsibilities of local, state, and federal educational agencies, (2) explain how SBAE programs fit in with general education at the local, state, and national levels, (3) identify the state agencies, organizations, and groups that have driven agricultural education policy and programs, and (4) describe how FFA has been structured, organized, and administered at the local, state, and national levels.

ROLES AND RESPONSIBILITIES OF EDUCATIONAL AGENCIES

Before learning how agricultural education and FFA have fit into the general education system, it is critical to understand how the U.S. educational system is organized and how its agencies at various levels have operated over time. The authority over the ***public school*** system in the U.S. is somewhat convoluted since it can incorporate several governmental entities at each of the federal, state, and local levels. With each of these governing bodies possessing some function in the public education system, it may sometimes seem confusing as to their roles and responsibilities. In this section, the current system of school governance in the U.S. was described, especially regarding to the specific roles and responsibilities of each educational entity at the federal, state, and local levels.

The United States Department of Education

As previously mentioned, education in the U.S. has primarily been a state and local responsibility. However, the ***United States Department of Education***, also referred to as the ED, has played an essential role in this hierarchical system. Located in Washington D.C. and led by the Secretary of Education, the ED was officially established as a cabinet-level agency of the U.S. government in 1980 (United States Department of Education, 2021). According to the U.S. Department of Education (2010), the mission of the ED has been to "promote student achievement and preparation for global competitiveness by fostering educational excellence and ensuring equal access" (para.1) by engaging in the following four activities: (1) establishing policies related to federal education funding, (2) collect educational data and research, (3) identification of educational issues in the U.S., and (4) enforcement of discrimination laws in programs that receive federal funding.

Policies Related to Federal Education Funding

The ED has established policy and procedures whenever legislation related to education has been enacted or reauthorized by the U.S. Congress. Congress also has been tasked with approving an annual federal budget with funds specifically allocated toward elementary and secondary education. However, the federal support of nationwide educational costs has been minimal, covering only about 8% of public-school operating costs requiring states and local entities to fund the remainder (United States Department of Education, 2021). Further, federal funding has had stipulations and regulations that states must abide by to receive them. Because states have ultimately been responsible for education, they may choose not to comply with federal educational regulations, but most have so that they could receive federal dollars.

Collect Educational Data and Research

Within the ED, the National Center for Education Statistics (NCES) has been an organization that has been the primary entity responsible for collecting and analyzing educational data at the federal level. Their research has been useful to identify the strengths and weaknesses in the U.S. public school system so educational leaders and policymakers have has access to data-backed evidence for educational reform. To view NCES publications and the types of data they collect visit their website at https://nces.ed.gov/.

Identification of Educational Issues

Often a result of data the NCES collects, the ED has identified and addressed significant issues and problems of the U.S. educational system. These issues may also be identified or brought forth by the various educational stakeholders, advisory groups, and organizations in which the department collaborates. Examples of these issues might be closing the achievement gap, providing funding for specific populations, or addressing the teacher shortage. These conversations may lead to legislation or policies to bring about federal education reform in the area(s) of concern.

Enforcement of Discrimination Laws

One of the most important responsibilities of the agency has been to ensure equal access to a high-quality education for everyone, regardless of gender, age, race, national origin, or disability. This has been accomplished by enforcing five federal civil rights statutes that prohibit discrimination in any educational program or activity that receives federal funding. Within the ED, the Office for Civil Rights (OCR) has been the unit responsible for ensuring these entities comply with federal laws.

State Education Agencies

The U.S. Constitution did not specifically mention the topic of education. However, the 10th Amendment provided states with any powers not explicitly delegated to the federal government. This provided a solid argument that the responsibility of organizing and administering the public school system belonged to state governments (Riley, 1997). As such, this decentralized system gave states direct oversight over most aspects of education. Unlike the U.S. Constitution, most state constitutions specifically included educational provisions, giving them authority to create a state educational agency and pass laws governing their public schools. According to the United States Department of Education (2008), states typically executed most fiscal, regulatory, and administrative functions related to education including:

- Providing funding for public education in their state
- Licensing faculty and staff in public schools (elementary and secondary)
- Providing oversight and support to local education agencies
- Overseeing educational services for special populations
- Licensing all private and public schools, including institutions of higher education
- Electing or appointing of state and/or local boards of control
- Creating policy related to state standards, curriculum, textbooks, assessments, etc. at the elementary and secondary levels

It should be further noted that all state governments have been different as to what they regulate and what authority they give to local school districts. For example, some states might set graduation requirements or determine the curriculum and/or textbooks schools will use for certain subjects, while others give that control to the local school district. To learn more about the organization and structure of a specific state's educational department or agency, visit their website. Each state's department or agency in charge of public education and their website addresses are in Table 1.

Local School Districts

At the core of the U.S. education system have been the local school districts. Most states will delegate the responsibility of operating their public schools to these Local Education Associations (LEAs). These local entities have been responsible for managing the school buildings and personnel in their districts, creating local policies and procedures, enforcing state educational mandates and laws, and raising funds to operate their schools (United States Department of Education, 2008). In 2018, federal and state governments provided funding for only 55% of school expenditures, requiring local entities to cover the remaining 45% (United States Department of Education, 2020). These funds are usually obtained through taxes or school bonds. According to data provided by the NCES, there are over 13,000 local school districts in the United States (Riser-Kositsky, 2019).

Each of these school districts is governed by a ***local school board***, a group of elected or appointed community members responsible for creating policy and making decisions. Some of these responsibilities may involve approving an annual budget, hiring administrators and faculty, and adopting curriculum, all while making sure their schools are following regulations and standards set forth by their state board of education. While they are heavily involved in the policymaking and administrative oversight within their districts, these boards are often not involved in the daily operation of the school district. Most hire a ***superintendent***, an administrator in charge of a school district, to carry out these tasks. Superintendents are often responsible for developing the district budget and managing their administrative staff, including building principals.

A ***principal*** is a school administrator in charge of managing the overall operations of an elementary, middle, or high school building. Building principals are the first instructional leaders responsible for working with faculty and staff to increase the academic achievement of the student body. This is all while implementing the policies and procedures set forth by the local school board, including overseeing the finances of their school. Depending on the size of their student population, they might also be responsible for supervising extracurricular or athletic events, managing student discipline, and hiring, supervising, and evaluating faculty and staff. Some schools employ additional building administrators, such as an assistant principal or a dean of students, to manage some of these responsibilities. Finally, at the heart of each school building are the individual classroom teachers. Classroom teachers are the individuals responsible for providing daily instruction and managing the learning environment for their students, among a multitude of other duties. The specific roles and responsibilities of the SBAE teacher are in Chapter 1.

Table 1. List of department or agency administering public education in each state and their website address

Name of Department or Agency	Website
Alabama Department of Education	https://www.alabamaachieves.org/
Alaska Department of Education & Early Development	https://education.alaska.gov/
Arizona Department of Education	https://www.azed.gov/
Arkansas Department of Education	https://dese.ade.arkansas.gov/
California Department of Education	https://www.cde.ca.gov/
Colorado Department of Education	http://www.cde.state.co.us/
Connecticut Department of Education	https://portal.ct.gov/SDE
Delaware Department of Education	https://www.doe.k12.de.us/
Florida Department of Education	https://www.fldoe.org/
Georgia Department of Education	https://www.gadoe.org/Pages/Home.aspx
Idaho State Department of Education	https://www.sde.idaho.gov/
Illinois State Board of Education	https://www.isbe.net/Pages/Home.aspx
Indiana Department of Education	https://www.doe.in.gov/
Iowa Department of Education	https://www.educateiowa.gov/
Kansas State Department of Education	https://www.ksde.org/
Kentucky Department of Education	https://education.ky.gov/
Louisiana Department of Education	https://www.louisianabelieves.com/
Maine Department of Education	https://www.maine.gov/doe/home
Maryland Department of Education	https://marylandpublicschools.org/
Massachusetts Department of Elementary & Secondary Education	https://www.mass.gov/orgs/executive-office-of-education
Michigan Department of Education	https://www.michigan.gov/mde
Minnesota Department of Education	https://education.mn.gov/MDE/index.html
Mississippi Department of Education	https://www.mdek12.org/
Missouri Department of Elementary & Secondary Education	https://dese.mo.gov/
Montana Office of Public Instruction	https://opi.mt.gov/
Nebraska Department of Education	https://www.education.ne.gov/
Nevada State Board of Education	http://www.doe.nv.gov
New Hampshire Department of Education	https://www.education.nh.gov/
New Jersey Department of Education	https://www.state.nj.us/education/
New Mexico Department of Education	https://webnew.ped.state.nm.us/
New York Department of Education	http://www.nysed.gov/
North Carolina Department of Public Instruction	https://www.dpi.nc.gov
North Dakota Department of Public Instruction	https://www.nd.gov/dpi/
Ohio Department of Education	https://education.ohio.gov/
Oklahoma Department of Education	https://sde.ok.gov/
Oregon Department of Education	https://www.oregon.gov/ODE/
Pennsylvania Department of Education	https://www.education.pa.gov/
Rhode Island Department of Elementary & Secondary Education	https://www.ride.ri.gov/
South Carolina Department of Education	https://ed.sc.gov/
South Dakota Department of Education	https://doe.sd.gov/
Tennessee Department of Education	https://www.tn.gov/education.html
Texas Education Agency	https://tea.texas.gov/
Utah State Office of Education	https://www.schools.utah.gov/
Vermont Agency of Education	https://education.vermont.gov/
Virginia Department of Education	https://www.doe.virginia.gov/
Washington Office of Superintendent of Public Instruction	https://www.k12.wa.us/
West Virginia Department of Education	https://wvde.us/
Wisconsin Department of Public Instruction	https://dpi.wi.gov/
Wyoming Department of Education	https://edu.wyoming.gov/

SBAE WITHIN THE EDUCATIONAL SYSTEM

SBAE has been situated with a structured, hierarchal system since its early beginnings. For example, SBAE has been considered a ***career and technical education*** (CTE) program that operates in a larger system in the U.S. Further, although the aims and purpose of SBAE are similar across the nation, its delivery differs dramatically from states and at the local level. Consequently, it is critical to understand how each of these systems functions and influences SBAE.

Career and Technical Education

CTE is an extensive system that has provided courses across a range of grade levels since its early beginnings. Nearly 98% of U.S. secondary schools offer credits in CTE (Gray & Lewis, 2018). Although CTE courses have been offered in a broad range of educational institution types, all CTE courses align within three major course types: (1) family and consumer sciences education, (2) specific market labor preparation, and (3) general labor (Gordon & Schultz, 2020). As a result, CTE courses attract a broad range of students with varying personal and career interests. The delivery and structure of CTE has evolved over time; however, the Association for Career and Technical Education (ACTE) (2021) represents seven major areas of study in CTE:

1. Agricultural Education – Courses include agricultural communications, agricultural business, agricultural power and technology, animal science, horticulture, leadership, and more.
2. Business Education – Courses include accounting, computer skills, grant writing, small business management, and more.
3. Marketing Education – Courses include advertising and promotion, sales techniques, marketing principles and problems, and more.
4. Family and Consumer Sciences Education – Courses include culinary arts, food science, hospitality, tourism, and more.
5. Trade and Industrial Education – Courses include carpentry and welding, cosmetology, automotive technology, and more.
6. Health Science Education – Courses include exercise science, healthcare diagnostics, nursing science, and more.
7. Engineering and Technology Education – Courses include energy and power foundations, technology applications, manufacturing systems, and more.

The Office of Career, Technical, and Adult Education (OCTAE) in the U.S. Department of Education is responsible for providing federal oversight of CTE programs. Specifically, the Division of Academic and Technical Education within OCTAE: (1) establishes a grant formula for funding CTE activities, (2) ensures program quality and accountability, and (3) creates initiatives to ensure CTE programs remain rigorous (Gordon & Schultz, 2020). Although CTE credits can be obtained as a post-secondary educational option, students must choose to earn credits in CTE courses rather than other elective options in high school. This is critical to CTE enrollment trends because some students may be interested in CTE courses but decide to enroll in more advanced coursework that they perceive better prepares them for college or their future career. As a result, some students may earn credits in a CTE course but never complete a curricular pathway or area of focus. Therefore, many CTE courses have no prerequisite requirements so that students can enroll in courses based on their interests. On this point, researchers have reported that as students completed more CTE credits, a significantly higher contribution was made to the U.S. labor market (Doughtery et al., 2019). Therefore, CTE coursework has a distinctly positive effect on students, especially those who have earned credits in SBAE.

Agricultural Education at the National Level

SBAE has been integral in the U.S. since its early formation. The first SBAE teacher was employed in Elyria, Ohio, in 1902 (Hamlin, 1956). However, SBAE was not formalized across the U.S. until the passage of the Smith-Hughes Act (1917), which provided federal funding for the teaching of agriculture, industrial trades, and home economics. Since that time, SBAE has consistently adapted to meet the needs of society. For example, SBAE was originally termed *vocational agriculture*; however, the National Research Council (1988) called for the discipline to change its name to *agricultural education* to promote innovations and opportunities that were available in the field through an expanded scope. As a result of this change, course offerings in SBAE began to expand to more advanced topics in animal science, biotechnology, horticulture, environmental systems, and more.

Although agricultural education has primarily been delivered in public school settings, there are also programs available across various contexts such as charter schools, home schools, magnet schools, and vocational institutions. Despite the various contexts in which agricultural education occurs, programs have been delivered systematically using an integrated three-circle model that includes learning experiences through (1) classroom and laboratory management, (2) supervised agricultural experiences (SAEs), and (3) the National FFA Organization (Croom, 2008). Classroom and laboratory management occur when SBAE instructors teach agricultural-related concepts in traditional classrooms as well as in other learning spaces such as agricultural mechanics laboratories, greenhouses, and food processing facilities. Meanwhile, SAEs occur when students engage in a structured experience in which they apply concepts learned in the classroom through project-based learning. Finally, the National FFA Organization provides opportunities for students to engage in personal and professional growth through career development events, leadership programming, and various award opportunities (Croom, 2008).

As a result of these learning experiences, over 1 million students were taught by more than 12,000 teachers in the U.S. in 2020 (National Association for Agricultural Education, 2020). In its delivery, SBAE fulfills a two-part mission: (1) the promotion of agricultural literacy, and (2) a skilled agricultural workforce (Roberts & Ball, 2009). As a result, the discipline has expanded considerably in its intent and scope (Roberts & Ball, 2009). To promote clarity regarding opportunities and expectations, the Agriculture, Food, and Natural Resources (AFNR) standards were created to promote rigor during the programmatic delivery of SBAE. The AFNR standards were organized by eight career pathways and provided standards for courses, units, and learning outcomes. The eight career pathways include (1) agribusiness systems, (2) animal systems, (3) biotechnology systems, (4) environmental service, (5) food products and processing, (6) natural resource systems, (7) plant systems, and (8) power, structural, and technical systems (National Council for Agricultural Education, 2015).

On a national level, SBAE is guided by the National Council for Agricultural Education (The Council). This group comprised of 14 directors provides leadership to the discipline on a range of issues. For example, The Council was critical in leading the effort for the development of AFNR Standards as well as a re-envisioning of SAEs through an initiative called SAE for All in 2017. Through these initiatives, SBAE has sought to increase the rigor of its programmatic delivery to ensure it is continuing to meet the needs of society. The Bureau of Labor Statistics (2018) projected that by 2026 SBAE teacher positions are expected rise by 8%. Further, jobs across agriculture are expected to increase in demand as the world grapples with providing adequate food, fiber, and fresh water for a growing population (Bureau of Labor Statistics, 2018). As a result of these changes, SBAE must not only be delivered effectively nationally, but also at the state level.

State Agricultural Education

Although the federal government provides funding for SBAE, they are not responsible for overseeing or evaluating its delivery. Instead, this responsibility was delegated to individual states through the Smith-Hughes Act (1917). Each state is responsible for ensuring that policies are upheld and delivered effectively. However, after the passage of the Perkins Vocational Education Act (1984), a shift occurred that began to change the way in which states approached programming. On this point, Phipps et al. (2008) explained that states began to seek to ". . .strengthe[n] the academic foundations of vocational education courses by applying mathematics and science principles in vocational education" (p. 31). This new priority was further stressed through the bill's reauthorization in 1990. With this new focus, state

Table 2. States with non-traditional agricultural education and FFA state supervision models

States with Executive Secretary not Affiliated with the Department of Education	**States with No Supervisor in the Department of Education**	**States with No Staff in the Department of Education**
Alaska	Alaska	Alaska
Colorado	Colorado	Colorado
Connecticut	Indiana	Louisiana
Delaware	Kentucky	Maryland
Florida	Louisiana	Massachusetts
Idaho	Maryland	New Hampshire
Illinois	Massachusetts	New Jersey
Indiana	Montana	New Mexico
Iowa	New Hampshire	New York
Louisiana	New Jersey	North Carolina
Maine	New Mexico	Pennsylvania
Maryland	New York	Rhode Island
Massachusetts	North Carolina	South Carolina
Michigan	Oklahoma	Vermont
Montana	Pennsylvania	Virgin Islands
Nevada	Rhode Island	Wyoming
New Hampshire	South Carolina	
New Jersey	South Dakota	
New York	Texas	
North Carolina	Vermont	
Ohio	Virgin Islands	
Oregon	Wyoming	
Pennsylvania		
Rhode Island		
South Carolina		
Texas		
Vermont		
Virgin Islands		
Virginia		
Wyoming		

Note. Reprinted with permission from "Reflections on reform: A historical narrative of the relocation of the Louisiana FFA Office" by C. A. LeJeune and R. Roberts, 2020, Journal of Agricultural Education, *61*(3), 277-278.

agricultural education leaders faced budget cuts in key areas that resulted in "…staff reduction and lost office space" (LeJeune & Roberts, 2020, p. 277).

As a result of these changes, some states removed agricultural education from their state Department of Education (DOE). For example, after losing funding, Louisiana and North Carolina came under the supervision of their state's 1862 land-grant university. These arrangements are not unique. In fact, 22 states have no state agricultural leader in their DOE. Further, 16 states have no state agricultural education staff in the DOE (LeJeune & Roberts, 2020). Table 2 provides a description of the states that have a non-traditional agricultural education and FFA supervision models.

Because of the unique delivery of supervision in agricultural education across each state, the local school districts have a prominent role in supervising SBAE. In the U.S., local school board and administrators have significant authority in drafting and upholding policy. Therefore, it is critical to understand the structure and role of local agricultural education programs.

Local Agricultural Education Programs

At the local level, SBAE varies among states and communities, especially regarding course offerings and learning experiences for students. This is because each SBAE program is designed to fulfill the needs of the communities they serve. Therefore, each local program must fulfill the expectations and policies mandated by the state agency, most commonly their department or state board of education, which oversees SBAE.

At the local level, SBAE programs are responsible for delivering SBAE's three-circle model as well as local, district, and state requirements. Because school sizes differ, the number of teachers in the program also varies considerably. For example, many rural schools only have one agricultural education teacher while suburban and urban spaces can have multiple individuals responsible for instruction and FFA advisement. Despite these differences, most school districts report to their local school administration regarding day-to-day functions. However, the local school board maintains legal authority to enact changes and make hiring decisions.

STATE LEADERSHIP IN AGRICULTURAL EDUCATION

Because of its complexity, it is critical to understand how leadership in SBAE has taken root and the ways in which its complexity has led to the diminished roles of some leadership positions over time. Further, this section will also describe how professional associations have helped the discipline navigate challenges inherent to changes in society and advocate for quality resources and opportunities for students.

State FFA Associations

Although FFA is a national organization, one of its strengths is its delivery at the state and local levels. At this level, the state FFA associations sponsor a range of activities, including (a) leadership development activities, (b) competitions, and (c) conventions. State FFA Associations are also responsible for ensuring that students have met the requirements to obtain their State FFA Degree. States differ regarding their programmatic delivery and opportunities; however, each state usually has at least one individual who has been designated as their leader for the state – the most common title for this individual is the

State FFA Advisor. In many states, another individual, often called the State FFA Executive Secretary, is usually responsible for providing guidance to the state FFA officer teams. They also provide leadership for FFA programming at the state level. However, it is important to note that the titles and duties of these individuals vary from state to state. Therefore, it may be helpful to explain the complicated history of state agricultural education staff.

Agricultural Education State Staff

With the passage of the Smith-Hughes Act (1917), the supervision of SBAE became mandated at the state level. As a result, state agricultural education staff job positions were created in many states to fulfill this role (Herring, 1999). Although the roles of state agricultural education staff have varied from state to state, the authority of these roles was particularly acute in the 1960s (Hillison, 1999). During this period, state agricultural education staff not only provided supervision to SBAE teachers but also inspected their quality of programs and teachers (Herring, 1999). The distinction between supervision and inspection has been critical in most states regarding the influence of state agricultural education staff.

In the context of SBAE, ***supervision*** has been defined as providing support to teachers to ensure their professional growth and success (Herring, 1999). Meanwhile, ***inspection*** referenced the power to evaluate SBAE programs and teachers regarding the degree to which they met established criteria and, in some cases, issue disciplinary action if standards were not met (Hillison, 1999). As a result of the dual roles of state agricultural education staff, Graham and Edwards (2018) explained that confusion emerged "among administrative and supervisory officers as to their authority" (p. 48). Consequently, SBAE teachers have historically been confused about whether they should follow the instructions of state agricultural education staff or their local school administration. This confusion has diminished the influence of state agricultural education staff in some states (Graham & Edwards, 2018).

Despite this, state agricultural staff continue to provide critical support for SBAE teachers throughout the U.S. For example, state agricultural staff are often responsible for sharing teaching resources, organizing professional development, and coordinating state-level activities for FFA. Although the positions and titles of state agricultural education staff differ among states, their roles remain critical to the success of SBAE in the U.S.

Agricultural Education Teacher Associations

To further support SBAE, agricultural education teacher associations emerged. The most prominent teacher association for the discipline is the National Association of Agricultural Educators (NAAE). NAAE has over 9,000 members across the U.S., and its members engage in SBAE at each level – from middle school teachers to faculty at post-secondary institutions (NAAE, n.d.). As part of its mission, NAAE "provide[s] professional development…[and also] works to recruit and retain agricultural educators in the profession" (NAAE, n.d., para. 2). Another unique aspect of this professional association is that members gain over $100,000 of professional liability insurance coverage. Further, they are also eligible for a $60,000 term life insurance policy (NAAE, n.d.).

NAAE is also often organized on the state level. For example, each state typically elects a slate of officers to fulfill the organization's mission. Further, most states hold an annual meeting in which members engage in professional development to discuss critical issues and create strategies to address them through advocacy, policy, or unique programming opportunities. The list of the state affiliates can be found at https://www.naae.org/affiliates/states/index.cfm.

Other Agricultural Education Groups

Besides NAAE, SBAE instructors can engage in a range of other professional organizations. These organizations have a range of intents and purposes. Below is an overview of the various professional organizations agricultural educators can belong to. Of these professional societies, six have the most prominent connection to agricultural education: (1) Association for Career and Technical Education, (2) American Association for Agricultural Education, (3) National Association of Supervisors for Agricultural Education, (4) National FFA Alumni and Supporters, and (5) the National Young Farmer Educational Association. Table 3 provides an overview of the intent of each professional association.

Organization	Intent
Association for Career and Technical Education (ACTE) – Agricultural Education Division	ACTE is the largest Career and Technical Education (CTE) professional association. It offers members a range of professional opportunities for educators who seek to prepare youth for future careers in CTE areas.
American Association for Agricultural Education	AAAE focuses on studying and promoting the teaching and learning of agriculture. The disciplines represented include (1) teacher education, (2) agricultural communications, (3) agricultural education, and (4) agricultural leadership.
National Association of Supervisors for Agricultural Education	NASAE is an organization primarily for state agricultural education staff to exchange ideas, debate relevant issues, and create new professional materials to support SBAE.
National FFA Alumni and Supporters	This organization seeks to promote the National FFA Organization through service and support. The National FFA Alumni and Supporters is open to individuals who have been past members, including collegiate students and others who support the organization's aims and causes.
National Young Farmer Educational Association	NYFEA provides leadership development, career education, and service opportunities for young professionals engaged in the agricultural industry.

Agricultural Education Teacher Educators

The earliest versions of teacher education began at Congressional District Agricultural Schools in Alabama (1889), Georgia (1906), and Virginia (1908). The intent of these institutions "was to provide instruction, in a comprehensive school setting for agricultural education and home economics education" (Hillison, 1998, p. 2). However, the Congressional District Schools also began to prepare teachers. After the passage of the Smith-Hughes Act (1917), teacher education programs began to appear more frequently in higher educational institutions, especially in the colleges of agriculture at land-grant universities (Brinkley, 1977).

Throughout history, the intent and purpose of teacher education in agriculture has been heavily debated. For example, some have advanced the notion that teacher educators should focus primarily on improving the technical knowledge of preservice teachers (Hillison, 1998). Meanwhile, others have argued that the central focus should be placed on the development of preservice teachers' pedagogical skills (Hillison, 1998). Today, many teacher educators have sought to blend both technical and pedagogical skill development. Therefore, the role of teacher educators is to provide instruction to preservice teachers regarding how to teach technical agricultural concepts effectively. Often, this involves exposing preservice teachers to a range of pedagogical approaches such as experiential learning, problem-solving, project-based instruction, and more. Teacher educators in agriculture are also often responsible for facilitating research and professional service to advance the discipline for future generations.

ORGANIZATION AND ADMINISTRATION OF THE FFA

Since its inception in 1928, the National FFA Organization has been the premier youth leadership organization for students enrolled in SBAE programs. As one of the three integral components of a SBAE program (with Classroom & Laboratory Instruction and Supervised Agricultural Experience programs being the other two), the FFA organization "helps members develop premier leadership, personal growth, and career success" (National FFA Organization, 2021c, p. 2). With over 700,000 members, the National FFA Organization is among the largest student-led youth organizations in the United States with more than 8,000 chapters, in all 50 states and two U.S. territories (National FFA Organization, 2021c).

The FFA organization is structured with three primary levels: (1) The National FFA Organization, (2) State FFA Associations, and (3) Local FFA Chapters (see Figure 1). Some states even have additional levels between the state and local levels. Just as state governments and local education agency's structure and execute their educational policies and programs differently, the ways in which state FFA associations and local FFA chapters organize and govern themselves are also quite unique. In this section, the organizational and administrative structure of the FFA at the national, state, and local levels are explored.

Figure 1. Structure of the FFA organization
Note. Reprinted with permission from the National FFA Organization.

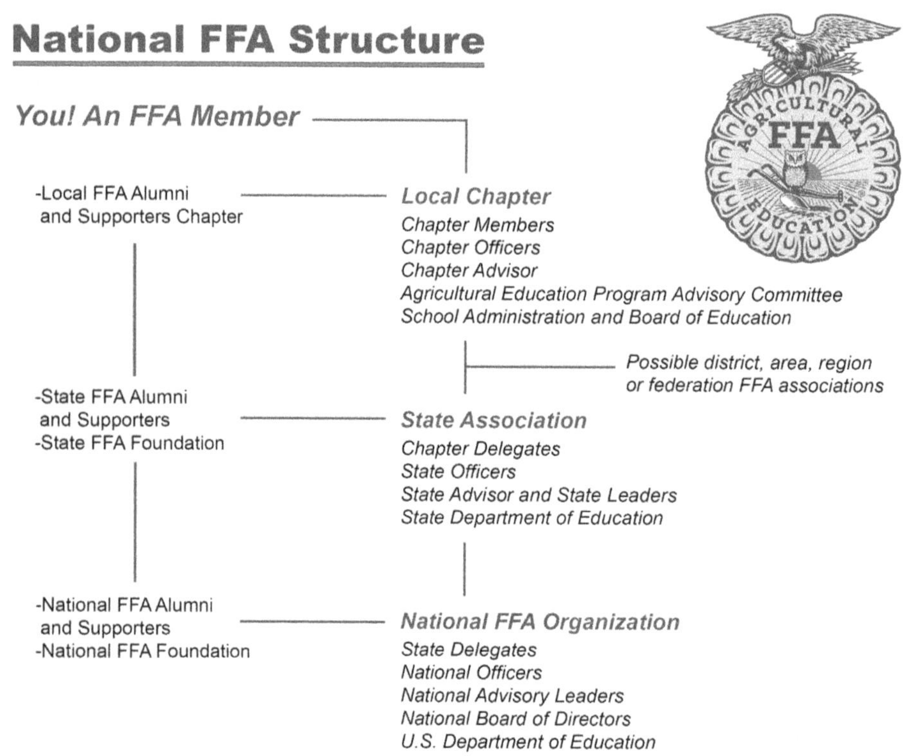

National FFA Organization

Following the guidelines of a federal charter and the policies and procedures set forth in the National FFA Constitution and By-laws, the National FFA Organization is led by a 14-member board of directors and the six student members of the National FFA Officer team (National FFA Organization, 2021a). The National FFA Board of Directors consists of the National FFA advisor, National FFA Executive Secretary, the National FFA Treasurer, and representatives from the following agricultural education stakeholder groups: (1) the United States Department of Education, (2) the National Council for Agricultural Education, (3) Agricultural Education State Staff, (4) Agricultural Education Teacher Educators, (5) Agricultural Education teachers, and (6) the agricultural industry (National FFA Organization, 2021b). The National FFA Officer team is composed of a President, Secretary, and four Vice Presidents representing each of the four FFA regions- Western, Southern, Central, and Eastern (National FFA Organization, 2021c). Beyond ensuring representation from different regions of the country for the National FFA Officers and several committees and boards, the National FFA regions serve no other organizational or administrative function. Collectively, these groups provide leadership and create policy initiatives for the National FFA organization.

The decisions and policies made by these entities are then implemented by the staff of the National FFA Center. Led by an executive management team, the staff of the National FFA Center administers and executes national-level programming, awards, and events. These national-level activities are supported by groups like the National FFA Foundation and the National FFA Alumni and Supporters organizations.

State FFA Associations

According to the National FFA Organization (2021c), the FFA at the national level strictly serves as a resource and support organization and does not "select, control or supervise state FFA association, local chapter or individual member activities" (p. 74). To receive a charter from National FFA Organization, a state association must be "composed of not less than four chapters ... and hav[e] a total student membership of not less than 125 members" (National FFA Organization, 2021c, p. 74). Individual state FFA associations operate independently from the National FFA Organization but must function within the guidelines of the National FFA Constitution and By-laws. A ***state FFA association*** is the state's basic organizational unit consisting of local FFA chapters. State associations are responsible for implementing their state's FFA programming such as a state convention, leadership camps and conferences, competitive events, and award programs. States are also responsible for handling FFA membership and chartering new chapters within their state associations. Similar to the national organization, state associations are often led by an executive staff and/or board of directors as well as a group of FFA officers.

The National FFA Organization Constitution indicates the student officers of a state FFA association shall be a president, vice president, secretary, treasurer, reporter, sentinel, and advisor, however, also granting each association the ability to elect officers as they deem necessary (National FFA Organization, 2021c). Further, the constitution signifies that each state supervisor for agricultural education shall assume the duties of the state FFA advisor (National FFA Organization, 2021c). While most states operate under this organizational structure, others elect to modify their structure to meet the needs of their state membership. Many states employ or elect a FFA executive secretary or director and/or other executive staff members to administer the programs and activities of their state associations. The duties of these individuals often include working with their state FFA officers and planning and executing activities of the association; however, specific duties of those individuals vary from state to state. Regardless, most will still report to some type of governing body, such as a state FFA board of directors.

Local FFA Chapters

Just as the local school district is at the core of the educational system, the local FFA chapter is the heart of the FFA organization. A local FFA chapter is the basic organizational unit of individual FFA members at the middle and high school levels. As previously mentioned, chartering of local FFA chapters is the responsibility of the state FFA associations. To obtain a charter, a school district must have an approved agricultural education program according to standards set forth by their state department of education. States may also possess additional requirements as set forth by their individual state associations. Once a local FFA chapter is established it must meet yearly requirements to remain in good standing with both the state and national organizations. These typically include paying dues and turning in required paperwork.

At the local level, FFA chapters are led by a group of student FFA officers. The six constitutional officers of the local FFA chapter are the president, vice president, secretary, treasurer, reporter, and the sentinel (National FFA Organization, 2021c). Just like the state associations, local FFA chapters have flexibility in the student officers they elect. At the local level, the agricultural education teacher(s) often serves as the local FFA advisor(s). The FFA advisor(s) works with both the chapter FFA officers and committees to execute the local chapter program of activities. A Program of Activities (POA) is the plan for all FFA activities they conduct during the year. It is within this working document that local FFA chapters develop goals and document results for their yearly activities.

Additional Structural Levels within States

Although state FFA associations are required to operate within guidelines set forth in the National FFA Constitution and Bylaws, they may also establish additional structural levels within their state for leadership, competition, award, and/or programming purposes (National FFA Organization, 2021d). These additional association levels have been positioned between the state and local levels and vary from state to state. They have often been referred to as districts, areas, regions, zones, or federations. Some states may only have one additional structural level. Examples of these states would include Alabama, Colorado, Connecticut, Idaho, Kansas, Minnesota, Montana, New Mexico, New York, North Dakota, Oregon, South Dakota, and Washington, which all use districts as their state-specified division title.

States with a larger geographic area, or higher membership numbers, may have additional structural levels beneath those. For example, the state of Illinois is divided into five districts, then further into 25 sections (five sections per district). Each of the 25 sections has historically conducted competitive events and are led by a section FFA officer team. Other states, such as Louisiana and Texas, have been organized into areas and further into districts. Mississippi and North Carolina have used the terms districts and federations to communicate their organizational structure. Meanwhile, California has used regions, they have been divided further into sections. Similar to Illinois, these sub-levels often have held competitions, and sometimes have been led by a group of FFA officers. It is up to each individual state association as to how they have chosen to organizationally structure. Regardless of what they are called, states often create these boundaries based on geographic area, while considering the number of chapters and the FFA membership in each of these structural levels. To see an example of the boundaries of two different states view Figure 2.

Figure 2. Map of organizational structural levels within the Illinois and Louisiana FFA associations
Note. Reprinted with permission from the Illinois FFA and Louisiana FFA Associations.

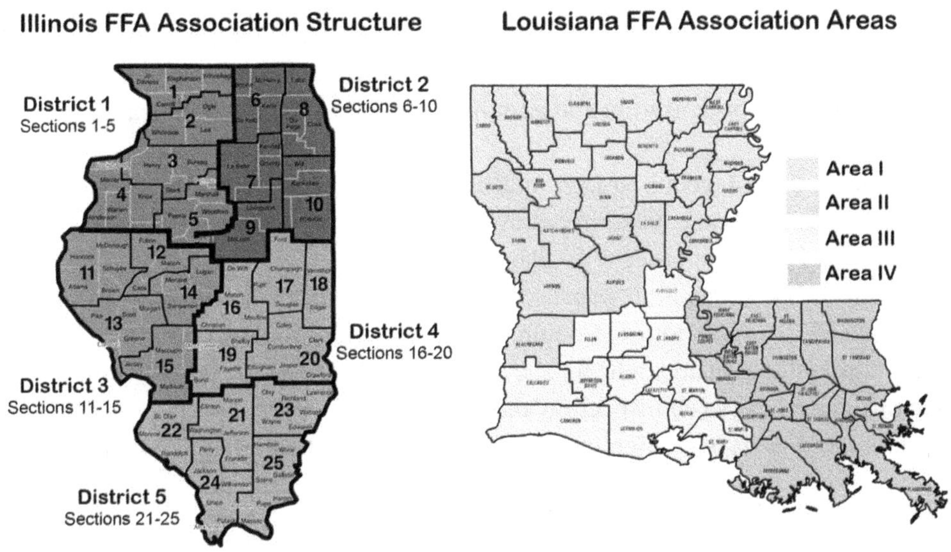

To learn more information about a specific organizational structure within a state, visit their websites by clicking on the following link: https://www.ffa.org/state-ffa-associations/

SUMMARY

Although the U.S. Constitution did not specifically mention the topic of education, the 10th Amendment provided states with any powers not explicitly delegated to the federal government. As such, although the federal government has provided oversight, each individual state uniquely delivers and assesses the effectiveness of teaching and learning. Similarly, although the National FFA organization and other national agricultural education entities have provided an effective, workable structure, the ways school-based agricultural education (SBAE) and FFA have been organized and administered in states and local spaces has varied considerably. These variations have allowed problems to emerge with the supervision of SBAE instructors. As a result, shifts occurred regarding the roles of state agricultural education staff in the U.S. Therefore, SBAE instructors and programs have often relied on support and guidance at the local level. Consequently, local agricultural programs have remained at the core of SBAE's continued success and progress.

REVIEW QUESTIONS

1. Explain the general roles and responsibilities of local, state, and federal governments as they pertain to developing and implementing educational policy and procedures.

2. How does school-based agricultural education fit into the public school system at the local, state, and national levels? How about within Career and Technical Education?
3. All states differ in how they structure and govern school-based agricultural education programs. Visit the website of your state's department or agency responsible for administering public education (see Table 1) and search for "agricultural education." What information and/or resources did you find regarding the organization and administration of agricultural education programs in your state?
4. Why might it be advantageous to have multiple state-level agencies, organizations, and groups responsible for agricultural education policy and program development in a state? Could there be any disadvantages to this approach? Explain.
5. Visit your state's FFA association website (https://www.ffa.org/state-ffa-associations/). Determine how your state FFA association is structured (including additional structural levels) and who provides leadership (student and adult) for the state organization.
6. Describe differences between inspection and supervision?
7. What benefits do SBAE instructors receive from joining professional organizations?

REFERENCES

Association for Career and Technical Education (ACTE). (2021). *What is CTE?* Author. https://www.acteonline.org/why-cte/what-is-cte/

Brinkley, H. R. (1977). Teacher education programs in agricultural education should be located in colleges of education. *The Journal of the American Association of Teachers in Agriculture*, *18*(3), 2–26. https://www.jae-online.org/attachments/article/1159/Binkley,H_Vol18_3_2,4-6.pdf

Bureau of Labor Statistics. (2018). *Occupational outlook handbook*. Author. https://www.bls.gov/ooh/#tab-6

Carl D. Perkins Vocational Education Act of 1984, *20 U.S.C. § 2301* et seq.

Croom, D. B. (2008). The development of the integrated three-component model of agricultural education. *Journal of Agricultural Education*, *49*(1), 110–120. doi:10.5032/jae.2008.01110

Dougherty, S. M., Gottfried, M. A., & Sublett, C. (2019). Does increasing career and technical education coursework in high school boost educational attainment and labor market outcomes? *Journal of Education Finance*, *44*(4), 423–447. https://muse.jhu.edu/article/738163/pdf

Gordon, H. R., & Schultz, D. (2020). *The history and growth of career and technical education in America*. Waveland Press.

Graham, C. M., & Edwards, M. C. (2018). Supervision of school-based, agricultural education: A historical review. *Journal of Research in Technical Careers*, *2*(1), 42–51. doi:10.9741/2578-2118.1030

Gray, L., & Lewis, L. (2018). *Career and technical education programs in public school districts: 2016-17. First look*. National Center for Education Statistics. https://files.eric.ed.gov/fulltext/ED582132.pdf

Hamlin, H. M. (1956). Fifty years of progress in agricultural education. *American Vocational Journal, 31*(9), 39–46.

Herring, D. R. (1999). Is state supervision of agricultural education needed in the 21st century? A teacher educator's perspective. *The Agricultural Education Magazine, 71*(6), 12–21.

Hillison, J. (1989). Congressional district schools: Forerunner of federally supported vocational agriculture. *Journal of Agricultural Education, 30*(4), 7–13. doi:10.5032/jae.1989.04007

Hillison, J. (1998). The role of the agricultural education teacher educator yesterday, today, and tomorrow. *Journal of Agricultural Education, 39*(1), 1–7. doi:10.5032/jae.1998.01001

Hillison, J. (1999). Whatever happened to the supervisor of the Smith-Hughes man? *Journal of Agricultural Education, 40*(2), 55–63. doi:10.5032/jae.1999.02055

LeJeune, C. A., & Roberts, R. (2020). Reflections on reform: A historical narrative of the relocation of the Louisiana FFA Office. *Journal of Agricultural Education, 61*(3), 276–293. doi:10.5032/jae.2020.03276

National F. F. A. Organization. (2021a). *Constitution, bylaws, and federal charter*. Author. https://www.ffa.org/constitution-bylaws/

National F. F. A. Organization. (2021b). *National FFA leadership*. Author. https://www.ffa.org/our-leadership/

National F. F. A. Organization. (2021c). *Official FFA manual*. Author. www.ffa.org/official-manual/

National F. F. A. Organization. (2021d). *State FFA associations*. Author. https://www.ffa.org/state-ffa-associations/

National Association for Agricultural Education. (n.d.). *NAAE membership benefits, service, and programs*. Author. https://www.naae.org/resources/index.cfm

National Council for Agricultural Education. (2015). *National AFNR content standards*. Author. https://thecouncil.ffa.org/afnr/

National Research Council. (1988). *Understanding agriculture: New directions for education*. Author.

Phipps, L. J., Osborne, E. W., Dyer, J. E., & Ball, A. (2008). *Handbook on agricultural education in public schools* (6th ed.). Thomson Delmar Learning.

Riley, R. (1997). The role of the federal government in education- Supporting a national desire for support for state and local education. *Saint Louis University Public Law Review, 17*(1), 29–54. https://scholarship.law.slu.edu/cgi/viewcontent.cgi?article=1479&context=plr

Riser-Kositsky, M. (2019, January). *Education statistics: Facts about American schools*. https://www.edweek.org/leadership/education-statistics-facts-about-american-schools/2019/01

Roberts, T. G., & Ball, A. L. (2009). Secondary agricultural science a content and context for teaching. *Journal of Agricultural Education, 50*(1), 81–91. doi:10.5032/jae.2009.01081

Smith-Hughes Act of 1917, 65 U.S.C. § 2301 *et seq.*

United States Department of Education. (2008, February). *Organization of U.S. education.* Https://Www2.Ed.Gov/About/Offices/List/Ous/International/Usnei/Us/Edlite-Org-Us.Html

United States Department of Education. (2010, February). *What we do.* https://www2.ed.gov/about/what-we-do.html

United States Department of Education. (2014, September). *10 facts about k-12 education funding.* https://www2.ed.gov/about/overview/fed/10facts/index.html

United States Department of Education. (2021, June). *The federal role in education.* https://www2.ed.gov/about/overview/fed/role.html

KEY TERMS AND DEFINITIONS

Career and Technical Education (CTE): Schools, programs, or educational institutions that focus on skills and knowledge for specific careers.

Inspection: The power of state agricultural education staff to evaluate SBAE programs and teachers regarding the degree to which they met established criteria and, in some cases, issue disciplinary action if standards were not met.

Local Education Agency (LEA): A public board of education or other public authority within a state that maintains administrative control or performs a service function for public schools.

Principal: An administrator in charge of managing the overall operations of an elementary, middle, or high school building.

Public School: A school supported by public funding, usually tax dollars.

School Board: The governing body of elected or appointed individuals responsible for oversight of a local school district.

State FFA Association: A state's basic organizational unit consisting of local FFA chapters.

Superintendent: An administrator in charge of a school district.

Supervision: Providing support to teachers to ensure their professional growth and success.

United States Department of Education (ED): Agency of the U.S. federal government responsible for establishing federal educational policy and coordinating federal funding for education.

Chapter 3
Promoting the Relationship Between School-Based Agricultural Education and the Community

Aaron J. McKim
Michigan State University, USA

Sallie McHugh
Abraham Baldwin Agricultural College, USA

ABSTRACT

This chapter explores the past, current, and future relationship between agricultural education programs and the communities in which programs operate. Throughout the chapter, opportunities to leverage communities to optimize the design and operation of an agricultural program are emphasized. Specifically, this chapter explores the development and utilization of a community profile, methods for identifying community needs and structuring a program to meet identified needs, state and national curriculum and program standards, and building and utilizing stakeholder groups. The chapter concludes with an opportunity to utilize the knowledge gained in the development of a program philosophy statement designed to guide the facilitation of an agricultural education program.

INTRODUCTION

In this chapter, readers will explore the role "community" plays within Agricultural Education including the three components of school based agricultural education- classroom/laboratory, supervised agricultural experience (SAE), and FFA. Before exploring the critical nature of community, investigating how this important concept has been operationalized is essential. In this chapter, community includes the area, individuals, businesses, industries, and opportunities within geographic proximity to the school. In addition, community represents the groups of individuals working in collaboration to support the agricultural

DOI: 10.4018/978-1-6684-3420-8.ch003

program, the academic department housing the program, and the school. Further, community includes local, state, and national groups of agricultural educators as well as groups of educators in associated disciplines (e.g., science educators, mathematics educators, etc.). Throughout the chapter, readers are introduced to these dimensions of community and how each dimension can positively impact an agricultural education program. Specifically, the objectives of this chapter are to (a) explore the historical perspective and importance of community-based agricultural education programs, (b) define the role and characteristics of community, (c) explore state and national curriculum and program standards, (d) investigate the role of relationships among teachers, school administration, and community in curriculum development, and (e) prepare a program philosophy statement for an agricultural education program.

HISTORICAL PERSPECTIVE AND IMPORTANCE OF COMMUNITY-BASED AGRICULTURAL EDUCATION PROGRAMS

Quick Review of the Evolution of American Education

To begin reviewing the evolution of agricultural education programs, readers must start with a quick review of how the educational system developed in the United States. In the early years of the United States, schooling existed for only a select few. Most children receiving educational services were white and were educated through a variety of methods including:

- Local schools organized by the community
- Schools requiring tuition to attend
- Boarding schools
- Dame schools- run by women in their homes
- Apprenticeships focused on work skills

Education was not free and open to all. The Founding Fathers recognized the importance of having an educated society to promote democracy. In the late 1780s, free public education initially began, but it took until the 1830s before the concept was fully developed across the nation.

In 1830, Horace Mann, a Massachusetts legislator and secretary of the state's board of education, began to champion the common school. Common schools focused on teaching the 3 R's (reading, writing, and arithmetic) as well as history, geography, grammar, and rhetoric. These schools were to be funded by the state, open to all children, and free of charge. The thought was that educating all children, from all socioeconomic levels, would help them to better themselves and their communities in the future. Initially, the idea of common schools was not universally accepted but, over time, gained acceptance from all.

In the 19th century, public schools (referenced previously as common schools), began to grow at a faster rate, especially in suburban areas. Over time, states began providing funding for universal public education. Figure 1 depicts how grade levels came online over time and the percentages of children who attended in the nation.

Figure 1. Development overview of elementary and high school education in the United States

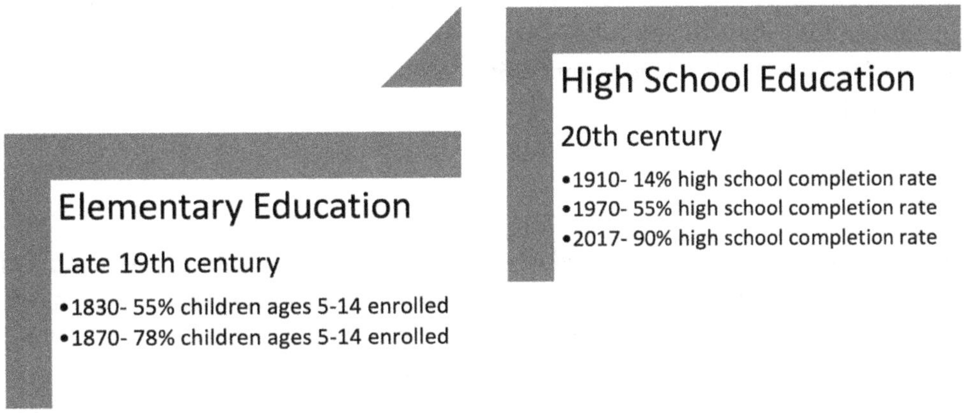

The Evolution of Agricultural Education

Morrill Act of 1862

The development of the educational system known today provides the backdrop to explore the genesis and growth of agricultural education. The first formal agricultural education teaching in the United States began in 1733 when Chief Tomochichi initiated daily conferences on agricultural practices for settlers. Through the passing of the Morrill Act in 1862, also known as the Land-Grant Act, federal support for agricultural education continued to grow and laid the foundation for this program area. First introduced in 1857 by Justin Morrill, this act wished to establish "colleges for the benefit of agriculture and mechanical arts" (Herren & Hillison, 1996). The final version of this bill gave 30,000 acres of public land for each member of Congress to use for their state's college. This land was often sold to provide funding for the establishment of these institutions. The Nelson Amendments were added to the Morrill Act in March 1907 to allow federal funds to be used by colleges of agriculture to prepare teachers to teach in the fields of agriculture and mechanical arts (Herren & Hillison, 1996). The need for secondary school agriculture education teachers was quickly growing in the nation.

Smith Hughes Act of 1917

With the onset of World War I, the United States realized its need to maintain a high position in the world market in relation to other industrialized countries. This brought an increased focus to vocational skills and the passage of the Smith Hughes National Vocational Education Act of 1917. The Smith Hughes Act provided funds to states with a matching requirement to pay salaries for teachers in the subjects of agriculture, home economics, and industrial education. This act helped to further define the roles of agricultural education in the United States and helped to provide even further teacher preparation services (Torres & Garton, 1991). The need for more agriculture teachers required the need for post-secondary institutions to prepare and produce these educators for the classroom. To meet this need, states already having public universities added a new agricultural and mechanical arts college (Kerr, 1987). In 1862, a total of 59 land grant institutions were developed (National Academy Press, 1995). In 1890, an additional Morrill Act was passed that required states to equalize opportunities for all land grant institutions and some states began removing race concerns from their admission criteria or established a separate land-grant institution for people of color. In 1994, the land grant system was expanded to include tribal colleges. As of 2021, there are 112 land grant institutions, 19 historically black, and 33 tribal (National Institute of Food and Agriculture, n.d.).

Public Law 740

Agricultural education continued to grow and develop. The FFA organization also continued to evolve with its official development in 1928 when it was established at the first national convention in Kansas City, Missouri. Another major milestone was the passing of a federal charter known as Public Law 81-740 by the United States Congress in 1950. The 2018 *FFA New Horizons* magazine stated:

This Charter serves as National FFA's articles of incorporation and as the organization's governance document. National FFA is the only charter organization that has stipulations within the charter mandating that the U.S. Department of Education selects or approves a majority of its members of the board of directors, including the national advisor, board chair and other board positions.

Vocational Education Acts

After the establishment of agriculture education in public schools, a series of Vocational Education Acts were passed to make greater strides for program growth and success. Find a quick review of these acts below.

- George Deen Act of 1936- added to Smith Hughes funding and loosened regulations for spending. Industrial training had slowed due to the Great Depression and lack of jobs for skilled workers.
- Vocational Education of 1946 (George Barden Act)- signed into law by President Harry Truman and provided additional funds for vocational education. Vocational education began growing again following the end of World War II.
- Vocational Education Act of 1963- continued to grow the vocational-technical education programs. These grants to states were to be used to maintain, improve, and develop vocational programs. This act helped to provide funds for work study programs.
- Vocational Education Act of 1968- continued to grow the vocational-technical education programs but transitioned focus from occupations to people. Advisory councils are required, and states must submit a yearly plan and complete an annual and five-year program plan. States must now meet requirements to receive federal funding.
- Vocational Education Act of 1976- continued to grow existing vocational education programs, develop new programs, and combat sex discrimination and sex stereotyping within programs.
- Carl D. Perkins Acts begins- vocational education acts took a name change in 1984 when the act changed to the Carl D. Perkins Act, named for long-time chairperson of the House Education and Labor Committee and a champion for vocational education. Below is a summary of the Perkins Act over the years.
    - *Perkins Act of 1984*- focused on improving job readiness, labor skills, and providing equal opportunities in vocational education programs. States were now required to have a State Council on Vocational Education as well as a National Council was now required. A rep-

resentative was appointed in each state to monitor sex equality within programs and eliminate sex bias and stereotyping. Also, this act linked vocational student organizations to the school-based course further supporting agricultural education and FFA.

- *Perkins Act of 1990-* a key focus of this Perkins Act was on technology preparation and integration of academic and vocational education studies. State grants were directed to special populations with emphasis on disadvantaged students. Programs were also developed to prepare students for transition from school to work. Program accountability continued to develop, and this Perkins Act was the largest amount of funding for vocational education to this date.
- *Perkins Vocational and Applied Technology Education Amendments of 1998 (Perkins III)-* focused on the continued development of high-quality programs that integrate academic and vocational education, present instruction on all industries, provide strong connections from secondary to post-secondary education, use technology, and meet needs of special populations.
- *Perkins Act of 2006 (Perkins IV)-* provided funding to secondary and postsecondary career and technical education programs to continue to develop more fully.
- *Perkins Act of 2018 (Perkins V)-* also known as the Strengthening Career and Technical Education for the 21st Century Act was signed into law by President Trump on July 13, 2018. It reauthorized the Perkins IV funding and added nearly $1.3 billion annually for career and technical education (CTE) across the nation. This act focuses on providing opportunities for every student to explore technical education programs and career pathways to earn credentials for professional use.

Education for All

Working with special needs children is a joy in the agriculture education classroom. Several acts and laws throughout history have helped to support and develop programs for these students. The Education for All Handicapped Children of 1975 was a major player by requiring public schools to provide free and appropriate education for *all* children ages 3-21. Funding was provided to agencies, on the state and local level, to support these students and programs. This act set to define the various handicaps such as vision impaired, special needs, etc. It is also stated that IEPs (Individualized Education Plans) should be developed jointly by teachers, parents, and other school representatives. The IEPs provide specific details on the student such as learning goals, objectives, services needed, date and length of services, and an evaluation process.

Another act that significantly impacted students with special needs was the Americans with Disabilities Act (ADA) of 1986. This act was not limited to the school sector but continued to further define requirements for special needs students. This act worked to provide equal opportunities for all. Schools were now required to have open and equal opportunities to students no matter their disability. This included laboratory areas and activities. Teachers must make sure all learning is fair and equal for all students.

Current Education Trends

In 2001 the Elementary and Secondary Education Act of 2001 also known as the No Child Left Behind (NCLB) Act was passed that implemented mandates for school performance and accountability at the state, district, and local levels. The goal was set for *every* student to show proficiency in reading and math by 2014. Schools were evaluated and required to meet a yearly benchmark known as Adequately Yearly Progress (AYP). AYP was based upon several factors including graduation rate (at the high school level) and test scores. Schools not meeting this benchmark for two consecutive years were placed on a needs improvement list and had to develop and apply an improvement plan. The thought was that by the year 2014, 100 percent of students would show proficiency in these content areas. NCLB also placed importance on teachers being highly qualified in their given area. This meant teachers must hold a bachelor's degree and pass the state's certification test for the subjects they teach.

The premise of NCLB was good, but unrealistic in that all students would achieve 100 percent proficiency by 2014. It is known that not all students learn the same neither is their growth the same. In 2015, Every Student Succeeds Act (ESSA) was signed by President Obama. This act still focused on student success but at the individual level. This act required accountability but granted flexibility to states regarding specific requirements. Major tenents of the act focused on closing achievement gaps, increased equity, improving the quality of instruction, and increased outcomes for all students.

Following the evolution of agricultural education programs throughout history, we can see the changes presented over time. Agriculture education is a major part of vocational education from the beginning years to now. Agriculture education has also realized the importance of staying in tune to societies' needs and being community focused. Readers will explore the importance of this relationship in the next section.

Importance of Agricultural Education programs fostering a positive relationship with the community

Agricultural educators develop and build relationships within the school and community. These relationships are vital to the success of the classroom, students, and program. From day one, agricultural educators must realize they are not alone in their efforts. As a beginning teacher, it is easy to sometimes feel isolated on an island, but relationships and support systems are available to help in the journey. Relationships to develop and build as an agriculture educator include:

- Students
- Parents/Guardians
- School staff, faculty, and administration
- Board of education and central office staff
- Local farmers and agriculturalists
- Agricultural organizations
- Other local agricultural teachers
- Community organizations
- State FFA staff
- Post-secondary contacts

In Table 1, a summary of each group is presented with a brief overview and examples.

Every relationship built in the community is important to an agriculture program. Upon entering a program and community, a new teacher should consider creating a directory with names and contact information for use throughout the year. Taking the time to create this will serve as a valuable resource. Relationships are vitaly important to the success of a program. Agricultural education teachers can

Table 1. Relationship types and details

Group	Relationship Dynamics	Examples	Ag Program Application
Students	Be their leader and teacher. Be friendly, but not a friend. Build a positive relationship and maintain professional distance.	Ag Teacher to FFA member	Most important relationship and foundational relationship to build. The agriculture teacher is the greatest recruiter for the program and then students go on to be the next best recruiters for the program.
Parents/Guardians	Vital partnership between students and the community. Can take on several roles within the program from alumni members, chaperones, guest speakers, volunteers, etc. Keep open lines of communication.	Parents or guardians of students	Direct tie to community. Win this group over and they will become the greatest cheerleaders for the program. Keep everyone informed, be transparent in dealings, consistent, and fair across the board.
School staff, faculty, and administration	This includes the people agriculture teachers work with daily within the school building. Work to educate all on the program and keep as supporters of the program.	Custodial staff, front office staff, maintenance department staff, Career Tech director, assistant principals, principals, fellow teachers	Support system behind the scenes. Help to keep room and lab clean, equipment repaired, bookkeeper, funding provider, and field trip approver.
Board of education and central office staff	Will see only a few times a year but need as supporters and they need to know the agriculture teacher. Attend board meetings to present successful students or to champion a large project for the program.	School Superintendent, Human Resource Representative, Head of Transportation	This group needs to know the good things being done in the program. As the agriculture teacher, have a working relationship and be able to call if needed. Transportation is also vital to the program's ability to travel.
Local Farmers and Agriculturalists	Community members likely to have a vested interest in the success of the program. Valuable resource for local agricultural knowledge that can be leveraged to enhance the program.	Farmer, Rancher, Gardener	Building positive relationships will increase opportunities for the program, including guest speakers, SAE locations, CDE preparation, equipment/land loans, program advocacy, and advisory board members.
Agricultural Organizations	Support groups within the community directly related to agriculture.	Ag credit associations, Commodity commissions, Marketing associations, Local Ag dealerships, Extension, Farm Service Agency	Make time to attend their meetings/ events, invite to program events, and they should know about the teacher and program. Huge supporters once relationship is developed!
Other local agricultural teachers	Competitors on some days, but also fellow agriculture teachers	Rival school FFA advisor	Show students how to be successful competitors and maintain professional relationships. Support system for the teacher as both can share materials and ideas.
Community organizations	A collective gathering of community leaders that agriculture teachers need to be present at throughout the year	Chamber of Commerce Rotary Club Lion's Club	Participate in some of their meetings and promote the program and students.
State Agricultural Education and FFA staff	Direct contact for FFA event information, event sign-up, and yearly evaluations (some states)	State FFA Executive Secretary State Director Regional Directors Content support teachers	Lifeline for all things FFA. Contact for event clarification, application help, and teaching content. They are only a phone call away and willing to help.
Post-secondary contacts	A content rich resource. Utilize as guest speakers/lecturers in courses. Work to help students be successful as an FFA member if completing dual enrollment.	Local technical school teachers, College or university professors	CDE team preparation, content expert, workshop host, college connection for students to continue education

never walk the aisles of the grocery store the same once they become an agriculture teacher. Teachers have connections and eyes everywhere. The same is true of the agricultural community. For example, an agriculture teacher who meets with a local farmer may quickly find out they have a family member working for the Farm Service Agency or the local agriculture sales company. These people then have connections to a state or national level entity. The connections span and the agriculture program and teacher want to be shared with a positive perspective. *Don't take this lightly!* Encourage students to see the value and importance in promoting the program's name and reputation. Treat every person as an asset to the program and a potential future resource.

ROLE AND CHARACTERISTICS OF COMMUNITY

Creating a Community Profile

One strategy for building positive relationships within the community is to explore and document potential relationships via a community profile. A community profile is a collection of information which details important features of the community relevant to the success of the agricultural education program. A community profile is critical to (a) increasing awareness of diverse groups within the community, (b) detailing services in the community (e.g., food banks, financial counseling, support groups), (c) building relationships with entities throughout the community, (d) communicating rationale for program support, (e) serving as a trusted resource within the community, and (f) informing long-term, strategic program planning efforts to meet the needs of the community. For these reasons, establishing a community profile early in an agriculture teacher's tenure within a community is essential to success.

Leveraging a community profile to enhance the quality of an agricultural education program involves three steps. First, identifying what community information needs to be collected. Second, planning how to collect the information. Third, utilizing the collected information to improve the quality of the agricultural education program *and* community. The following sections details how to successfully accomplish each of these steps.

Step One: Identifying Important Information

Step one of a community profile is to develop a list of information to collect about the community. Certainly, there is a seemingly endless amount of information one can learn about a community; therefore, agricultural educators must be strategic about selecting information that will be most useful. To help prioritize, collect information within three categories: (a) *conditions* – data which inform the priorities and trends of the community; (b) *features* – resources within the community; and (c) *norms* – representations of the non-formal operations of the community.

Importantly, each community is different; therefore, some of the examples provided may be irrelevant for a given community. Further, there may be important elements to a particular community not represented on the list. As the reader moves forward with building a community profile, use the list provided as a starting point to adapt to fit the specific needs of the community.

Table 2. Example conditions, features, and norms within a community profile

Conditions	Features	Norms
Community Population	Major Industries	Community Values
Population Trends	Agricultural Support Organizations	Community Attitudes
Income Statistics	Religious or Spiritual Centers	Community Issues
Agricultural Statistics	Additional Agricultural Education	Information Networks
Community Type	Activity Centers	Decision Making Processes
School Type	Cultural and Aesthetic Resources	Problem Resolution Methods
School Size	Natural Resource Areas/Parks	
School Schedule	Historic Districts and Structures	
School Facilities	Prominent Neighborhoods	
Academic Performance	Local Landmarks	
School Funding	Community Gatherings and Events	

Step Two: Collecting Information

The second step of a community profile is to plan how to collect the information identified as important in step one. A valuable distinction when collecting information regarding a community is the difference between primary and secondary data. Primary data refers to information collected from an original source. For example, conducting a personal interview with a community member or surveying members of the community falls within primary data collection methods. Secondary data, on the other hand, are pieces of information collected by others about the community. Examples of secondary data include newspaper articles and census data. Using a combination of primary and secondary data is essential to gathering the diversity of data needed to assemble the community profile.

Collecting primary data is critical to building a community profile while also building relationships throughout the community. Review the three categories of data identified in Table 1, primary data collection is required to identify the community norms, many of the community features, and a few of the community conditions; therefore, it is important to consider *how* to collect primary data. The first step is to identify sources of primary data. When considering who to collect data from, start by reflecting upon who are the agricultural education program stakeholders. Stakeholders are individuals who are directly or indirectly impacted by the program (e.g., students, school personnel, school administration, parents, local business operators, program alumni, community leaders, agricultural educators in neighboring communities, etc.). The most effective way to gain primary data from these stakeholders is through personal interviews.

When conducting a personal interview with a program stakeholder, prepare a list of relevant questions beforehand but be amenable to the stakeholder guiding the conversation in a different direction. The interviewer shoud listen most of the time so they can fully understand the perspectives of the stakeholder. One challenge to conducting personal interviews many new members of a community face is identifying who to interview. When a challenge, consider using the "snowball sampling method" in which each interviewee is asked who else should be interviewed. This sampling approach will exponentially increase the list of potential stakeholders. Importantly, however, if the community includes populations of minoritized groups, additional effort is often required to gain access; however, this effort is essential to ensuring the community profile is representative of all the voices within the community.

Many of the conditions and features found in Table 1, though, are best answered through secondary data sources. Secondary sources of data agricultural educators may consider exploring include past newspaper articles; local, state, and national population census data; local, state, and national agricultural census data; school demographic reports; school performance reports; and community directories. Fortunately, internet technology has made many of these secondary data sources easily accessible; therefore, educators are encouraged to explore governmental and non-governmental resources to collect community profile information. When utilizing secondary sources, individuals should ensure the data being used are the most recent data available.

Step Three: Using the Information

Developing a community profile should be viewed as a continual process in which the information is updated frequently to reflect changes within the community. As a continually updated collection of information, the community profile can be utilized to ensure the agricultural education program remains attentive to the community in which it resides. Specifically, the community profile helps to identify *resources* the agricultural education program can leverage within the community to achieve programmatic goals. The community profile also illuminates potential *relationships* between educators, students, and community members essential to expanding opportunities for learning and skill development through the agricultural education program. Building upon resources and relationships, a community profile pinpoints opportunities for the agricultural education program to participate in bettering the community through *revitalization* efforts to address specified community needs. Finally, the community profile details the *routes* required to create positive change within the community. In these ways, the community profile exists as a key feature to developing and maintaining a community-based agricultural education program.

Evaluating Community Needs

The Value an Agricultural Education Program adds to the Community

An agricultural education program which evaluates and addresses community needs becomes an indispensable feature of the community. An indispensable agricultural education program is both sustainable and supported through its symbiotic relationship with the community. In this way, both the program and community increase their social capital – i.e., scope of positive relationships which yield mutual benefit. The expanded social capital of both program and community is pivotal for continued growth and impact. Later in the chapter, readers will unpack ways in which the community can serve the agricultural education program; however, in this section, the focus is on ways in which agricultural education programs can serve their community. The most important role the agricultural education program plays within a community is preparing the next generation of community members to be leaders working to create a better community. Through engagement in classroom/laboratory instruction, supervised agricultural experiences, and leadership development via the National FFA Organization, students are empowered to enhance their community as both professionals and members. In addition to preparing the next generation of leaders, the agricultural education program provides an important connection to the school for community members. Fostering a relationship with the community enhances understanding of the challenges, opportunities, and successes of younger community members. Finally, the agricultural education program contributes to the community by responding to the needs of the community in the form of service projects, outreach, and community engagement. To be responsive, however, the agricultural education program must have an established approach to evaluating the needs of its community, a topic discussed in the next section.

Considering Economic, Cultural, Environmental Factors

Evaluating the needs of a community is done primarily through continuous updating and utilization of the community profile. As the agricultural education program utilizes this collection of information, attention should be paid to three dimensions of the community in which to identify and address needs. These three areas are economic, cultural, and environmental. *Economic* refers to the financial health of the community. Specifically, a community which, at a minimum, meets the basic needs of its members while avoiding a level of economic productivity which deteriorates ecological and cultural assets within the community. *Cultural* refers to the human dimensions of community; specifically, ensuring all members of the community are afforded the opportunity to participate equally within a community that acknowledges the challenges and successes of their past, present, and future. *Environmental* focuses on the natural dimension of community. Specifically, encouraging all members of the community to improve the health of the ecosystem services (e.g., air purification, water filtration, natural beauty) found throughout their community. Economic, cultural, and environmental dimensions are viewed as the three pillars to a sustainable community and should, therefore, be the lenses through which an agricultural education program continually evaluates its community to identify, and respond to, needs. In this way, the agricultural education program plays a timely and important role in bettering its community both today and tomorrow. Additional community needs evaluations (e.g., evaluating job/professional outlooks within the community) are discussed within the program evaluation chapter of this textbook.

STATE AND NATIONAL CURRICULUM AND PROGRAM STANDARDS

When considering the word standard, one usually thinks of a set way of doing something. Standard operating procedure comes to mind; the process for completing a task detailed step by step. In education, standards have been developed for agriculture programs and courses. The standards help lay a foundational framework for programs and curriculum. They also work to ensure all students across the nation are receiving similar instruction in agricultural education courses and the standard agriculture program experience. Below these standards are discussed in more detail.

National Agricultural Content Standards

The agriculture, food, and natural resources (AFNR) industry is constantly changing and becoming more and more technological. To prepare students for a career in this changing industry, the National Council for Agricultural Education (The Council) worked to update and revise the AFNR Career Cluster Content Standards. These standards were originally developed in 2003 as part of the United States Department of Education (USDE) Career Clusters Project. The focus of these standards is to "provide state agricultural education leaders and educators with a high-quality, rigorous set of standards to guide what students should know and be able to do after completing a program of study" in an AFNR Career Pathway (The National Council for Agricultural Education, 2015). The standard content areas include:

- Agribusiness Systems
- Animal Systems
- Biotechnology Systems
- Environmental Science Systems
- Food Products and Processing Systems
- Natural Resource Systems
- Plant Systems
- Power, Structural, and Technical Systems

These standards are to serve as guides to developing quality programs and curriculum. The standards cover all components of the agricultural model including classroom/laboratory, supervised agricultural experience, and FFA. Since the standards' beginning, they have been reviewed and revised in 2009 and 2015. These revisions have led the AFNR Standards to align to the Common Career Technical Core (CCTC) Standards for Career Ready Practices and the AFNR Career Cluster. They have also been cross walked to some Common Core Academic Standards and National Standards for Financial Literacy. These standards help students become introduced and find their area of interest in the agriculture industry. For a complete review of the AFNR Content Standards and their organization visit AFNR Standards through The National Council for Agricultural Education at https://thecouncil.ffa.org/afnr/.

Career Pathways within the United States

Existing within the umbrella of Career and Technical Education, it is important for agricultural educators to prepare individuals for diverse careers in agriculture. Each of the AFNR standard content areas correlate to a Career Pathway. Students work to determine the desired content area of interest in high school and tailor their coursework to complete the given pathway. Each pathway offers career skills and content knowledge to students. To become a pathway completer, students must accomplish the sequence of outlined courses and complete the certification assessment. The outlined courses vary by state. For example, in Georgia, if a student wishes to complete the plant and floral design systems pathway they must take:

1. Basic Agricultural Science
2. General Horticulture and Plant Science
3. Floral Design and Management

After successful completion of these courses, students will then sit for the national assessment, which is the Precision Floriculture and Greenhouse Management Exam. Students must meet or exceed the given cut score for each assessment. Upon successful completion of the courses and the national assessment, students are then determined to be a pathway completer. Pathway completion is recognized on the student's academic transcript, and they typically will receive a special distinction during the graduation ceremony. Throughout the process of pathway completion, students gain valuable skills and knowledge in their AFNR area and are prepared with career skills for their future. Pathway completion numbers can also be used in a school's overall evaluation process to determine annual performance levels.

Figure 2. National AFNR content standards (The National Council for Agricultural Education, 2015)

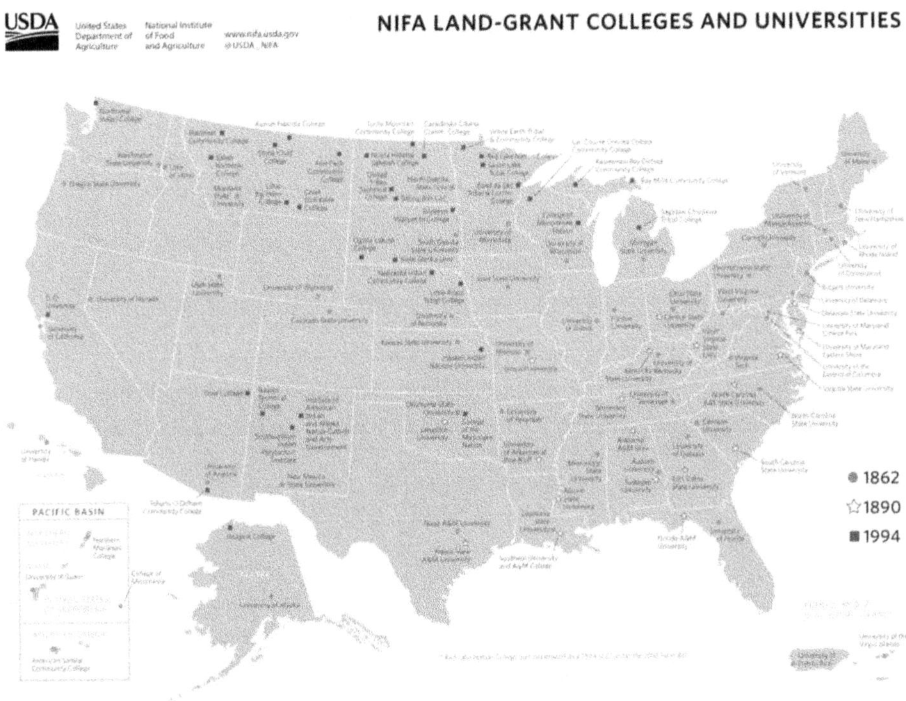

Academic Standards

From the program perspective, the next level of standards relates to the daily classroom courses. Some states now have a developed curriculum for each agriculture course. These standards provide the overall framework and guidance for the educator. Once hired and courses are determined, the first step should be to access the standards through the state association. The content standards can help determine units to teach along with daily activities and lessons. Below is an example (Figure 3) of content standards from the Indiana Department of Education.

Again, the focus of standards is to ensure all students are receiving similar content and instruction. Teachers teach differently, but standards ensure a baseline and content focus for all.

In addition to AFNR Standards, agricultural educators should be aware of, and build connections to, core academic standards by designing interdisciplinary learning experiences. Creating interdisciplinary learning environments, for example, aligns agricultural education to broader trends within the education community like the increased importance of science, technology, engineering, and mathematics (STEM) learning. Furthermore, teaching core academic content within agricultural education better prepares students for the interdisciplinary realities of AFNR systems. Therefore, two sets of commonly-adopted standards are introduced which agricultural educators should consider when designing learning experiences within their program.

Figure 3. Example Indiana AFNR standards

Content Standards

Domain - Classroom and Tool Safety

Core Standard 1 Students analyze and implement safe work practices which apply to agricultural mechanics.

Standards

APST-1.1	Explain the importance of safety in agricultural mechanics
APST-1.2	Identify and differentiate between safe and unsafe shop and work safety practices
APST-1.3	Describe the methods utilized to implement safe work and proper use of safety equipment practices
APST-1.4	Identify and explain the purpose of signals and symbols in agricultural safety
APST-1.5	Explain the importance and function of an operator's manual
APST-1.6	Identify and explain the role that various agencies play in regulating shop safety
APST-1.7	Locate and demonstrate the proper uses of the first aid and emergency equipment found in an agricultural shop
APST-1.8	Develop proper safety skills to use for hand and power tools

Common Core

Buzz terms in education today include common core standards. Common Core State Standards (CCSS) were developed in 2009 when state leaders from 48 states felt the need to "[recognize] the value of consistent, real-world learning goals and launched this effort to ensure all students, regardless of where they live, are graduating high school prepared for college, career, and life" (Common Core State Standards Initiative, n.d.). As of 2015, 42 states had adopted the CCSS in English/language arts/literacy and math. To learn more about common core standards and to view the standards visit www.corestandards.org.

Next Generation Science Standards

The Next Generation Science Standards (NGSS) represent a dramatic shift in the structure of science education. The NGSS were developed to transform the outcomes of science education from the memorization of facts to engaging in the practice of being a scientist. In this pursuit, science education was restructured into a three-dimensional approach, including practices, crosscutting concepts, and disciplinary core ideas. Practices include the actions and behaviors of scientist that should be represented in science learning spaces. Crosscutting concepts refer to principles which apply across all aspects of science (e.g., patterns, sytems, change) which should be addressed throughout science classrooms. Disciplinary core ideas are essential elements of content categorized into physical sciences; life sciences; earth and space sciences; and engineering, technology, and applications of science. In addition to establishing the importance of scientific behaviors and ways of the thinking, NGSS highlight the importance of integrating authentic phenomena within science learning experiences to increase authenticity and applicability. In this way, the science education community is poised to leverage agriculture, food, and natural resource contexts to engage learners in science education. To learn more about NGSS and to view the standards visit www.nextgenscience.org.

RELATIONSHIPS AMONG TEACHER, SCHOOL ADMINISTRATION, AND COMMUNITY IN CURRICULUM DEVELOPMENT

In simplest form, fostering a community-based agricultural education program relies on relationships formed through information gathering, overlapping goals, transparency, and mutual benefit. In this section, readers will explore agricultural educators' management of the diverse relationships important to maintaining a quality program.

Addressing Needs through Programmatic Structure

Selecting Experiences which Address Community Needs

One element which strengthens the relationship between an agricultural education program and its community is program efforts which directly address the needs of the community. These efforts not only improve the community, they also instill among students the importance of community improvement efforts and their capacity to lead initiatives to better their community throughout their life.

Identifying community experiences for the agricultural education program to engage in occurs as a component of the community profile development process. As teachers become more aware of the assets within the community, they will also uncover areas of need. These areas should be seen as opportunities to revitalize the community for a better future. In addition to illuminating areas of need, the community profile will uncover routes to achieving community change. Specifically, information relating to key community decision makers, systematic barriers, and community norms should be considered when preparing to undertake a community change effort.

There are many instances, however, in which the development of a community profile uncovers myriad community needs. For example, a community may be struggling with road littering, lack of recreation for youth, recruiting community volunteers at the senior center, abandoned lots in need of clearing, trail maintenance required at the community park, and so on. Addressing every need within a community is untenable; therefore, the agricultural education program must decide which opportunities to pursue. Importantly, the final decision must be made by students; however, educators can guide their decision by reviewing these three criteria for selecting a community intervention: (a) *priorities* – which project most closely aligns with the goals, interests, and strengths of program members, (b) *people* – which project has the likelihood of significantly impacting the most members of the community *or* the members of the community who are most in need, and (c) *potential* – which project are members able to accomplish given identified constraints (e.g., money, time, equipment, etc.). Using a combination of these decision-making criteria and information gathered through the development of a community profile should yield community-based projects beneficial to the community and students alike.

Identifying Partnerships to Optimize Programmatic Experiences

Just as the identification of community projects required evaluation of the community, the identification of partnerships to optimize programmatic experiences requires evaluation of the program. Evaluating the program, a process discussed in more depth throughout this textbook, will yield opportunities to expand or reinforce the program to better meet the needs of students. As these programmatic needs are identified, consult the community profile, advisory council, alumni, friends, and chapter to match the program need to a resource within the community that may be able to address that need. As an example, a teacher may identify students in the program that want to compete in the land judging contest; however, the teacher may not yet be knowledgeable about the skills required to succeed in the contest. To address the programmatic need, the teacher may discover there is a retired Extension agent in the community that specialized in soil management that can teach the teacher and students about the skills and knowledge required to be successful in the contest. As the example details, the origin on the partnership was identifying a direct need within the program that would benefit student learning.

The program-initiated process for identifying partnerships is straightforward and simple, but how do teachers manage a situation in which a community member articulates an interest in supporting the program without the teacher's request? First, recognize these instances are indicative of the program having a positive image in the community. Community members are most likely to offer their time and effort when they know it will be within a program achieving success. Importantly, however, teachers do not have to say "yes" to every offer of support and assistance from the community. In fact, there are instances in which an offer of support can be detrimental to the program; therefore, exercise caution when making decisions regarding what types of community support to accept. Here are some things to consider when making these decisions: (a) do I have the time and energy required to complete my role in this task; (b) does their offer relate to an identified programmatic need that will increase student learning and success; (c) can I modify their offer to better align to the needs of the program, and (d) will accepting this offer obligate the program to reciprocate via a future commitment? If the teacher decides to accept an offer of support, they should frame their acceptance as an initial "trial period" to provide an easy out should the support not be in alignment with expectations.

One additional strategy for agricultural educators to adopt to help utilize community partnerships is to have an updated programmatic "wish list." This list should include equipment, projects, and opportunities the teacher believes are important elements of programmatic growth; however, are not pursuing due to time and energy limitations. Example items could include an aquaculture tank, automatic watering system for the greenhouse, land lab, microscopes, coach for an Envirothon team, etc. Once this list is created, when individuals or organizations indicate resources are available to support the program (e.g., school administrator shares, "we have $5,000 to spend by Friday"), the teacher will be ready to provide specific requests that align with the program vision and objectives.

Advisory Council

Another essential relationship an agriculture teacher must manage is their relationship to a program advisory council. An advisory council is a collection of knowledgeable individuals who meet regularly with the role of offering advice and guidance regarding the facilitation of the agricultural education program. Advisory councils differ by school district, region, and state; therefore, specifics regarding the development, composition, and meeting frequency should be made in accordance with localized policies. In this section, however, readers will explore general principles for developing and utilizing an advisory council.

Advisory councils are an important element of program development, maintenance, and expansion. Comprised of experienced professionals in agriculture, food, and natural resource systems, advisory councils provide valuable insights into emerging industry trends and innovations to incorporate within the curriculum. In addition, an advisory council is useful in securing resources for the agricultural education program given council members' direct connection to industry. Further, members of the advisory council exist within networks of industry professionals they can call upon for support, expertise, and effort to aid the agricultural education program. Members of the advisory board are also critically important advocates for the program to the community, industry, and school. In each of these roles, the advisory board supports and extends student learning and program success.

The makeup of the advisory council is instrumental in its ability to successfully enhance student learning and program success. First, the advisory council should mirror the makeup of the community regarding diversity of genders, races, industries, ages, ability levels, etc. Establishing a diverse advisory council is essential to ensuring the breadth of voices within the community are represented in the guidance of the program. In addition, diverse teams are more successful in achieving innovative and impactful outcomes. To increase the diversity of representation, it is recommended to have individuals rotate on and off the advisory council. Importantly, structure member rotations so only a fraction of the members rotate off every year, increasing institutional knowledge amongst the committee.

In addition to representation amongst committee members, intentionality regarding the qualities of members should be considered. Recruiting individuals who bring insights, experience, connections, time, passion, and a collaborative spirit will result in a more cohesive and effective advisory council. Once identified, schedule an initial meeting with new members to orient them to the norms, goals, and structure of the advisory council so they are ready to contribute from the onset. Finally, if an agriculture teacher is in the situation of starting an advisory council, they should draw upon their local school administration, program alumni (if available), nearby agricultural educators, and community profile process to develop a diverse list of potential advisory council members to invite.

In addition to the information provided above, here are some general tips for facilitating an advisory council:

- Include 9-12 members on the advisory council.
- Meet as an advisory council at least twice a year.
- Be considerate of the agricultural season when scheduling meeting dates.
- A chair should be elected or appointed who is not the agricultural educator.
- Have current students (i.e., 2-4 members) actively engaged on the advisory council (e.g., making presentations, giving tours, and sharing requests).
- Provide ideas or artifacts for the advisory board to respond to instead of undirected ideation.
- Identify new advisory council members with access to resources or expertise aligned to program growth areas.
- Assign advisory council members a job (e.g., "program needs this piece of equipment, can you help us get access?").

Using Community to Enhance Student Learning

Relationships often include mutual benefit. In this section, the benefits a community can offer an agricultural education program are examined. Specifically, attention is given to the educational opportunities which exist within the community (e.g., land-based learning, field trips, guest speakers, SAE connections) and how agricultural educators can utilize these opportunities to enhance student learning.

Land-Based Learning

Learning experiences which address complex, national, or international challenges are important, however, they often fail to foster student confidence in their ability to address the problem. For example, lessons on global climate change are important but often leave students feeling powerless to address the immense scale of the problem. The emergence of place-based education was a direct response to the challenges associated with teaching globalized curriculum. Place-based education emphasizes the educational value of learning experiences which occur within a learner's community. Advocates of place-based education suggest the connection learners have to their community makes these localized experiences more relevant and impactful.

In agricultural education, place-based education has been adapted into an approach called land-based learning (McKim et al., 2019). Land-based learning engages students in a four-stage process which starts with identifying a local phenomenon and community partners (e.g., local farm and owners/employees of the farm). Then, students and the teacher learn about the local phenomenon via field trips and in-class presentations by relevant community members. The second stage of the land-based learning process, called understanding, includes identifying sustainability challenges within the local phenomenon relating to ecological, social, or economic systems (e.g., the ecological challenge of soil erosion on the farm). Following the understanding stage, students engage in the third phase of land-based learning, intervention. In this stage, students work with community members and educators to implement an intervention to increase the sustainability of the farm (e.g., using cover crops). In the final stage, students engage in evaluation in which they explore the impact of their intervention on the ecological, social, and economic systems of the farm and community.

In addition to building technical agriculture knowledge, problem solving, collaboration, and leadership skills, land-based learning has the potential to increase student confidence in their ability to positively impact their community. In this way, land-based learning serves as an exemplary method for fostering mutual benefit, a key component of relationships, as students learn through their land-based learning experiences and the sustainability of the community increases through the intervention and the development of individuals who recognize their potential to impact their community. Additionally, land-based learning provides valuable opportunities for the educator, students, and community members to expand or deepen their relationships through collaboration.

Field Trips within the Community

Another approach to utilizing community resources to increase learning is a field trip. Field trips provide students with opportunities to experience a locally relevant example of content while building relationships with professionals in their community. Given the time and financial costs of field trips, however, educators must be intentional about the types of field trip experiences in which they engage students. To aid in the selection of field trips, educators should consider value, venue, and validity. *Value* refers to an assessment of the learning that will occur compared to the time and financial costs associated with embarking on the field trip. In addition to value, consider the *venue* where the field trip will take place. Specifically, the space, set-up, and experience must be conducive to all students having equal opportunities to learn. The final element to consider is *validity*. A field trip is valid if the experience is the most efficient and effective way to achieve an identified learning outcome. To measure value, venue, and validity, educators should visit the field trip location and discuss the learning experiences with hosts before taking their students.

Community Connections in the Classroom

In many instances, the costs of a field trip preclude teachers for using these learning experiences as often as they prefer. A common substitute for a field trip is to have community members come to the classroom to share their knowledge or demonstrate a particular skill. These experiences provide tremendous educational value as community members bring relevant experience, current knowledge, and industry connections for students. As is the case with all educational experiences, intentionality is required when utilizing a guest speaker. To facilitate a guest speaker's engagement, consider these five steps. First, identify the learning outcomes to be addressed by a guest speaker. Second, based on the learning outcomes, identify a guest speaker from the community who can address those learning outcomes in a public speaking format. Step two is a great time to engage the advisory council for speakers or recommendations. Third, prepare students beforehand by developing questions they will ask and reviewing appropriate and professional expectations. Fourth, facilitate the guest speaker experience. Fifth, follow-up with a thank you (e.g., student-signed thank you card) and, if appropriate, invitation to support the program in future capacities.

Supervised Agricultural Experiences and the Community

The final dimension of leveraging the community to support student learning is, potentially, the most important. Supervised agricultural experiences (SAEs) often have students completing work, service, or research opportunities within the community. SAEs are explored in more depth throughout this textbook; therefore, the discussion of SAEs in this section is brief. Educators often play an important role in making students aware of the diversity of SAE opportunities within the community. To increase knowledge of these opportunities, rely on the community profile, advisory council, alumni and friends chapter, and fellow educators and administrators within the school. In addition, educators' role will be to prepare students to be respectful and professional in their SAE opportunity so they can continue the positive relationship between community entity and program. Further, the reader should continue reflecting on the importance of leveraging community relationships to identify SAE opportunities *and* the importance of fostering positive relationships with community members by preparing students to engage in their SAE as representatives of the agricultural education program.

Communicating with Advisory Council, Alumni, and School Administration

Successful relationships consist of structured and meaningful conversations. Therefore, to enhance the relationship with the advisory council, alumni, school administration, and other stakeholders, teachers must enhance the quality of communication with these individuals. To do so requires effort and intentionality.

In the book *Leadership & the Art of Conversation*, author Kim Krisco (1997) highlights conversations occur in three domains: description, action, and possibility. The most common type of conversation is within the domain of description, which are conversations focused exclusively on events that have occurred in the past. Alternatively, conversations in the domain of action are conversations about the present, usually involving some request for action. Finally, conversations within the domain of possibility are conversations exclusively about the future. Krisco notes that while conversations occurring within the domain of description are common and useful, they don't necessarily result in relationship building and collaborative efforts due to their existence discussing previous events. One of the main action items emerging from Krisco's book is that effective relationship building and leadership requires shifting conversations out of the domain of description and into the domain of possibility and action.

Insights into the types of conversations informs how to structure conversations with programmatic stakeholders. Certainly, educators need to provide frequent and meaningful communication with advisory council members, alumni, school administrators, and other stakeholders detailing the achievements of the program, including teacher(s) and students. However, these conversations are within the domain of description. To enact change and build relationships, accompany consistent and meaningful communication about achievements with consistent and meaningful communication about future possibilities: program goals, new directions, and possibilities. Shifting communication in this way provides stakeholders with a call to act, support, and innovate.

As an example, take communicating with administration regarding students in the program. If teachers only communicate with their administration to share things that have already occurred, administrators are not likely to view them as proactive. If, instead, teachers integrated their vision for how a situation could be addressed (domain of possibility) and the actions being taken to achieve this vision (domain of action), administrators will see the teacher as proactively shaping the culture of the program. Further,

Table 3. Eleven elements of marketing an agricultural education program

Concept	Description	Application
Price	Cost of the service, product, or experience.	Illustrating the low cost to benefit ratio while seeking opportunities to lower the cost for students and the school district.
Product	Characteristics of the service, product, or experience.	First, ensuring a quality program and then, communicating the positive impacts of the experience.
Promotion	Communication of information about the service, product, or experience.	Identifying how the agricultural education program addresses specific student and community needs.
Place	Location and timeliness of promotional materials.	Emphasizing proactive communication about the program (e.g., program engagement with elementary students).
Personal Relationships	Interactions between consumer and seller.	Building and maintaining relationships with current and future students and community supporters via effective communication.
Position	The context or environment in which the service, product, or experience is offered.	Articulating the unique opportunities within agricultural education as compared to other opportunities available within the school.
Performance	Traditions of success associated with the service, product, or experience.	Communicating a pattern of success and how individuals can build on that success through their engagement.
Philosophy	The core values embedded within the service, product, or experience.	Identifying the core principles of the agricultural education program (e.g., service, career preparation, leadership).
Physical Evidence	Tangible examples illustrating a quality service (e.g., case study).	Qualitative and quantitative evidence of the impact of agricultural education, including empirical research and student stories.
People	Individuals who are marketed a service, product, or experience.	Adapting communication to meet the population; including ensuring messages are inclusive to underrepresented communities.
Process	Processes within the system which influence the experience.	Including engagement in classroom/ laboratory learning, work-based learning, and leadership engagement for all students.

this new communication approach invites administrators to be co-developers of the future learning environments of students, a more energizing role than serving to address things that have already occurred.

Promoting and Marketing the Agricultural Education Program

As readers have uncovered, expanding the impact of agricultural education relies of leveraging positive relationships with individuals throughout the community. Inherent within the development of relationships is quality communication; therefore, in this section readers explore specific strategies for marketing the agricultural education program to community members and students using effective marketing communication.

Historically, marketing communication has relied on the "four p's of marketing," including: price, product, promotion, and place as essential to consider during any marketing effort. However, more recently, the list of essential elements to consider has expanded to eleven, which include personal relationships, position, performance, philosophy, physical evidence, people, and process. In the following table, readers will explore these eleven elements and their interpretation when marketing an agricultural education program (see Table 3).

Marketing an agricultural education program is a system-wide effort which includes considerations for the quality of the learning experience, communication approach, relationships, and diversity of potential students and supporters. Beyond these program-level elements of the system, marketing agricultural education is a task requiring all individuals within the discipline work in collaboration to empower educators at the local level to expand their program. For example, National FFA must work to ensure student experiences (e.g., conventions, contests) are inclusive to all participants; scholars must conduct research highlighting the benefits of student engagement in agricultural education *and* barriers to engagement among underrepresented populations; teacher educators and university faculty must expand recruitment efforts for future educators to include underrepresented populations; and state leaders in agricultural education must ensure curriculum standards and educator resources cultivate learning experiences which prepare individuals with knowledge of current and emerging technologies. These examples are just a few ways in which the discipline must work in concert to ensure the entire system is supporting inclusive and impactful marketing efforts.

PROGRAM PHILOSOPHY STATEMENT FOR AN AGRICULTURAL EDUCATION PROGRAM

In an effort to market a program, one of the best foundational pieces is to develop a program philosophy. Throughout teacher education courses, the word philosophy is mentioned many times. Philosophy in Greek means the "love of wisdom." When applying philosophy to the agricultural education program, educators are trying to convey their thoughts or wisdom on how to effectively run their program to the community and stakeholders. Think of the program philosophy as incorporating attitudes and beliefs and reflecting this into the program. When creating a philosophy, keep in mind the audience. The philosophy is often shared with parents, school administration, and used in an interview setting. Be sure to explain acronyms and give enough content information for the reader to understand the agriculture education program. The writer is sometimes teaching others about agriculture education through their philosophy. Possible questions to consider when creating the philosophy should be:

- What is the purpose of agricultural education?
- What is the role of the teacher and advisor?
- How should the teacher teach? What educational philosophy aligns and supports a preferred teaching style?
- What is the role of the student?
- How should the student-teacher relationship look?

To make the philosophy a solid statement, include a clear discussion of beliefs on running an FFA chapter and supervising supervised agricultural experience (SAE) programs. A personal piece is also nice to intertwine because this highlights the writer's beliefs and thoughts. Lastly, don't forget to include the community view. How should these relationships look? Below is a possible outline of how to approach the development of an agricultural education philosophy.

1. Personal journey finding agricultural education
2. Introduction to the three components of the agriculatural education program and discussion on the purpose of agricultural education
3. Class/lab
    a. Include aligned theory or philosophy to support thoughts
    b. Role of the students
    c. Dynamics of the student-teacher relationship
4. FFA
    a. Role of the advisor
    b. Role of the students
    c. Discussion of activities such as Career Development Events, Proficiency Applications, Degrees, etc.
5. Supervised agricultural experience
    a. Explain what it is
    b. Role as the teacher working with students
    c. Role of the students in this project
    d. Vision for students
6. Community connections
    a. Thoughts on ways to engage the community within program
    b. Community resources available
    c. View of current and future community relationships
7. Conclusion
    a. Bring all thoughts home
    b. Discussion of trends and changes in education and agriculture industry
    c. Powerful ending statement

Know that a philosophy is never a finalized document. The philosophy will continue to evolve and change as educators grow. The philosophy should be reviewed and updated each year. It should be professional in appearance and proofread.

SUMMARY

In this chapter, readers explored the development, facilitation, and value of the variety of communities in which agricultural educators operate. At the crux of these conversations was the importance of fostering positive relationships based on effective communication while upholding established standards for the discipline. Moving forward in an exciting and impactful career in agricultural education, hold closely the principles of community and relationships explored in this chapter as these elements are footholds to expanding student growth and development. Further, seek opportunities to lead the communities explored in this chapter through service to the profession and community. Those in leadership position themselves to enhance the diverse communities impacted by, and impacting, agricultural education.

REVIEW QUESTIONS

1. What key milestones in the development of the agricultural education discipline are most important? Why do these milestones stand out?
2. In this chapter, a diversity of individuals and entities in which agricultural educators develop and maintain a relationship were explored. Of those explored, which will be most and least challenging to cultivate?
3. In addition to the items described in this chapter, what other pieces of information could be useful to include in a community profile?
4. Why is it important to identify and address community needs within an agricultural education program?
5. In what ways can agricultural and core academic standards be leveraged to increase student learning and success in school?
6. Consider a familiar community, who are individuals who could be on the agricultural education advisory council? Are there any identities within the community not represented on the list?
7. In this chapter, land-based learning, field trips, guest speakers, and SAEs were explored as opportunities to utlize community resources to enhance student learning. Which of these examples are most exciting when considering their implementation in an agricultural education program?
8. As explored, communication can exist within the domain of description, action, or possibility. Take a moment to reflect on the conversations from today, in which domain did these conversations exist?
9. What are innovative approaches to marketing implemented by agricultural education programs and how do these approaches align to the eleven principles of marketing explored in this chapter?
10. How should somone plan to revisit and revise their program philosophy statement throughout their career as an agricultural educator?

REFERENCES

Common Core State Standards Initiative. (n.d.). *Development process*. Common core state standards initiative: Preparing America's students for college & career. http://www.corestandards.org/about-the-standards/development-process/

Herren, R. V., & Hillison, J. (1996). Agricultural education and the 1862 land-grant institutions: The rest of the story. *Journal of Agricultural Education*, *37*(3), 26–32. doi:10.5032/jae.1996.03026

Kerr,H.T.(1987).Withaviewtothefuture.*SAGEJournals*,*44*(2),164–164.10.1177%2F004057368704400202

Krisco, K. H. (1997). *Leadership and the art of conversation*. Prima Publishing.

McKim, A. J., Raven, M. R., Palmer, A., & McFarland, A. (2019). Community as context and content: A land-based learning primer for agriculture, food, and natural resources education. *Journal of Agricultural Education*, *60*(1), 172–185. doi:10.5032/jae.2019.01172

National Academy Press. (1995). *Colleges of agriculture at the land grant universities: A profile*., doi:10.17226/4980

National Institute of Food and Agriculture. (n.d.). *College Partners Directory*. National Institute of Food and Agriculture. https://nifa.usda.gov/land-grant-colleges-and-universities-partner-website-directory?field_u_s_states_and_territories_target_id=All&field_map_filter_value=All

New Horizons, F. F. A. (2018). *Frequently asked questions: National FFA charter revision*. National FFA Organization. https://www.ffa.org/the-feed/frequently-asked-questions-national-ffa-charter-revision/

The National Council for Agricultural Education. (2015). *Agriculture, food, and natural resources (AFNR) career cluster content standards*. https://thecouncil.ffa.org/afnr/

Torres, R. M., & Garton, B. L. (1991). *A program for the preparation of preservice teachers of agricultural education*. https://files.eric.ed.gov/fulltext/ED347382.pdf

United States Department of Agriculture. (2019). *1994 land-grant colleges and universities map*. National Institute of Food and Agriculture. https://nifa.usda.gov/resource/1994-land-grant-colleges-and-universities-map

KEY TERMS AND DEFINITIONS

Academic Standard: Statements of expected outcomes for learners within a given discipline.

Career Pathway: Collection of courses which meet established outcomes for initial success within an identified career.

Community: Collection of individuals with a shared identity (e.g., geographic region, profession, employment location).

Community Engagement: Immersed participation which address the needs and strengths of a community.

Community Profile: A living document describing the conditions, features, and norms of an identified geographic area.

Program Standard: Statements of expected outcomes achieved by, or structures present within, an educational program.

Stakeholder: Individual with an interest in the success of a program.

Chapter 4
Utilizing What We Know About How People Learn

Jeremy M. Falk
University of Idaho, USA

ABSTRACT

Teaching and learning are skills that can be improved that begin at birth. Formal education in schools has employed strategies for teachers to help students acquire new information, process that information, and transfer it to new situations. The prior experiences of students help to provide a foundation for all future learning. Problem-solving and inquiry-based instruction can be strategies to help students engage in their learning. The principles of teaching and learning are outlined in this chapter that guide teachers in their understanding of how students best learn. Helping all students succeed is critical, and teachers can engage with the community to provide resources to students, including formal programming and individualized education plans.

INTRODUCTION

Learning is done with or without a teacher. In fact, we've all learned *what not to do* because of how observant we are as humans. Teachers can make an incredible impact on the guidance of how people learn, but we should always remember that teachers are not the only person who can direct learning. Family and others who are close to a student can also direct learning. The comfort and relevance that family can bring to learning can be major factors to increasing a student's desire to learn new skills and grow. Students will also guide themselves in skills that they want to learn, especially with the accessibility of technology for learning. Videos and social media are often easily available for learning new skills and many people are willing to share their skills.

As teachers, we should incorporate all the senses and experiences that can help with learning to provide opportunities for students to make meaning of their world. Certainly, we can consider giving students opportunities to learn through direct experiences and vicarious experiences. Direct experiences are those in which a student can interact with a situation and be immediately involved in responding to people or a stimulus. Vicarious experiences are those that they can observe other people having the interaction

DOI: 10.4018/978-1-6684-3420-8.ch004

with people or stimuli. The social interaction of technology, including short videos of people sharing their lives, allows for a lot of vicarious experiences that help people learn and develop a way to interact.

The National Research Council (National Research Council, 2000) has collected findings from many studies around schools, students, teaching, and cognition. A new emphasis on how learning occurs has transformed pre-conceived notions about children and students and how they make meaningful connections to the interactions and experiences in the world. We will need to uncover more about the differences between understanding facts and having useable knowledge. Information is readily available to many Americans through internet searches, social media, and videos of people sharing tutorials. Useable knowledge is the key to producing students who can navigate how to solve problems and make decisions.

Schools can become large systems that condition students on how to function throughout their day. The structure of schools can result in seemingly universal behaviors, like raising your hand to ask a question, that continue throughout our lives in all situations that resemble school. We know that some behaviors that are needed in the workforce can be taught in schools, and we rely on schools to also teach students how to approach problems and find solutions. There continues to be research on how the brain works, but also on how education works, and some of the most influential research comes from teachers who observe patterns of learning with their students.

Objectives of this chapter:

1. Describe learning theories for teachers and students
2. Describe approaches to teaching and learning
3. Apply the principles of teaching and learning
4. Apply teacher behaviors that influence student achievement
5. Describe student needs and individual differences

DESCRIBE LEARNING THEORIES FOR TEACHERS AND STUDENTS

We will begin with an introduction to some of the theories on how children learn. As teachers, we should consider how students learn and that there is often a complex system of how students are processing information. Teaching is never as easy as simply telling students the right way to do something because the students need to be ready to learn. This section will briefly introduce four common theories that are used in schools to help students and teachers (Cohen & Waite-Stupiansky, 2017). These theories are incorporated by the many teachers that may see a student in a day and can be used to guide learner development.

Behaviorist Theory

Behaviorist Learning Theory, or Behaviorism, is rooted in the work of B. F. Skinner. This learning theory states that learning is done through a series of rewards and punishments and that people are constantly learning by finding the outcome to their actions. In classrooms today, Behaviorist Theory can be seen in the memorization of facts and repetition of skills, as well as a tool for classroom management.

According to Skinner, rewards will increase the likelihood that a behavior will be repeated, and punishments will decrease the likelihood that a behavior will be repeated. Rewards and punishments can be provided by adults and children but can also be provided by the environment. In Behaviorism, rewards and punishments can be provided and withheld, meaning we can provide a reward or punishment, or we can withdraw or hold back a reward or punishment. Behaviorists will refer to these as positive and negative, where positive means that we will add a reward or punishment, and negative means that we will hold back a reward or punishment. As we think of classrooms, a positive reward could be when students receive points or praise for answering questions correctly and a positive punishment is when students are required to do more cleaning because they made a mess.

In the agricultural classroom, we can quickly see opportunities to provide structure for students to learn using the Behaviorist Learning Theory. Teachers can work with students to learn the expectations and rules for the classrooms and lab spaces, and then be held accountable for their behavior. If students learn that there are consequences to their actions, then they are more likely to learn from that experience. Setting rules is important in a classroom, but this theory also relies on an objective response to when the rules are broken.

As teachers, we need to plan ahead and consider the behaviors that we want to have repeated, and which behaviors we want to eliminate. A quick understanding of Behaviorism lets us realize that consistency matters and that a plan will let us know what actions to take. As we think about a lab environment, if we want students to wear safety glasses, then there needs to be consistent rewards or punishments for the students. Providing a stimulus and a consistent response will help students find the path that is most appealing to them. For the FFA, we can use the rewards of additional experiences and trips, plaques and banners, and other recognitions for students at all levels to feel motivated to learn.

Cognitive Theory

Cognitive Learning Theory is based on the work of Jean Piaget, who believed that children do more than simply react to a stimulus or their environment. He believed that children develop their ability to think and process and make sense of the stimulus. Cognitive Learning Theory uses Piaget's 4 Stages of Cognitive Development to describe a person's ability to understand abstract and complex concepts.

Piaget's stages are:

Sensorimotor: birth to 2 years

- Infants explore their world through movement and learn by listening, grasping, and watching
- Infants learn that things continue to exist even when out of sight (object permanence)
- There are separate and distinct objects
- Their actions can influence how other things react

Preoperational: ages 2-7

- Children begin to think symbolically and will use words and pictures
- Preoperational children struggle with seeing the perspective of others
- Children in this stage think in very concrete and real terms

Concrete Operational: age 7-11

- Children begin to think about linear steps to concrete events
- They begin to understand conservation, that a quantity of water doesn't change when it is transferred to a new container
- They begin to reason and will generalize to others from a specific event that they experienced

Formal Operational: age 12 and up

- Young adults will begin to think abstractly and can think in hypothetical terms
- Teens begin to think about moral, philosophical, and social issues that require reasoning
- Begin to reason from a general principle to a specific situation

Piaget concluded that children are not less intelligent than adults, but they simply think differently. In his observations, he found that children will scientifically interact with their environment and make their own observations about a stimulus and a response. Early cognitive development is based on actions, while later stages will implement more mental operations and processing.

As teachers in agricultural education, we should consider that our students are familiar with observing and learn through their interactions as well as thinking. Our students will benefit from seeing how other people think and interact. The career experts in our field are great resources for helping explain to our students how they think through their work. Supervised Agricultural Experiences and internships will allow our students to continue to make their own observations and learn about application to other situations. Cognitive Learning Theory in an agricultural classroom can help bring concrete experiences to the abstract concepts that they have learned in other classrooms. As teachers, we can use geometry to solve problems and allow students to develop their processing of knowledge they gained in other curricular areas.

Constructivist Theory

Constructivist Learning Theory was developed by Lev Vygotsky and focuses on how a learner is the constructor of knowledge (Steffe & Gale, 2012). New learning is shaped by existing schemas that the student has built with their previous experiences and learning. Vygotsky believed that learning is a process and requires social interaction to grow connections between what we knew and what we are learning. According to Vygotsky, students are best able to learn when they work collaboratively with others who are already able to perform that task. This is a scaffolded approach where the skills to be learned are divided into manageable pieces for students and then they are guided from one step to the next by a facilitator. Constructivists recognize that there is a process to learning and that we learn best when we're ready to learn the next step.

In Agricultural Education, we can present students with problems that will allow them to interact in teams and rely on mastering skills. As we consider engaging our local community in the learning process, we can identify an end goal and then divide the material to be learned into manageable pieces and help students to reach each goal. Experienced students can be engaged in teaching novice students and the collaboration can help everyone learn. As we consider projects, we can help put students into teams and groups if the members are close in their understanding of the topic. A large gap in knowledge can be frustrating and can lead to students not engaging in their learning.

Humanist Theory

Humanist Learning Theory is an approach that is based on Maslow's Hierarchy of Needs (Cohen & Waite-Stupiansky, 2017). The individual student is central to this theory and focuses on the whole person, including cognitive, physical, and emotional development. In this learning theory, teachers guide students to set their own expectations and goals and to proceed when they are ready.

Figure 1. Maslow's Hierarchy of Needs

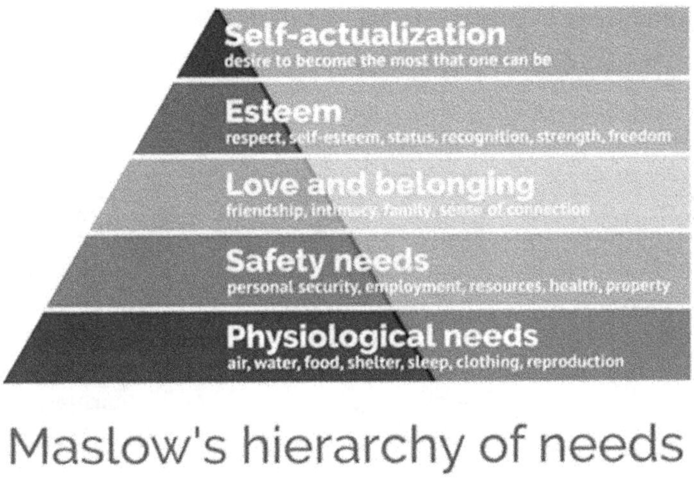

As we consider each student as an individual in the Humanist Learning Theory, teachers need to be prepared for students to be at many different levels and to progress at their own pace. This flexibility will allow some students to rapidly achieve their goals and perform to high levels, while they may need more time to progress with other goals. The teacher will need to help students realize their potential and what the appropriate expectations could be for each student. Students will need to make decisions on their goals and how they will evaluate their progress. Each student in a classroom could have different goals and measurable outcomes.

In Agricultural Education, we can certainly help students by providing a menu of assessments as options to show their mastery of a skill. The students could also set their own goals for progression toward a chapter degree. The benchmarks that students attain could be provided and adjusted by each student as a starting point. The entry level for students to learn is that they arrive ready to learn. For some students, this may mean that they have more options on what time of the day they learn best, and where they learn best. Online learning can afford students the ability to learn at their own pace and teachers empower students to engage.

The four learning theories that are provided in this section are often used in classrooms and selected for the appropriate situation. Each theory has strengths and weaknesses as we consider the school environment and outcomes for teachers. As you consider these theories, you will balance expectations that your school and community have for you to reach standards for the curriculum as well as the well-being of the students. Teachers can use these theories to help set plans and curriculum in an evolving learning environment.

DESCRIBE APPROACHES TO TEACHING AND LEARNING

There is a continuum of instructional strategies and at one end, it identifies teacher-centered teaching, and the other end is student-centered teaching. This continuum is helpful to understand as a future teacher because each of us have probably experienced some form of purely teacher-centered teaching, student-centered teaching, or somewhere in between (Banchi & Bell, 2008).

Figure 2. Teacher and student-centered instruction

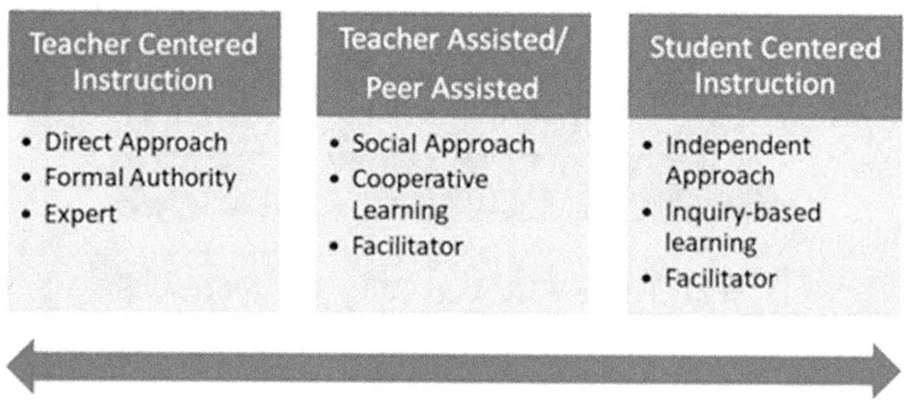

The approach of teacher-centered or student-centered can depend on many variables, including:

- Time available
- Resources
- Content to be delivered
- Age and maturity of student
- Preference of the teacher

Teacher-centered teaching relies heavily on the teacher being the expert and delivering content to students. This may be one of the images of school that comes to mind when we think of a teacher delivering a lecture to an auditorium full of students. This approach can be effective for delivering information to a large group of people in a short amount of time. The teacher can test or quiz students to help check for comprehension, and often the materials to be learned are memorization of facts and procedures.

Student-centered teaching provides more expectation for the student to be prepared to learn and work toward uncovering what they need to know. This approach requires teachers to provide guidance and ask additional questions to facilitate learning. Students will choose goals that they will learn and will often be able to use daily goals to complete a project.

The importance of solving problems should be highlighted in this chapter. As a species, humans have continued to progress and adapt because of their ability to solve problems. Problems occur everywhere and all the time, and at all magnitudes. Schools are a place where we can focus on developing skills and allowing students to solve guided problems to help prepare them to solve problems outside of school.

In an Agricultural Education program, we are fortunate to have real examples to solve and have community members to help us all grow. The problems that we may choose to solve can be planned in a general sense, like a packaging and preservation technique in food science, or a problem with fences in a livestock program. The specific problem or scenario will allow us to customize the learning and make it authentic. The community partnerships in class may turn into jobs and careers for the students who are involved in the problem solving.

We should also consider Supervised Agricultural Experiences as a great source of authentic problems for students to solve. The motivation for a student to solve their own problems could be strong and allow the student to learn much more than just what's going to be covered on an exam.

As we guide students, we should be aware that the learning process can be illustrated as being similar to the scientific method:

1. Experiencing a provocative situation
2. Defining a problem
3. Seeking data and information
4. Formulating possible solutions
5. Testing proposed solutions
6. Evaluating results

The systematic delivery of instruction may be varied by topic and by teacher, but this structure to teaching can help students process and understand the content. Allowing students to engage in the problem and determine solutions will take time and will provide value for everyone if the learning opportunity is well-constructed.

In addition, Kolb (1984) helps us understand that students can enter at any point the experiential learning cycle. His theory highlights concrete experiences and abstract concepts, as well as reflective observations and active experimentation. As we think about problems in agriculture, we can think of situations where this model can directly impact how students learn. For example, in a greenhouse setting, our students can propagate new plants by seed and experience how water pressure affects the germination of their new flats of plants. A teacher can then lead the students through reflective observations and uncover what happened. Students can then be guided through abstract conceptualization and use existing knowledge to think about how to address any problems. When the teacher guides the students back to the greenhouse for another try to propagate plants, there are opportunities for active experimentation to hone the ability to water a flat of new plants.

Figure 3. Kolb's Model of Experiential Learning

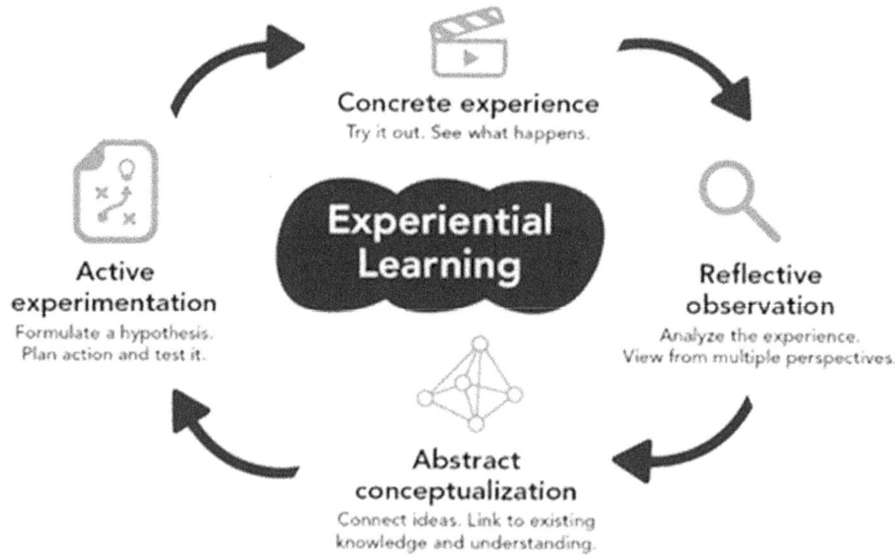

a. Define problem-based learning, decision-making, and inquiry-based instruction

As we learn about incorporating strategies to teach using problems in our agriculture programs, it is important to define a few commonly used terms in education:

Inquiry Based Instruction is when a teacher sets up experiences for students to ask questions and determine what they need to learn. With inquiry-based instruction, there are many questions that a teacher should help guide students to ask. Another hallmark of Inquiry Based Instruction is that teachers will help students to discover their own answers rather than providing the answers. The scientific method is used to help students systematically answer questions.

Banchi & Bell (2008) outline the four types of inquiry that teachers can use:

1. **Confirmation Inquiry:** Students confirm a principle through an activity when the results are known in advance.
2. **Structured Inquiry:** Students investigate a teacher-presented question through a prescribed procedure.
3. **Guided Inquiry:** Students investigate a teacher-presented question using student designed or selected procedures.
4. **Open Inquiry:** Students investigate questions that are student formulated through student designed or selected procedures.

As we consider the role that Inquiry Based Instruction can play in agriculture programs, we may find quick connections for Supervised Agricultural Experiences (SAE) in guided inquiry and open inquiry. Those real experiences can help students ask questions that have direct implications for their projects. Student organization opportunities through the FFA may include service learning, Career Development

Events, and team-building experiences. Many of the FFA activities could be guided using confirmation inquiry and structured inquiry. Again, the teacher will lead the students to ask their own questions and help them to discover their own answers.

Problem-Based Learning (PBL) is a student-focused approach to teaching that uses real problems as the content of the lesson and encourages students to develop problem-solving skills. A main concern is to identify the problem and then determine solutions and make decisions.

Problem-Based Learning has been used in agricultural education to help improve production and processing. Often, teachers can help students identify common problems in class and then encourage students to consider how to solve a similar problem when it arises in their SAE project or in a career. Problem-Based Learning can be used in all four types of Inquiry-Based Instruction, however, PBL is often associated with guided inquiry when it is conducted in the classroom. One distinction for PBL is that it emphasizes decision-making strategies, which need to be taught to students to implement.

Decision-Making Strategies

There are several strategies that teachers can use to help students make decisions, including (Newcomb, et. al, 2003):

Forked Road, Pro/Con List – student identifies two or more possible solutions and then lists all possible advantages and disadvantages for each solution. A decision can be made after reviewing the lists.

Steps and Key Points – student will list out the process for resolving a problem and identify major steps. For each major step, the student will consider options and factors that may include pricing, time available, and practicality.

Situation to be improved – students are presented with a specific situation and then find opportunities to increase efficiency or effectiveness, or other factors. This strategy encourages students in look into how other people do things.

Given the Effect – find the cause – students can dismantle and evaluate a series of events. This backward modeling may help find solutions from a new perspective.

Learning Styles

Novice teachers, and students, can create dissatisfaction in the classroom when they assume that all students learn the same way, and that the preference of the teacher is always the best way for all students to learn. We should consider that students have preferences on how they learn and that teachers have preferences on how they teach.

The idea that students learn best when teaching methods and school activities match their learning styles, strengths, and preferences has been a popular concept since the 1970s and 80s. The research and evidence that has been reported to connect personal learning preferences and actual learning outcomes has had inconsistent results, meaning that matching teaching methods to a student's learning style may not improve their learning.

Nevertheless, it has been valuable for teachers to understand learning styles so that they can add variability to their lessons and to possibly increase learner satisfaction. Students may have increased motivation when there is variability in the lesson, and they are happy with the delivery of the content.

A popular model for learning styles has been the VARK model, created by Neil Fleming in the later 1980s. Using the VARK model, learners are identified by whether they prefer:

- Visual learning (pictures, movies, diagrams)
- Auditory learning (music, discussion, lectures)
- Reading and writing (making lists, reading textbooks, taking notes)
- Kinesthetic learning (movement, experiments, hands-on activities)

An instrument was created that asked students to self-report how they learn best. With each question, they would be asked in which way they'd prefer to learn. There are several instruments that are available today to customize questions for specific content areas. Students tend to score higher in one area, and that area would then be considered their learning style. Students who prefer more than one single way of learning could be considered multimodal in their learning style.

Figure 4. The VARK Model of Learning Styles

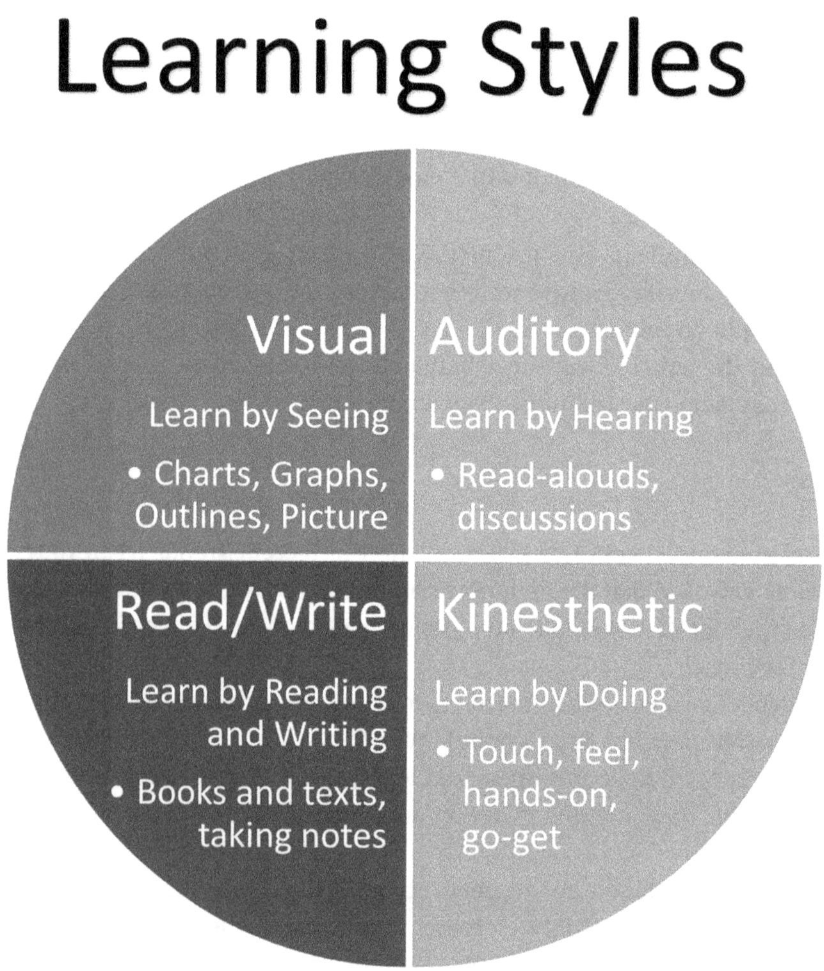

Teachers may choose to vary their instruction to help students find a style that they prefer. Similarly, teachers may help students stretch and strengthen all learning styles and offer many ways of learning that could be less preferred, but perhaps equally or more beneficial. As we consider Agricultural Education, some of the preferred learning styles may be more effective when teaching students how to weld, germinate seeds, or construct a fence. The awareness of a teacher about learning styles and preferences can help that teacher prepare students to feel confident and supported in their education. Teachers can remind students that there are many ways to learn and can offer students time to see, hear, write, and move.

Multiple Intelligences

Additionally, in his theory of multiple intelligences, Gardner (1991) proposed the existence of seven relatively autonomous intelligences: linguistic, logical, musical, spatial, bodily kinesthetic, interpersonal, and intrapersonal. Recently, Gardner (1997) proposed an eighth intelligence, "naturalistic." The first two intelligences are familiar to many people because they are often what we think about when given a test or think about a traditional school. The theory of multiple intelligences is a psychological theory and has been brought into education for teachers to consider that their students may have varying levels of intelligence in these eight areas.

Figure 5. Gardner's Eight Multiple Intelligences

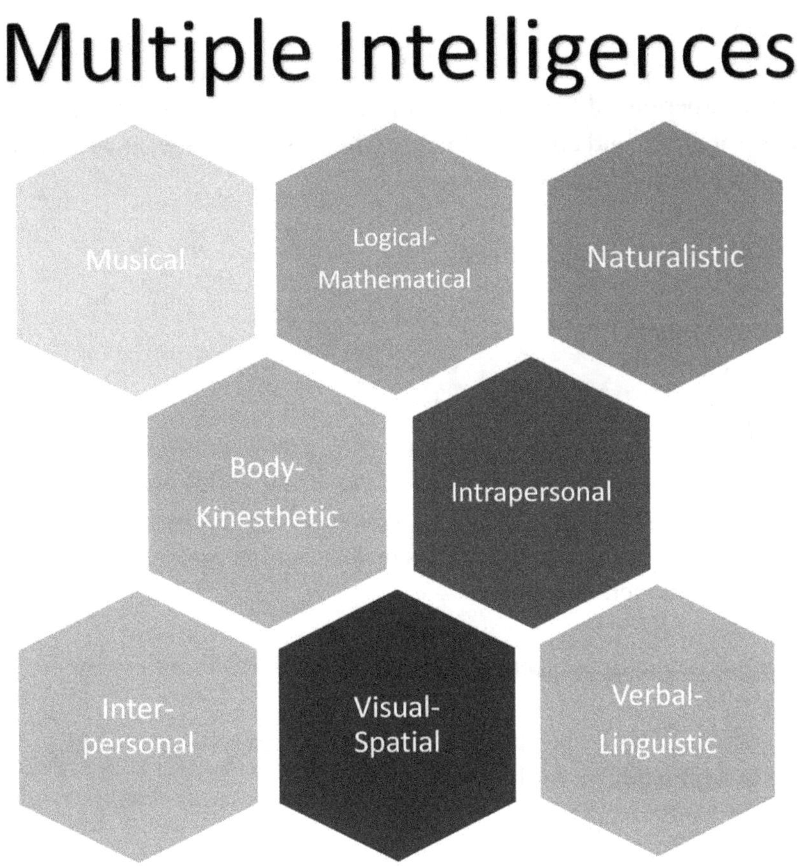

Teachers have applied the theory of multiple intelligences mainly in one of two ways. One option is that the teacher will try to stretch and grow all eight intelligences in each of the students. There is a belief that we should challenge and support the growth in all intelligence areas to prepare students for unknown situations in the future. The other option is that teachers will allow choice and help students to further strengthen the areas in which they are most intelligent. The skills and objectives could still be met for all students in the class, but with each student selecting their own way to learn the material and to perform different learning activities and assessments. There are strengths and weaknesses to each approach. Continued research will help us understand if either of these options are the best, or if there are other approaches that are better to help students learn.

Education has adapted in some classrooms and programs to modify existing curriculum to provide students the choice on how they want to learn and be assessed. The idea of metacognition and that students are aware of their learning has helped transform schools to be increasingly student-focused and we will continue to see the impact on our workforce.

APPLY THE PRINCIPLES OF TEACHING AND LEARNING

There are 16 Principles of Teaching and Learning in Agriculture, as provided by Newcomb et al (2003). They are the foundation for student learning and are a critical guide for teachers to use as they prepare and deliver lessons. As you discover the Principles of Teaching and Learning, you may wonder if they focus more on the content or on student management. The principles are effective in both realms and help keep learning content and student engagement together.

These principles can be intuitive for novice teachers, and they build a strong foundation for guiding and teaching students. Experienced teachers can improve their existing lessons and curriculum when they rediscover these principles and continue to make their lessons relevant to their students.

The principles fall within five main categories:

1. Organization and structure of subject matter
2. Motivation
3. Reward and Reinforcement
4. Techniques of Teaching
5. Transfer of Learning

The Principles of Teaching and Learning (Newcomb et. al, 2003) are based in the psychology of teaching and allow for teachers to start with a firm understanding that the 16 principles can be effective when present in their teaching. In fact, the principles make teaching easier because the students are more engaged and have meaning in their education.

1. Organization and structure of subject matter

    a. The subject matter should be structured and organized so that it is clear to the students
    b. Readiness is a prerequisite for learning

Discussion on Organization: I can clearly remember a teacher who posted the learning objectives for an entire semester on the wall in the classroom. The students were introduced to the outcomes of the course and were then walked through the skills that they'd learn to meet the final outcomes. It was a landscaping course and students needed to plant trees, maintain a yard, and operate equipment. The teacher was flawless in easing concerns of the students that they couldn't operate a chainsaw because he pointed to the steps that they would take so that each student could be successful. The course was prepared in a way that allowed students to see where they were headed and could see what needed to happen. The amount of planning for this course was impactful for students to know what skills they would have at the end.

2. Motivation

   a. Students must be motivated and engaged in the subject matter.
   b. Students are motivated by being involved in setting goals and planning learning activities
   c. Success is a strong motivating force
   d. Students are motivated when they attempt tasks that fall in a range of challenge where success is perceived but not certain

Discussion on Motivation: Some of our students will be competitive with themselves and with others, however, not all students are competitive with every topic, or even in academic goals. Finding what motivates students can take time and need rapport to be built but can encourage a student to learn more than the minimum. An encouraging teacher will help students find their motivation by offering many ways to be successful. A congratulatory call home when a student earned a "B" on a test was enough motivation for a student to complete all assignments from that point on in the course.

3. Reward and Reinforcement

   a. Students learn better when they understand the progress in their learning
   b. Behaviors that are reinforced are more likely to be learned
   c. Rewards or reinforcement should immediately follow the desired behavior and be clearly connected to the behavior

Discussion on Reward and Reinforcement: Accurate feedback for students is important because we want them to have accurate expectations for their behavior and performance. As we build a class environment, we should guide students along the course objectives and let them know where they can perform better and what is working well. As we talk about managing a classroom and shared space, we should consider the work-skills that employers will want in our community. Be prepared to have accolades for students when they're performing the way you want them to.

4. Techniques of Teaching

   a. Direct learning is more effective than undirected learning
   b. Students should *inquire into* rather than be *instructed in* the subject matter
   c. Students learn what they practice
   d. Supervised practice allows students to have real experience with the content to be learned and should be provided within the instruction

Discussion on Techniques of Teaching: Novice teachers can sometimes have the desire to talk for an entire class session and expect the students to listen. Teachers should be able to guide students to learn and provide opportunities for students to engage in the content of the lesson. If it is an agribusiness lesson, what opportunities are there for student engagement? If it is a plant propagation lesson, what opportunities are there for student engagement? Teachers will help students learn when the opportunities are local and relevant in their communities.

5. Transfer of Learning

   a. Learning is most likely to be transferred from the educational experience to a real situation when the learning immediately precedes the time when students need the knowledge or skill
   b. Learning is likely to be transferred when it is generalized for other situations or has a formula to follow
   c. Students should be instructed on how to transfer learning from an educational experience to other life and work environments

Discussion on Transfer of Learning: Agricultural Education programs are fortunate to have many other contents (math, English, history, etc.) come to life in the real application of the skills they learn. Transferring what students learn in class to a new environment, like a job, is impactful and needed for students to get good jobs. Teachers must connect with employers to help solidify skills in the classroom and then help students see the application to other areas of their life.

Table 1. Categories of the Principles of Teaching and Learning

Categories of Principles	Applications
Organization and structure of subject matter	Program goals and course objectives are clear to students. Units have measurable outcomes that students work toward. Lessons build upon what students learned the previous day.
Motivation	Students can use their learned skills to get jobs and earn money. Students find the curriculum challenging and rewarding. Students set goals and accomplish goals through their learning.
Reward and Reinforcement	Students can see their progress in skill acquisition over time on a board or chart. Students know the rewards and consequences of their actions and inactions from the start of the course.
Techniques of Teaching	Students are directed in what they will learn with opportunities to guide their own pace. Experiences are real and provide value to the students' life.
Transfer of Learning	Students can work on equipment like they'll encounter in a career. Students can see meet professionals who have benefitted from the experiences the students are having.

APPLY TEACHER BEHAVIORS THAT INFLUENCE STUDENT ACHIEVEMENT

Teacher behaviors can directly impact how students learn in our classrooms and agricultural education programs. Teachers of any experience level can question what they can do to motivate students, get the students to like the content, and to have learning be enjoyable. Other teachers may realize that the process of teaching can vary and that the results, or student outcomes, of teaching are important. Certainly, we need to be aware that teaching and learning are inseparable and that there must be a good match between the teacher and the student.

Since 1971, a meta-analysis of research in education was conducted by Rosenshine and Furst, and they found variables that a would increase student learning as the teacher exhibited them more often (Barrick & Thoron, 2019). Using these variables in your teaching can positively affect how the students perform. There variables can be central when developing lessons, and a lack of any of them may help explain why a lesson didn't go as well at it could have gone. The five variables that had the most correlation to student outcomes are:

Businesslike Behavior – This is the degree to which a teacher is task-oriented. The teacher is concerned that the students learn something rather than that the students enjoy themselves.

Clarity – Cognitive clarity of a teacher's presentation. Teacher explains concepts clearly and logically. There is transparent organization of the lesson.

Opportunity to Learn Criterion Material – Students are given the chance to engage and learn the lesson as well as perform the task for review. Students know that they will be expected to demonstrate that they learned.

Variability – A teacher uses a variety of instructional materials, teaching devices, and assessments. The students are stretched in the activities in which they participate, and the teaching is novel rather than boring.

Enthusiasm – The teacher can be perceived as being enthused by the material and the chance to work with the students. This enthusiasm can be seen in engagement and interest in the topic, but also includes movement, voice inflection and gestures.

One of the best ways to find examples of these variables is to watch and experienced teacher who is proficient at teaching, and then observe a teacher who is not finding success with the students. The art of teaching can include a lot of variation between teachers, but these characteristics are important for helping students be successful.

1. Business-like Behavior. This variable can perhaps be observed both in the professionalism that a teacher has toward the organization of the learning space, and also in that the teacher actively tries to get students to become learners.

   As we consider the professional disposition of the teacher, we may see care in the teaching materials and the lab spaces that emphasize that the students are part of a larger community and are gaining skills that can be seen outside of the school. For example, the pride in keeping tools clean and organized can demonstrate that the tools and workspace are important. The skills learned in the laboratory or classroom can be modeled after workplaces in the community so that students can directly see the connection to the skills they're learning and the businesses.

The second component is that the teacher acts in a way that encourages students to be learners. The priority is not to be liked by students, but that the teacher emphasizes student learning and student outcomes. An exemplary teacher can be observed by being friendly with students, but also guides students to focus on the task and develop skills.

Business-like behavior can also be seen in the form of routines in the classroom that encourage autonomous behavior. Teachers can set proactive routines, like providing assignment submission locations or binder checks, that students will understand their job and can complete it. Teachers will be able to provide structure and follow the structure to help student performance.

2. Clarity. The notion of clarity in the classroom focuses on the teacher in what is said, written, and demonstrated to students. Just as important, the students need to perceive the teacher as being clear and not just that the teacher thinks that they are being clear.

   Teachers who are clear in their presentation often have a lesson plan and are prepared with their delivery of the lesson. Regardless of the experience level of the teacher, clarity can be achieved when a plan is created and practiced by the teacher. Here are some strategies to become clearer in your teaching:

   a. Create a lesson plan for the material that is specific to your students.
   b. Prepare key questions and word them to be clear and concise.
   c. Write key concepts to be learned and keep them to 3-7 main points.
   d. Avoid using vague words, like "some," "of course," and "a lot of people."
   e. Practice the lesson and gauge time and ability to monitor student progress.
   f. Ask the students. Questioning will need to go beyond "Do you understand?" and into questions that require a specific answer to what was taught that day.

3. Opportunity to Learn Criterion Material. Significant, positive, and consistent correlations have been found between the opportunity to learn and student achievement. This makes sense when we state that the more we teach, the more students will learn. The criterion material is a standard that is measurable by the teacher, and a way to let students know if they've met the standard.

   Some ideas to support student learning in this variable are:

   a. Time on task. Teachers should begin and end class on time and engage students for the entire class period. Students should know how to be productive in the class and work toward meeting the daily objectives.
   b. Out-of-class time. Students should engage with learning course material out of the classroom. Time can be spent in group projects, gathering materials for an upcoming lab, studying rote memorization facts to be used in class, or applying the skills they've learned.
   c. Readiness for learning. Teachers need to provide the appropriate level of instruction to the students. The students can be ready to learn if the teacher will assess their current skill with a pre-test or find ways to know if the students are ready to learn the material.
   d. Teach to the test. Teachers must ensure that course content is directed toward the criteria to be used as a measure of student success. There is nothing wrong with "teaching to the test" if the test is an accurate representation of the criterion material.

4. Variability. Providing a variety of ways for students to be engaged in the classroom can increase motivation and engagement. Teachers should have a variety of instructional materials, teaching techniques, types of assessments, and a variety of levels of student tasks.

   Variability may be one of the easiest areas for teachers to increase, but it can also be one of the first to lose as we find routine and comfort. Here are a few strategies to increase variability in teaching:

a. Change seating arrangements in the classroom and direct learning to different walls in the classroom.
b. Provide different types of media for the students, including music, video, drawings, and smells. Engage their senses and strive for them to be intrigued.
c. Vary the types of assessments and types of questions on tests. Multiple choice and short answer questions may be common, but add variety with projects, completions, presentations, and skits.
d. Change the decorations and media in the room, including bulletin boards, posters, and displays. Showcase the content that you are teaching and showcase student work.
e. Watch others teach and find what works well in other classrooms.

5. Enthusiasm. Enthusiasm of the teacher can be difficult to measure, and also difficult to discuss. When Rosenshine and Furst discussed enthusiasm, they found that the three traits associated with it are movement, gesture, and voice inflection. As this variable relates to student achievement, we need to remind ourselves that movement, gesture, and voice inflection must be noticeable to the students and seen as enthusiastic around the content and the students. The personality of the teacher can play a role in natural enthusiasm, but it has been shown that teachers can increase enthusiasm by actively trying to increase a level of excitement for the material. Above all else, a genuine interest in teaching and in being around students is the key to being effective in showing enthusiasm.

DESCRIBE STUDENT NEEDS AND INDIVIDUAL DIFFERENCES

1. IEPs and 504 Plans

Students with delayed skills development or other learning disabilities may be eligible for special services that provide for individualized support initiatives in public schools. These programs are free of charge to families and are subject to two important federal mandates. Programs are designed to help ensure that children's' special needs are met. The Individuals with Disabilities Education Act (IDEA) of 1990 and the U. S. Rehabilitation Act of 1973 provide guidance for educators and provide funding for schools to conduct programs.

2. Individual Education Programs (IEPs)

The most recent enactment of the IDEA legislation makes teachers responsible for planning, implementing, and monitoring educational plans to help students be successful in school. A well-planned IEP describes the goals set for the student for the school year and outlines any special support the students need to achieve those goals. States are required, by law, to provide public education for all students with disabilities ages 3 to 21, regardless of the severity of the disabilities.

Students who are struggling in school may qualify for support services, allowing them to be taught in a special way. Reasons for qualifying include such attributes as:

- learning and/or physical disabilities

- attention deficit hyperactivity disorder (ADHD)
- emotional disorders
- cognitive challenges
- autism
- hearing and/or visual impairment
- speech or language impairment
- developmental delay

Teachers should be able to recognize the signs of students with a learning disability:

- getting poor grades in spite of significant effort
- needing continual guidance for tasks
- inability to remember problem-solving steps due to the lack of understanding the tasks or the logic involved
- having poor memory of spoken or written material
- difficulty in mastering tasks or transferring skills to other tasks
- inability to remember skills and facts over time
- having strong general knowledge abilities but being unable to rea, write or do math at that level
- having difficulty with communication process
- being extremely frustrated with school and homework

Services outlined in an IEP are typically provided in regular classroom settings. However, some IEP services might be delivered in separate resource classrooms or separate schools, depending on the students' needs. Some students may have an IEP for one subject area only, while others may have one for all academic subjects. Services should be provided in the least restrictive setting. The least restrictive setting is the regular classroom. Students with IEPs usually join regular education classes for special subject areas such as science, social studies, art, music, library, gym, and health. Classroom teachers need to be familiar with the services and monitoring that are required in the plan.

The next least restrictive setting is a resource or learning support classroom where groups of students with similar needs are brought together for small-group instruction. A certified special education teacher is the instructor and other school personnel (aides or support teachers) assist with teaching. Students who need intense intervention, however, may be taught in a special school environment. These schools have fewer students per teacher, allowing for more individualized attention. Teachers in these schools usually have specific training in helping students with specific special educational needs.

Initiating the IEP process generally begins when a teacher, parent, or doctor is concerned that a child may be having trouble in the classroom, based on specific data regarding the student's progress or academic problems. This may be done through a conference with parents, a conference with the student, observations of the student, and an analysis of the student's performance (attention, behavior, work completion, tests, classwork, homework, etc.). Then strategies specific to the student could be used to help the child become more successful in school prior to any formal testing. If this doesn't work, the child would be tested for a specific learning disability or other impairment to help determine qualification for special services.

The presence of a disability does not automatically guarantee a child will receive services. To be eligible, the disability must affect functioning at school as determined by teachers, psychologists, therapists, and other specialists. To determine eligibility, a multidisciplinary team of professionals will evaluate the child based on their observations, the child's performance on standardized tests, and daily work such as tests, quizzes, classwork, and homework.

The classroom teacher is central to offering suggestions about how the plan can help the child's progress in the standard education curriculum and how it can be used in a regular classroom setting. The IEP outlines the related services and supports students will receive and how often they will be provided. These can include many different things; for example, transportation; speech-language pathology and audiology services; psychological services; physical and occupational therapy; recreation, including therapeutic recreation; social work services; and medical services (for diagnostic and evaluation purposes only). Other services may be provided on a consultative basis. In these cases, the professional consults with the teacher to come up with strategies to help the child but doesn't offer any hands-on instruction. Other services can be delivered right in the classroom, so the child's day is not interrupted by therapy. When deciding how and where services are offered, the child's comfort and dignity should be a top priority. Some services may be offered outside the regular education classroom, with students getting small-group instruction in a particular subject area.

The IEP should be reviewed annually to update the goals and ensure the levels of service meet the student's needs. During the school year, progress monitoring will be done often to make sure the student is achieving goals set in the IEP. IEPs can be changed at any time on an as-needed basis. If parents disagree with any part of the IEP, mediation and hearings are options. It is important for teachers to understand the IEP process and their role in delivering instruction to students who have an IEP.

3. 504 Plans

Section 504 of the Rehabilitation Act of 1973 is designed to help parents work with educators to design educational plans for students with physical or mental impairments, legally ensuring that students will be treated fairly at school. Students can qualify for a 504 plan if they have impairments that affect their abilities to:

- walk, breathe, eat or sleep
- communicate, see. Hear, or speak
- read, concentrate, think, or learn
- stand, bend, lift, or work

A 504 plan is different from an IEP. The main difference is that a 504 plan modifies a student's regular education program in a regular classroom setting. A 504 plan is monitored by classroom teachers. IEP programs are delivered and monitored by additional school support staff. Also, parental approval and involvement is required for an IEP, but not for a 504 plan. Full parental participation in the 504 plan process, however, is important for the student's academic success.

The goal of the 504 plan is for students to attend regular classrooms along with the services, accommodations, or educational aids they need. Examples of accommodations that may be in a 504 plan include:

- preferential seating
- extended time on tests and assignments
- reduced homework or classwork

- verbal, visual, or technology aids
- modified textbooks or audio-video materials
- behavior management support
- adjusted class schedules or grading
- verbal testing
- excused lateness, absence, or missed classwork
- pre-approved nurse's office visits and accompaniment to visits
- occupational or physical therapy

A 504 plan should be considered when a student is not benefiting from instruction due to a physical or mental impairment. The issue can be raised by a parent or legal guardian, teacher, physician, or therapist. A 504 plan can help when a student returns to school after a serious injury or illness, or when a student is not eligible for special education services or an IEP, but still needs extra services to succeed academically. Once an educational concern is raised, the school principal or other academic advisor sets up a meeting of a 504 planning team, usually consisting of parents, the principal, classroom teachers, and other school personnel (such as the school nurse, guidance counselor, psychologist, or social worker). Unlike an IEP, there are no set rules for what a 504 plan should include, and written plans are not required. Schools only are required to have written policies regarding their 504 plans.

The overall intent of the U. S. Rehabilitation Act of 1973 is to protect students from disability discrimination by any school that receives federal funding. The Act is a part of civil rights legislation.

In summary, parents have the right to choose where and how their children will be educated. The utilization of an IEP and/or 504 plan helps ensure that students will be treated fairly and equitably and be given a reasonable chance to succeed in school.

SUMMARY

The foundations of how people learn are based in being inquisitive and wanting to know more. Teachers have been able to use the Principles of Teaching and Learning to guide the way that students experience new information and provide the new information in an understandable way. The use of Inquiry Based Instruction and Problem-Based Learning allows students to inquire into problems and make decisions that will make smart solutions in their lives. As teachers consider that their behaviors can affect student performance, we should always act with purpose and make education about offering learner-appropriate experiences. Students each have different backgrounds and rely on their previous experiences to build new knowledge. Teachers should learn about their students and use resources to construct opportunities for students to engage in their own learning and set expectations that are reasonable for them to attain new skills and knowledge.

REVIEW QUESTIONS

1. Describe how the scientific method can be used in designing lessons.
2. Select a topic area in agriculture. How does the Experiential Learning Model from Kolb relate to learning that topic?

3. What are the four levels of inquiry-based instruction?
4. Select a topic area in agriculture. How might a teacher use the decision-making strategies to help students determine how they will proceed?
5. According to the VARK Model, what are the four learning styles?
6. How could an agriculture teacher help students determine their learning style and grow their other learning styles?
7. Which of the eight multiple intelligences are most common in a high school? Which are least common?
8. Provide an example for each of the five main categories of the Principles of Teaching and Learning when the principles were used well. Then, provide an example for each of the five main categories of a time when the principles were not used well.
9. How are the Variables of Effective Teaching (Rosenshine & Furst) similar to the Principles of Teaching and Learning?
10. Describe how an IEP can be used to support student learning.
11. What are the differences between an IEP and a 504 Plan?

REFERENCES

Banchi, H., & Bell, R. (2008). The many levels of inquiry. *Science and Children*, *46*(2), 26–29.

Barrick, R. K., & Thoron, A. C. (2019). *Principles of Teaching and Learning*. Retrieved from University of Florida website: https://edis.ifas.ufl.edu/pdf/WC/WC24300.pdf

Barrick, R. K., & Thoron, A. C. (2019). *Teaching Behavior and Student Achievement*. Retrieved from University of Florida website: https://edis.ifas.ufl.edu/pdf/WC/WC24400.pdf

Cohen, L. E., & Waite-Stupiansky, S. (2017). *Early Childhood Education: Developmental, Behaviorist and Critical*. Routledge. doi:10.4324/9781315641560

Gardner, H. (1991). *The unschooled mind: How children think and how schools should teach*. Basic Books.

Kids Health. (n.d.). *504 Education Plans: Tips for Teachers*. Jacksonville, FL: The Nemours Foundation. www.kidshealth.org

Kids Health. (n.d.). *Individualized Education Programs: Tips for Teachers*. Jacksonville, FL: The Nemours Foundation. www.kidshealth.org

Kolb, D. A. (1984). *Experiential Learning: Experience as the source of learning and development*. Prentice-Hall.

National Research Council. (2000). *How People Learn: Brain, Mind, Experience, and School: Expanded Edition*. The National Academies Press., doi:10.17226/9853

Newcomb, L. H., McCracken, J. D., Warmbrod, J. R., & Whittington, M. S. (2003). *Methods of Teaching Agriculture* (3rd ed.). Pearson Prentice Hall.

Steffe, L. P., & Gale, J. (2012). *Constructivism in Education*. Routledge. doi:10.4324/9780203052600

U. S. Department of Education. (n.d.). *Individuals with Disabilities Education Act*. Washington, DC: Author. www.sites.ed.gov/idea

U.S. Equal Employment Opportunity Commission. (n.d.). *The Rehabilitation Act of 1973 Sections 501 and 505*. Washington, DC: Author. www.eeoc.gov/statutes/rehabilitation-act-1973

KEY TERMS AND DEFINITIONS

504 Plan: An individualized and formal plan to help students in a regular classroom to get the support they need, often with accommodations for additional time or resources.

IEP: A legal document that outlines special education accommodations for a child that is created with a team of teachers, parents, and staff in a school.

Inquiry-Based Instruction: Is when a teacher sets up experiences for students to ask questions and determine what they need to learn.

Learning: A change in cognitive, psychomotor, or affective behaviors that is brought on by an experience.

Learning Styles: Generally referred to as the areas in which people prefer to learn. The VARK Model describes these preferred areas as visual, auditory, reading, and kinesthetic.

Problem-Based Learning (PBL): Is a student-focused approach to teaching that uses real problems as the content of the lesson and encourages students to develop problem-solving skills.

Teaching: The process of guiding other people and helping them learn something and go beyond what's provided.

Chapter 5
Designing the Curriculum and Developing a Course

Catherine A. DiBenedetto
https://orcid.org/0000-0003-1802-0854
Clemson University, USA

ABSTRACT

The purpose of this chapter is to describe various types of school-based agricultural education (SBAE) programs and explain the process to develop a comprehensive agriscience course in a career pathway of a SBAE program. Current issues and challenges associated with curricular decision making are posed and curriculum is clearly defined. Strategies to develop a school instructional program plan, facility use calendar, and a curriculum map are presented as steps to build the scope and sequence and align content with the course objectives. The principles of the understanding by design curricular model are demonstrated in three stages (identify the desired results, determine acceptable evidence, plan learning experiences and instruction). The chapter concludes by identifying the components of a quality course syllabus to arrange relevant information to design a document that can be utilized to describe the course, communicate expectations, and outline guidelines for learner success.

INTRODUCTION

Chapter 5 will identify and describe types of School-based Agricultural Education (SBAE) Programs while introducing the process to develop an agriscience course. Strategies will be presented to design curriculum and develop an effective course outline to provide a framework for a complete SBAE program. The objectives of this chapter are to 1.) explain SBAE programs and the course development process, 2.) examine strategies to develop an effective SBAE course outline, and 3.) develop a comprehensive course for a SBAE Program.

DOI: 10.4018/978-1-6684-3420-8.ch005

TYPES OF SCHOOL-BASED AGRICULTURAL EDUCATION (SBAE) PROGRAMS

How are SBAE Programs Described?

There are various types of School-based agricultural education (SBAE) programs which include P-5 (elementary) school, middle school, comprehensive high school, Career and Technical Education (CTE) Centers and adult education programs. When the Smith-Hughes Act was passed by the United States Congress in 1917, agricultural education joined the public education system. According to the National FFA Organization, "over 800,000 students participate in formal agricultural education instructional programs offered in grades seven through adult throughout the 50 states and three U. S. territories" where SBAE programs "prepare students for successful careers and a lifetime of informed choices in the global agriculture, food, fiber and natural resources systems. Students are provided opportunities for leadership development, personal growth and career success" (FFA.org, n.d.).

There are three main components through which SBAE is delivered:

1. Classroom/Laboratory instruction (contextual learning);
2. Supervised Agricultural Experience programs (work-based learning);
3. Student leadership organizations (National FFA Organization, National Young Farmer Educational Association and National Post-secondary Agricultural Student Organization).

Over 13,250 agricultural education teachers share their passion, knowledge and technical expertise to teach students in nearly 8,500 agricultural education programs throughout the United States (Foster et al., 2020). SBAE programs are accessible for students in middle school programs, comprehensive high school programs, and Career and Technical Education (CTE) Centers.

Elementary agricultural education programs (K-5) are beginning to form as evidenced by a pilot test in the state of Georgia when Senate Bill 330, titled the Quality Basic Education Act was signed into state law during the 90th Annual Georgia FFA Convention on April 27, 2018. To provide agricultural education in elementary schools in the state of Georgia, the three-year pilot program began in 26 schools throughout 22 counties during the 2019-2020 school year. The goal of the pilot program was to utilize results from data collected over the three-year timeframe to determine if and how to implement state wide elementary agricultural education programs. Standards and curricula for the elementary agricultural education program following the National FFA three-component model were developed with input from Georgia agricultural education teachers and approved by the Georgia State Board of Education in June of 2019 (Georgia Senate Bill 330, Quality Basic Education Act, n.d.).

The purpose and main focus of a middle school (grades 6-8) program is for students to explore career opportunities within the broad agriculture industry. Students participate in activities to introduce them to leadership and personal skill development while helping them develop agricultural literacy skills, as well as global competence. In comprehensive high school programs (grades 9-12), students are further introduced to the agriculture industry more specifically through career pathways. A **career pathway** is a focused area of study where students complete a sequence of coursework that aligns content and skills to industry certifications, and preparation for post-secondary college majors affiliated with careers in agriculture, food, and natural resource sectors. Career and Technical Education Centers operate within public school systems to specialize in teaching students skilled trades, applied sciences, modern technologies, and career preparation. Ultimately the goal for any SBAE program is to enhance leadership and personal skills while ensuring students are agriculturally literate citizens, competent and prepared to enter the workforce and compete in the global economy.

Adult agricultural education programs also exist in various states throughout the nation. Adult education programs address distinct sets of needs and goals of individuals, of the institution, and of society (Knowles, 1980). The mission of an adult agricultural education program is to provide the opportunity to educate adults for entry into agricultural occupations, and to improve career skills that support them to be more productive and efficient in their current positions or entrepreneurial enterprises (Weller et al., 2013). The adult agricultural education program is an extension of the local, public, county SBAE program. In some states, a specific teacher is hired to serve as the adult agricultural education instructor. In other states the SBAE teacher serves the role of adult educator in addition to their other roles and responsibilities. The major goal of the adult agricultural educational program is to organize instruction for agriculturists and consumers in the local community. Meetings are conducted to expand current and highlight progressive agricultural significance. Instructors serve to enrich understanding and improve communication between the general consumer and the agricultural community. Programming reflects suitable agricultural practices for community members involved in both production and non-production agriculture careers to enrich the local economy and expand the technical skills for members to become more resourceful and industrious in their current position or business (Weller et al., 2013).

The National Young Farmer Education Association (NYFEA) is a professional organization that also serves the adult agricultural education population. The mission of the NYFEA is to "promote the personal and professional growth of all people involved in agriculture" (NYFEA, 2013). The target audience includes young producers, 4H and FFA members, agricultural education students, instructors, mentors and leaders.

Current issues and Challenges associated with SBAE Program Curricular Development

Two issues and challenges associated with SBAE program curricular development are defining the curriculum and its purpose and identifying factors that influence curricular decisions. When developing curricula for SBAE programs several factors influence the decisions that occur due to issues and challenges that can be associated with program development. Conducting a school and community profile is an important first step in designing a SBAE program. Once the school and community profile is completed, data can be utilized to help design each course within the associated career pathway which should be determined by and aligned with the school and community needs and available resources.

Curricula can be defined through broad, prescriptive and descriptive language. The meaning of curricula has evolved since 1902 when John Dewey stated a prescriptive definition as, *"Curriculum is a continuous reconstruction, moving from the child's present experience out into that represented by the organized bodies of truth that we call studies...the various studies...are themselves experience—they are that of the race"* (pp. 11–12). Harold O. Rugg (1927) affirmed *"[The curriculum is] a succession of experiences and enterprises having a maximum lifelikeness for the learner...giving the learner that development most helpful in meeting and controlling life situations"* (p.8). In 1957, Ralph Tyler expounded "*[The curriculum is] all the learning experiences planned and directed by the school to attain its educational goals* (p. 79). A decade later Robert Gagne (1967) *declared "Curriculum is a sequence of content units arranged in such a way that the learning of each unit may be accomplished as a single act, provided the capabilities described by specified prior units (in the sequence) have already been* *mastered by the learner"* (p. 23). McBrien and Brandt (1997) maintained *"[Curriculum] refers to a written plan outlining what students will be taught (a course of study). Curriculum may refer to all the courses offered at a given school, or all the courses offered at a school in a particular area of study"* (as cited in McBrien, 2019, p.4). In a quest to summarize the previous definitions of **curriculum** to closely align terms evolved throughout time and intended for study today, curriculum can be expressed as, learning experiences that are planned in a sequence of units outlining the content and skills learners will be taught to attain educational goals in a specified area of study.

Now that a working definition for curriculum has been outlined, it is important to mention the factors that can influence the decision-making process when developing a course for a SBAE program. The learners, resource materials and facilities, the teacher, the school environment, historical, ideological, cultural, social, political, economic, theoretical and pragmatic influences can all impact curricular decisions (Livingston et.al, 2015). While it is not the intent of the content in this chapter to deeply explore these factors, it is important for curriculum designers to be conscious of and sensitive to the influences each of these factors may enact on the overall development and planning process for the broad SBAE program and therefore overall individual course development.

Reflect upon the following factors that may influence how a course is developed for a SBAE program.

- What types of facilities and resources are available?
- What can be learned from conducting a school and community profile?
- What are the community members perceptions of agriculture and the SBAE program?
- What is the socioeconomic status of the students who will enroll in the SBAE program?
- What are the school administrators perceptions of agriculture and the SBAE program?
- What philosophical, theoretical and historical views influence a teachers' personal philosophy of education?

STRATEGIES TO DEVELOP AN EFFECTIVE SBAE COURSE OUTLINE

School Instructional Program Plan

Developing a school instructional program plan will help the teacher determine course sequence by career pathway offering, align classroom/laboratory, SAE and FFA components, develop a facility use calendar to include an inventory of facilities and resources, and coordinate facility use between courses and multiple teachers. The teacher also develops a rationale for the instructional program plan to align with the learning objectives for the course.

The **school instructional program plan** provides an opportunity to assess the SBAE program and categorize the career pathway(s) into a sequence of courses. While the school instructional program plan is a valuable tool to assist with the development of each course in the career pathway sequence, the plan may also be used to design a marketing tool for the SBAE program. Courses should be sequenced to increase the knowledge and skills of learner as they progress through the career pathway. Rigor and relevance are both essential to build a hierarchy of learning among the courses in the pathway. The goal of a successful school instructional program plan outlines the course standards for how learners will achieve proficiency based on the course objectives. It is important for the course objectives to align with the classroom/laboratory, Supervised Agricultural Experience (SAE) program and the student leadership organization (FFA) components addressed throughout the overall SBAE program.

A general description of the course should be written based on the content outlined by the curriculum. Course topics can then be identified and a series of five to seven **course objectives** should be written to help guide instruction. The intended outcomes for what the students will learn in the course should be stated as the course objectives. Mager (1984) defined a behavioral or performance objective, which can also be called an instructional or learning objective as "an intent communicated by a statement describing a proposed change in a learner- a statement of what the learner is to be like when [they have] successfully completed a learning experience" (p. 3). The course objectives should align with the big ideas, enduring understandings, and class topics to outline what the teacher will teach, what the students will do and how learning will be assessed. Course objectives serve as the general road map for instruction. Course objectives also indicate learning expectations for the student, assist the teacher in determining appropriate learning activities, and provide the basis for evaluation and assessment of student learning.

Lesson objectives convey specific behaviors, outline the condition for the behavior and establish the criterion for the behavior. Instructional or behavioral objectives include three domains - cognitive, affective, and psychomotor. The cognitive domain relates to the development of the learners' intellectual abilities, knowledge, and skills. The affective domain relates to the learners' emotions and feelings to assist with the development of attitudes, beliefs, and values. The psychomotor domain is the coordination of physical movements and performance that assesses what the learner will be able to do.

Enduring understandings are identified as the concepts and information learners should retain deeply to later transfer and apply to situations as a result of what they know, understand, and can do (Wiggins & McTighe, 2005). Clearly outlining the course objectives based on the course description, centered around the intended performance objectives, will help the teacher guide their learners to successful mastery. Course objectives should be measurable and define the competencies learners will achieve as a result of completing the course. Domains of Bloom's Taxonomy (Krathwohl, 2002) should be aligned to the knowledge, skills, or attitudes the learner is expected to know, do and understand as a result of instruction throughout completion of the course.

Reflect upon the following questions to help write course objectives.

1. What is the central purpose of the course?
2. How does the course correspond to the career pathway sequence?
    a. Does the course establish the introductory level content for the learner to begin to explore the subject?
    b. Is this the second course in the sequence for the career pathway that builds upon the prerequisite knowledge, skills and attitudes expected of the learner in level one?
    c. Is this course the final course in the career pathway where the goals intend the learner to transfer and apply the prerequisite knowledge, skills and attitudes gained from the level one and two courses?
3. How does the content of the course align with school-wide goals, the intracurricular activities related to the FFA organization, Supervised Agricultural Experience Projects, and Career Development Events?
4. How does the content of the course align with college and career preparation?

Developing a **facility use calendar** that provides an inventory of facilities and resources accessible for program use is the next step in completing a school instructional program plan. The facility use calendar can be very instrumental in coordinating timing and sharing of resources between courses, multiple career pathways, multiple teachers and FFA activities and events within a SBAE program. A calendar for the program should list all activities and events with dates aligned with the school calendar and FFA Chapter, county 4H Club and for the Adult Education Program as applicable. The calendar will help the teacher(s) be more organized and provide a valuable resource to share the programming efforts with all students, parents, administrators and stakeholders.

Reflect upon the following guidelines to help develop a facility use calendar and provide rationale for the instructional program plan designed.

- The facility use calendar is the result of the program development process.
- Creating a calendar is the logical next step following completion of the Program of Activities (POA) for the FFA Chapter.
- A calendar defines the program and organizes goals and events. All curricular and programmatic activities should be outlined in the calendar.
- Avoid scheduling major curricular, program or FFA chapter events on other important dates outlined on the master school calendar.
- Add major FFA chapter activities to the master school/district calendar.
- Use the calendar to help outline the courses within each career pathway.
- Use the calendar and the instructional program plan for recruitment, marketing of the program and sustaining fund raising efforts.
- Consider the best method to assure proper use of the instructional program plan and identify widespread distribution of the facility use calendar. Reference school policy for guidelines and social media/website posting.

HOW TO DEVELOP A COMPREHENSIVE COURSE

Alignment to Curricular Standards

Curricular decisions are typically the responsibility of administrators and teachers in the local school, district, or state. In 2015, the National Council for Agricultural Education published a revised version of content standards for Agriculture, Food and Natural Resource (AFNR) career pathways to serve as a benchmark curricula for states to utilize when developing courses for SBAE programs. Curricular standards outlined by SBAE programs may also be cross walked with core academic standards (NRC, 2012). An example for cross walking standards was offered by a panel of experts who identified disciplinary core ideas recommended for inclusion in secondary school agriscience programs to integrate the AFNR Career Pathways and the Next Generation Science Standards (Barrick, et al., 2019). The structure that was developed provided a framework around practices, crosscutting concepts and disciplinary core ideas for cross-curricular collaboration which focused on highlighting science, technology, engineering, and mathematics concepts and aligned with recommendations published by the National Research Council (2012). Agriscience teachers should utilize research-based practices and state standards documents recognized within respective states as a starting point for course development.

Once the appropriate standards documents are identified, a process to unpack the concepts begins. Unpacking can be described as deconstructing curricular standards to make sense of the meaning to support students potential to meet the learning targets. Learning targets focus on the essential knowledge and technical skills students need to know, understand and be able to do at the conclusion of the course. Unpacking the standards as a starting point to design curriculum enables teachers to present the content they are teaching in a manner which can be understood by their students (Wiggins & McTighe, 2005).

Outlining a Course

It is important to identify the components of a quality course outline. The process to develop the course outline can be structured by first developing a curriculum map that will assist in determine course overarching goals and objectives. Next, units of instruction within a course should be determined. Then, factors for sequencing the units in a course should be analyzed. Finally, the timeframe required to teach each unit of instruction should be identified.

Once the teacher has explored how to conduct a school and community profile and has the instructional program plan and facility use calendar developed, next it is time to outline a course by developing a curriculum map. A quality course outline starts by selecting the course in the career pathway sequence, determining overarching course goals and objectives framed by the identified curricular standards, scaffolding units of instruction within the course, analyzing factors to determine the sequence of each unit within the course and identifying an accurate timeframe required to teach each unit of instruction. A well-developed curriculum map will help both the teacher and the learners achieve the goals and objectives of the course. Course goals and objectives closely align with standards to determine the overall distribution of content for the course outline. Goals provide a sense of direction for the course, while standards are derived from goals and identify what must be accomplished or learned by completion of the course.

A **curriculum map** serves as a guide for instruction throughout the course. The first stage in the process to develop a curriculum map begins by outlining the scope and sequence for the course. The scope and sequence of the content within the course is scaffolded into logical units, designated by the number of days required to teach each unit, and determining the order in which the units would be taught. Generally, there are several steps to follow to unpack standards. The first step requires identifying key concepts and skills described by the standard. Next, types of learning targets should be identified, followed by determining the big ideas the standard represents (Wiggins & McTighe, 2005). Curricular decisions are made on the teachers' behalf that also factor in the school instructional program plan and the facility use calendar. Table 1 provides a suggested outline to follow to determine the course scope and sequence to assist with developing the curriculum map for the agriscience courses in the SBAE program.

Once the scope and sequence of the course has been outlined using the above recommendations, the second stage in the process to develop a curriculum map can be activated. A curriculum map organizes information about the content in a course into a single document to determine the basic flow and timeline of instruction. The main purpose of a curriculum map is to outline the scope and sequence of a course to identify what must be taught to all students to achieve the course goals and objectives as outlined by the state and/or national standards.

After the many factors that account for the rationale of the decisions that are made when developing the curriculum map are determined, the teacher will then align each lesson topic to the unit essential question and then to the lesson essential question(s) and instructional objectives planned for each lesson. Essential questions are written to encourage engagement in the content outlined by the standard. An es-

Table 1. Determining the course scope and sequence

How to Determine the Scope and Sequence for an Agriscience Course		
Step in the Process	**Rationale for Decision**	**Example**
Determine the total number of instructional weeks in the school year.	How long is the course? Is the course taught over the entire year or by semesters?	School Year (36 Weeks), Semester (18 Weeks), Quarter (9 Weeks)
Determine loss of instructional time due to school activities/events, meetings	How much instructional time will be lost due to activities on the schoolwide calendar?	Standardized Testing = 2 Weeks Holidays = 1.5 Weeks School Assemblies = 3 days Teacher Inservice Meetings = 5 days/1 week Other
Scaffold the course into logical units of instruction based on the topic and organized by the content.	What is the course of focus? Where does the course align within the career pathway?	Introduction to Agriscience Horticulture I, II, III, Animal Science I, II, III, Agricultural Mechanics I, II, III, etc.
Unpack the curricular standards related to units in the course.	What state and/or national standards are required to be addressed in the course?	Identify and outline State Standards and/or National AFNR Standards to meet course objectives as recommended by the local school, district, or state.
Determine how much time to spend on each unit.	How should a course be scaffolded into daily lessons? What are the student, teacher and community interests and needs? What is within the teacher's ability and what is the availability of resources/ facilities?	Course = 36 Weeks, Unit = 1 to 5 Weeks, Lesson = 1 to 2 Days Depth of content versus breadth of content Refer to the facility use calendar
Determine the order in which to teach units.	How should the units be scaffolded into a logical order? What prerequisite concepts are needed?	Time of year Overall logic based on growing season and weather Timing of FFA activities and Regional or State CDE, Fair Dates, Access to facilities Review and demonstration of proper safety procedures before working in lab

sential question helps frame the course or unit and is written as an open-ended, provocative question that focuses instruction and helps elicit inquiry (McTighe & Wiggins, 2013). As standards are unpacked and the essential questions are developed, the teacher can also align formative and summative assessments to provide evidence of student achievement towards the learning targets.

Table 2 provides an example of how to develop one unit of instruction for an introductory Agriscience and Technology course. Create one unit of instruction using the format outlined for each unit identified for the course. Typically each unit of instruction will account for a minimum of one and no more than three weeks of instruction.

Organizing the Course Content

Once the curriculum map is completed, it is time to move deeper into organizing the content of the course. The **Understanding by Design (UbD)** curricular model developed by Wiggins and McTighe (2005) encourages the curricular designer to "begin with the end in mind". In doing so, priorities for content are established based upon course enduring understandings, knowledge and skills that are im-

Table 2. Example unit in curriculum map for agricultural science and technology course

<table>
<tr><th colspan="4">Course: Agriscience and Technology</th></tr>
<tr><td>Number of School Days: 180</td><td colspan="2">Number of Days for School Activities: 15</td><td>Number of Instructional Days: 165</td></tr>
<tr><td colspan="3">Unit of Instruction: Using Agriscience Tools and Equipment</td><td>Total # of Days in Unit: 9</td></tr>
<tr><td colspan="4">Curricular Standard: CS.03. Examine and summarize the importance of health, safety and environmental management systems in AFNR workplace
Performance Indicator: CS.03.04. Use appropriate protective equipment and demonstrate safe and proper use of AFNR tools and equipment.</td></tr>
<tr><td colspan="4">Unit Essential Question: How do I properly identify, use, and maintain agriscience tools and equipment?</td></tr>
<tr><td>Lesson Topic:</td><td>Lesson Essential Question:</td><td>Instructional Objectives derived from Curriculum Standards
The student will be able to:</td><td># of Days</td></tr>
<tr><td>Practicing Safety in the Lab</td><td>Why is laboratory safety important?</td><td>1. Discuss the meaning of safety
2. Analyze hazards found in agriscience laboratories
3. Identify personal protective equipment (PPE)
4. Demonstrate how to properly use personal protective equipment (PPE)
5. Describe safety practices with machinery and tools
6. Demonstrate safety in agriscience laboratories</td><td>3</td></tr>
<tr><td>Identification of Tools and Equipment in the Agriscience Laboratory</td><td>What types of tools and equipment are used in agriscience laboratories?</td><td>1. Discuss the importance of tools and equipment in agriscience.
2. Identify common agriscience equipment.
3. Demonstrate how to maintain agriscience equipment.</td><td>2</td></tr>
<tr><td>Using the Microscope</td><td>What skills are required to use a microscope?</td><td>1. Explain the meaning and importance of microscopes in agriscience.
2. Identify the major parts of a microscope.
3. Demonstrate the proper use and care of a microscope.
4. Demonstrate how to make a temporary wet-mount slide.</td><td>3</td></tr>
<tr><td>Summative Assessment</td><td colspan="2">Unit Test: Students will be assessed on their ability to properly identify, use and maintain agriscience tools and equipment. Note: 1) Minimum passing score of 85% is required on the safety exam for students to participate in the laboratory. Students will be required to retake the exam until a 100% passing score is achieved. 2) Students may not participate in laboratory activities until the safety contract is signed and on file. (Alternative assignments will be provided). Safety contracts must be signed by student and parent/guardian and on file at school by September 30th.</td><td>1</td></tr>
</table>

portant for learners to know and do, and knowledge the learner should be familiar with to achieve the goals and objectives of the course. There are three stages of the UbD backward design model. First, identify the desired results. Second, determine acceptable evidence, and third, plan learning experiences and instruction that will actively engage students to show if they met the lesson objective(s) (Wiggins & McTighe, 2005).

In the first stage, the curricular designer takes into consideration the goals for the course and the standards to identify the desired results expected for what learners will be able to know, understand, and do at the completion of the course. Normally the amount of content to be taught exceeds the instructional time allotted within the academic semester or year. Therefore, decisions must be rendered to prioritize the content. It is important to determine the enduring understandings that are anticipated for the learner to transfer and apply as a result of instruction. The enduring understandings are fundamental to the course.

Standards are then unpacked to prioritize knowledge concepts and content that are important for learners to know and do, and are assessed by performance tasks and projects, versus prioritizing information that is worth being familiar with and assessed by traditional exams and quizzes (Wiggins & McTighe, 2005).

Stage two involves determining how evidence of student learning will be collected to determine if they have met the standards. Evidence of student learning is collected in the form of assessments (quiz, test, project) and must be aligned to the desired results previously outlined in stage one. The assessment evidence that is needed to document mastery of the unit or course is designed before instruction occurs and specifies a variety of assessment methods that directly align with the instructional experiences and activities (Wiggins & McTighe, 2005).

The third stage of the UbD model (Wiggins & McTighe, 2005) utilizes the evidence determined in stage two and aligns with planning of instructional experiences and activities. The instructional activities create a linkage to the previously defined instructional goals and connect the content directly to the assessments designed in stage two. Only content that is taught can be tested. In this stage enabling knowledge and skills needed for students to achieve the desired results as outlined in stage one are planned. Activities are created to prepare students with the required skills to master the content. Resources and materials are identified to assist with accomplishing the goals for the course.

When the UbD model for curricular design (Wiggins & McTighe, 2005) is followed, course structure and organization of content can be achieved to promote clear expectations for student achievement and mastery of the learning objectives. Curriculum design can be challenging. Using the UbD model (Wiggins & McTighe, 2005) specifies a method of purposefully thinking about curriculum planning, helps develop and deepen student ability to make meaning of and transfer learning, offers an opportunity for the teacher to unpack standards to develop desired results and select appropriate assessments for their learners, allows regular reviews of units and the curriculum against standards to improve overall quality, and is a continuous-improvement approach to achievement.

Reflect upon the following questions when using the UbD curricular model.

- How can a teacher look at curriculum like goal setting in order to provide clear direction for where the instructor and learners are going?
- What are the core performance tasks related to the course content?
- What should learners know and be able to do?
- What evidence will show the learner understands?
- What are the expectations for learning?
    - Learning can be in the form of performance tasks, other evidence such as tests, quizzes, work samples, projects, peer assessment and cooperative learning, learner self-assessment or reflection.
- What are the needs of the individual learners?
    - Consider learner readiness, interests, and preferred learning styles.

Presenting the Course

To present the agriscience course to students, parents, guardians, and administrators, a course syllabus should be designed and shared. Presenting the course by designing the syllabus allows the teacher to set the tone for the manner intended to conduct the overall management of the classroom environment and goals for instruction, to engage and challenge the learners access to educational activities and experiences aligned with the curriculum. A **syllabus** is a document that includes course policies, rules and regulations, required text books, grading policy, course expectations, outlines or summarizes topics to be included in a course, and helps students be successful in the course. The purpose of the course syllabus is to provide a descriptive outline of the course expectations, while framing how students should prepare and engage in the course. The course syllabus serves as a contract and communication tool between the teacher, the student, and the parent/guardian. The components of the course syllabus will guide the course for the semester or year, whichever the case may be, dependent upon the school schedule.

The three broad topics to consider when designing a course syllabus include 1) curriculum design and planning, 2) expectations and procedures outlined for students to follow, and 3) interventions to indicate how the policies adopted by the teacher will be monitored to evaluate and assess student learning in the course. The components of a quality course syllabus can include a combination of the items listed below.

- Course information:
    - Course name
    - Course description
    - Course goals
    - Required texts/resources, readings
    - Required materials, supplies and technology
- Course administration:
    - Teacher name
    - Teacher contact information- office phone number, email address, office hours
    - Course location, room number
    - Course schedule/meeting days and times
- Course Student Learning Outcomes
- Course Requirements
- Schedule of assignments
- Evaluation of student performance, grading policy
- Requirements for written and/or typed assignments
- Policy for assignment due dates, missing, late work
- Test/Exam policy
- Course calendar
- School resources
- Other important information regarding FFA, SAE expectations – dues, field trips etc.
- Policies and guidelines required by the school/district
- Signed, contractual agreement

Many of the components listed above may be outlined by school policies that should be reviewed prior to establishing any course syllabus. Grading and all other policies should adhere to school/district guidelines.

Reflect upon the following questions to design a course syllabus.

- How does a personal philosophy of teaching influence the tone a teacher sets for the course?
- What logistics are important for students to know to be prepared for learning and achieve the course goals?

- What are the expectations and responsibilities required of students to be successful in the course?
- What resources, materials, supplies and technology are needed for students to be actively participate in the course?
- What expectations should be outlined for FFA, SAE and other school-related activities and events?
- What policies and guidelines need to be outlined for the syllabus to serve as a learner contract and communication tool?

SUMMARY

This chapter described various types of SBAE programs and provided a detailed description with approaches and strategies to develop an effective SBAE course outline for an agriscience course in a career pathway. Developing a comprehensive course for a SBAE Program can be challenging and compels the curricular designer to think and plan with the end in mind. By recognizing factors that influence curricular decisions, the curricular designer becomes equipped with the knowledge and skills essential to develop a rationale for the instructional program plan and facility use calendar that will lead to the creation of a curriculum map. A curriculum map guides instruction and organizes the scope and sequence of the content within the course. Course objectives serve as a road map for instruction and are written to outline intended outcomes and the enduring understandings for what the students will know, understand, and do by the completion of the course. The conceptual model in Figure 1 depicts a visual summary of the steps to develop a course for a SBAE Program.

Figure 1. Developing an agriscience course

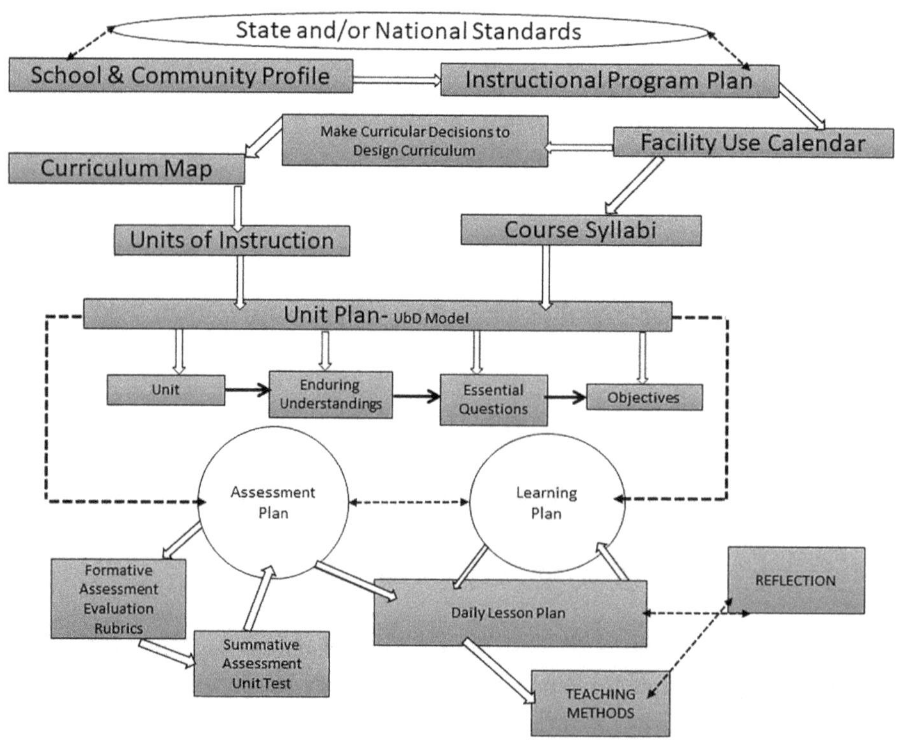

REVIEW QUESTIONS

1. How can a SBAE program be described to a stakeholder in the school community?
2. How is curriculum defined?
3. What issues and challenges influence the curricular decisions required to develop an agriscience course?
4. What is the purpose of a school instructional program plan?
5. How does a curriculum map help align all components of a quality course outline?
6. What are the three stages of the backward design process recommended to organize course content?
7. Why is it important to design a quality syllabus to present an agriscience course?

REFERENCES

Agricultural Education About FFA. (n.d.). https://www.ffa.org/agricultural-education/

Barrick, R. K., Heinert, S. B., Myers, B. E., Thoron, A. C., & Stofer, K. (2018). Integrating disciplinary core ideas, the agriculture, food and natural resources career pathways and Next Generation Science Standards. *Career and Technical Education Research*, *43*(1), 41–56. doi:10.5328/cter43.1.41

Bloom, B. S. (1956). Taxonomy of educational objectives. Vol. 1: Cognitive domain. *McKay*, *20*(24), 1.

Dewey, J. (1902). *The child and the curriculum.* University of Chicago Press.

Foster, D. D., Lawver, R. G., & Smith, A. R. (2020). *National Agricultural Education Supply and Demand Study, 2019 Executive Summary.* http://aaaeonline.org/Resources/Documents/NSD2019Summary.pdf

Gagne, R. W. (1967). Curriculum research and the pro-motion of learning. In R. W. Tyler, R. M. Gagne, & M. Scriven (Eds.), *Perspectives of curricular evaluation.* Rand McNally.

Georgia Senate Bill 330, Quality Basic Education Act. (n.d.). https://tinyurl.com/2dntpjpj

Knowles, M. S. (1980). The modern practice of adult education: From pedagogy to andragogy (revised and updated). Englewood Cliffs, NJ: Cambridge Adult Education.

Krathwohl, D. R. (2002). A revision of Bloom's taxonomy: An overview. *Theory into Practice*, *41*(4), 212–218. doi:10.120715430421tip4104_2

Livingston, K., Hayward, L., Higgins, S., & Wyse, D. (2015). Multiple influences on curriculum decisions in a supercomplex world. *Curriculum Journal*, *26*(4), 515–517. doi:10.1080/09585176.2015.1095015

Mager, R. F. (1984). *Preparing instructional objectives* (2nd ed.). David S. Lake.

McBrien, J. L. (2019). *The Nature of Curriculum.* SAGE Publications, Inc. www.sagepub.com. https://pdf4pro.com/view/chapter-1-the-nature-of-curriculum-sage-b784c.html

McBrien, J. L., & Brandt, R. (Eds.). (1997). *The language of learning: A guide to educational terms.* ASCD.

McTighe, J., & Wiggins, G. (2013). *Essential questions: Opening doors to student understanding*. ASCD.

National Council for Agricultural Education. (2015). *Agriculture, food, and natural resources (AFNR) career cluster content standards.* https://www.ffa.org/SiteCollectionDocuments/council_afnr_career_cluster_content_standards.pdf

National Research Council. (2012). *A framework for K-12 science education: Practices, crosscutting concepts, and core ideas*. National Academies Press., doi:10.17226/13165

National Young Farmer Educational Association. (2013). http://www.nyfea.org/educational-overview.html

Rugg, H. O. (Ed.). (1927). The foundations of curriculum-making (Twenty-Sixth Yearbook of the National Society for the Study of Education, Part II). Bloomington, IL: Public School Publishing.

Tyler, R. W. (1957). The curriculum then and now. In *Proceedings of the 1956 Invitational Conference on Testing Problems*. Princeton, NJ: Educational Testing Service.

Weller, C., Rewind, C., & Huber, B. (2013). *A Guide for Adult Agricultural Education and Young Farmer Programs in Pennsylvania.* A project of Vision for Pennsylvania Agricultural Education in Cooperation with Pennsylvania Department of Education Pennsylvania State University Department of Agricultural and Extension Education.

Wiggins, G. P., & McTighe, J. (2005). *Understanding by design*. ASCD.

KEY TERMS AND DEFINITIONS

Career Pathway: A focused area of study where students complete a sequence of coursework that aligns content and skills to industry certifications, and preparation for post-secondary college majors affiliated with careers in agriculture, food, and natural resource sectors.

Course Objective: Intended outcomes written to help guide instruction for what the students will learn in the course.

Curriculum: Learning experiences that are planned in a sequence of units outlining the content and skills learners will be taught to attain educational goals in a specified area of study.

Curriculum Map: A document used to organize information about the content in a course to determine the basic flow and timeline of instruction.

Enduring Understandings: The concepts and information learners should retain deeply to later transfer and apply to situations as a result of what they know, understand and can do.

Essential Question: An open-ended, provocative question that focuses instruction and helps elicit inquiry.

Facility Use Calendar: An inventory of facilities and resources accessible for program/course development over a predetermined time period.

School Instructional Program Plan: Organization of the SBAE program to categorize the career pathway (s) into a sequence of courses. While the school instructional program plan is a valuable tool to assist with the development of each course in the career pathway sequence, the plan may also be used to design a marketing tool for the SBAE program.

Syllabus: A contract and communication tool between the teacher, the student, and the parent.

Understanding by Design (UbD): Curricular model developed by Wiggins and McTighe (2005), encourages the curricular designer to begin with the end in mind.

Chapter 6
Planning for Teaching

Rebekah Barnes Epps
The University of Kentucky, USA

ABSTRACT

This chapter consists of a clear and direct path to assist educators in creating specific unit and lesson plans. The chapter clearly outlines the differences among unit plans, lesson plans, the use of the three-circle model in agricultural education, and how to evaluate classroom resources. There are templates and examples available for the educator to utilize in planning for their curriculum, program, and resources. If an educator utilizes this chapter effectively, many problems with student engagement, pedagogical understanding, and planning will be eliminated.

INTRODUCTION

An old British Military adage, sometimes known as the 6 Ps, states "Proper Planning and Preparation Prevents Poor Performance". No educator desires to deal with poor performances of their students and should model high achievement and performance. This is done best through planning and preparation. This chapter will assist the educator in the planning process. Being able to plan lessons that are engaging, inclusive, rigorous, and relevant is extremely important for learning to occur. Educators will be able to develop unit plans that utilize the problem-solving and inquiry-based methods to meet the needs of all learners. Educators will integrate the problem-solving method and inquiry-based learning into the classroom, laboratory, Supervised Agricultural Experiences, and FFA. Educators will also learn to evaluate and assess curriculum, instructional materials, and equipment to be utilized in the classroom.

DOI: 10.4018/978-1-6684-3420-8.ch006

UNIT AND LESSON PLANS USING PROBLEM-SOLVING AND INQUIRY-BASED METHODS

Differentiate Between Unit and Daily Lesson Plans

Unit plans are longer in length and last between 15-20 days. Unit plans are typically larger in the depth of standards and contain broad concepts. These broad concepts align with national standards and usually help to guide unit plans. In utilizing unit plans, student learning is typically assessed through summative means, or assessment of learning, at the end of the unit. Formative assessment is used throughout the 15-20 days to assist the teacher in forming the lesson content, procedures, or activities (Epps, 2021).

Daily lesson plans are between 1-3 days in length and contain specific and measurable goals for the learner. Learning is usually measured through formative assessments, or assessment for learning, including entry and exit slips, gaming quizzes and polls, kinesthetic activities, and drawings.

Example of a Unit Plan containing Daily Lesson Plans

The individual unit, lasting between 15-20 days, may be entitled Plant Propagation. Plant propagation could be delineated between Asexual and Sexual Propagation. Daily lessons on asexual propagation methods would include 1) Cuttings, 2) Air Layering, 3) Plant Division, 4) Grafting, and 5) Budding. Daily lesson plans on sexual propagation methods would include 1) seed germination, 2) breaking seed dormancy, and 3) transplanting seedlings. The author has included a template to utilize for unit plans later in this chapter along with an example of a detailed unit plan. Please see Table 1, Table 2, and Table 3 (Epps, 2021).

Key Principles of Unit Lesson Planning

As an educator, it is important to align or crosswalk the unit with proper national, state, and local standards. This ensures a standard of accountability within the course and can easily show how electives such as art, humanities, and career tech can work to enhance student knowledge just as math, science, English, or social studies. Standards should include Next Generation Science Standards (NGSS), International Society for Technology in Education Standards for Students (ISTE), Common Core Math Standards, National Agricultural Content Standards, State Standards, and Local School Initiative Standards if applicable.

The educator should choose a variety of teaching methods, teaching techniques, and teaching strategies to meet the needs of all learners. This should include diversity through Socio-Economic Status (SES), race, ethnicity, gender identity, learning levels and content matter. Examples of various teaching methods include: The Socratic Method, Problem-Solving Method, Inquiry-Based Learning, Direct Instruction, and Indirect Instruction (Epps, 2021).

Describe the Steps in the Lesson Planning Process

Choose a sequence of steps that are clear and concise. The steps listed below are available to assist the educator in planning a lesson.

a. Objectives/Standards

The educator must determine why the lesson learned is important and identify what standards will be met and aligned to within the unit. Clear descriptions of how learning of the unit goals will be assessed must be included.

b. Mental Set
The mental set, sometimes known as the anticipatory set or the interest approach, prepares the learners for what is to be taught/learned that day or for that lesson. It is used to pique the learner's interest in what is yet to be taught.

c. Critical Content Material
Critical content material is what the learners need to know before moving forward with a support activity. Educators must decide what previous knowledge the learners have already attained to determine if scaffolding is needed, or the requisite knowledge is enough to proceed in gaining of new knowledge.

d. Teaching Strategy
Choosing an appropriate teaching strategy is especially important in the planning process. This could be a demonstration of learning, evaluation of a case-study, worksheets, laboratory, dissection, etc. A major fallacy, or pitfall, for young teachers is planning the "cool activity" with no real objective or correlation to learning.
For example: The soil in a cup activity to teach soil layers is fun and everyone enjoys a delicious dessert at the end. However, taking a true soil sample, using water to separate the layers is easier to connect for learners and much more affordable for the educator as soil tubes/jars are reusable and natural soil tends to be cheaper than Oreo cookies and pudding.

e. Assessment
Assessment for, and of, learning is needed in the planning process. When determining what type of assessment, remember that formative assessment forms how the teacher may form lessons based upon the student's assessment for learning. Assessment of learning, or summative assessment, is utilized to gauge learning larger concepts. This type of assessment is done mainly at the end of the unit in the form of an exam, project, or presentation.

f. Teacher Reflection
After using the assessment, the educator must reflect if students learned the material. If learning occurred at a satisfactory level, the educator should move forward to the next construct/concept/standard in the course. If not, the educator must decide what type of remediation is necessary and what changes need to occur in the teaching strategies for learning to occur.

Creating a Unit Plan

Many beginning educators are sometimes intimidated by the idea of planning an entire course or unit. It is not a task that should be taken lightly, and it takes practice, forethought, and planning for it to occur successfully. One way to combat the intimidation factor is the use of a template. The author has utilized some version of this template for many years as a teacher educator and finds most beginning educators continue to use some form of the template as they meet with principals and superintendents for evaluations while in service teaching. The use of pathways is one system which is discussed in Chapter 4 which some utilize in creating curriculum.

Table 1. Unit plan template

UNIT 1:		Length:	
Essential Question(s):			
Key Terms: *(not limited to the following)*			
Connection to Standards:			
Subjects/Daily Lessons	**Objectives**	**Assessment**	**Skills Acquired**
Section 1: **Section 2:** **Section 3:** **Section 4:** **Section 5:**	Upon completion of this unit, students will be able to:		

Each Section of the unit correlates to 55-minute class periods for instruction.

Table 2. Example of a beef cattle unit plan for an animal science course

UNIT 1: Beef Cattle	Length: 16 Days
Essential Question(s):	**SECTION 1: Functions of Anatomy** What are the external parts of the beef animal? What are the internal parts of the beef cow? What is the anatomy of the reproductive anatomy? What are the parts of the ruminant digestive system? What is the function of the ruminant digestive system, and how does it differ from the dairy industry in terms of reasons for production? **SECTION 2: Breeds** What are the major production breeds for the beef industry? What are the benefits to each of the major breeds of beef cattle? What are the populations of each of the major breeds? Where did major breeds originate from? What purpose do major breeds serve in the food production system? **SECTION 3: Production Systems** What products are provided through the beef? What is the variance in products within the beef industry? What parts of the beef animal give which retail cuts? What is the difference in muscle of the beef animal? **SECTION 4: Housing Facilities** What does a cow/calf operation consist of? What does a stocker operation consist of? Where in the US are most beef cattle produced? What are feedlots? **SECTION 5: Reproduction** What is the purpose of reproduction in the beef industry? What does the reproductive cycle in the beef cow look like? How long is the gestation length for a beef cow? What are the reproductive hormones being produced during gestation? What purpose do the reproductive hormones serve? What is an EPD? How can EPD'S help farmers predict the value of offspring? What are the desirable characteristics in beef cattle? How can farmers evaluate their herd? What reproductive technologies are available for beef producers? **SECTION 6: Careers** What are the careers available in the dairy industry? What degrees are required to find these potential jobs? How can I be involved in the dairy industry? How can I be an educated consumer?

continued on following page

Table 2. Continued

UNIT 1: Beef Cattle		Length: 16 Days
Key Terms: *(not limited to the following)*	External anatomy, skeletal, internal, organs, functions, four-compartment, omasum, abomasum, reticulum, rumen, pH, acid, stomach, digestive, reproductive, technology, CIDR, Cervix, vagina, uterus, uterine horn, artificial insemination, natural cover, bull, cow, calf, straw, udder, poll, back, loin, rump, hooks and pins, dewclaw, brisket, hoof, horns, polled, ears, nose, rostrum, barrel, ribs, spine, tail, docked, switch, knee, mammary, rear udder attachment, beef breeds, angus, Hereford, Gelbvieh, Chinana, Simmental	
Connection to State Standards:	OF3 Identify the major skeletal and external parts of production and companion species OF2 Differentiate major animal breeds within the industry and their production strengths OF4 Identify wholesale and retail cuts of production species OG4 Evaluate preventative measures for controlling and limiting the spread of diseases, parasites, and disorders among animals OE4 Evaluate measures to maintain the quality of the environment in areas used to raise and house OB1 Describe characteristics of effective animal care facilities OD1 Describe the functions of major reproductive organs and hormones in the male and female reproductive systems OD2 Contrast the processes of natural and artificial breeding methods OD3 Infer how age, size, life cycle, maturity level, gestation, and health status affect the reproductive efficiency of male and female animals OD4 Evaluate the use of quantitative breeding values in the selection of genetically superior breeding stock (e.g., EPD's) OD5 Discuss the importance of efficient and economic reproduction in animals OC1 Describe the anatomy and physiology of digestive systems (monogastric, ruminant, hind gut fermenters, avian) OC2 Discuss the function of the six classes of nutrients regarding animal nutrition OC3 Compare common types of feedstuffs and the role they play in the diets of animals OC4 Explain the purpose of feed additives and growth promotants in animal production OC5 Formulate animal feeds based on nutritional requirements using feed ingredients for maximum nutrition and optimal economic production OF5 Evaluate desirable anatomical and physiological characteristics of animals within and between species OG1 Characterize common disorders, parasites, and vaccinated diseases of both production and companion animal species OG3 Select proper medical dosages and delivery methods EM2 Research and identify emerging technologies for specific careers, EL4 Discuss the importance of flexible career planning and career self-management EK2 Define jobs associated with a specific career path or profession	

continued on following page

Table 2. Continued

Subjects	Objectives	Assessment	Skills Acquired
Section 1: Functions of Anatomy (4 Days) a. External Parts b. Internal Parts c. Reproductive Anatomy d. Digestive Anatomy e. Functions of the Digestive Anatomy f. Differences between Beef and Dairy **Section 2:** Breeds (2 Days) a. Angus, Simmental, Charolais, Hereford, etc. a. Production Differences in breeds. **Section 3**: Production Systems (2 Days) a. Processing Standards, humane slaughter act, Temple Grandin? b. Retail Cuts c. Differences in Production across U.S **Section 4:** Housing/facilities (2 Days) a. Operation types, cow calf, stocker, order buyer, Stockyards? b. Grazing operations c. Geographic populations **Section 5:** Reproduction (4 Days) a. Reproductive Cycle b. Gestation Length c. Gestation and Lactating Hormones d. Reproductive tools and technologies. **Section 6:** Careers (2 Days) a. Careers in Animal Science b. How to be an Educated Consumer	Upon completion of this unit, students will be able to: **SECTION 1:** -Identify the internal and external parts of a beef cow through an interactive notebook. -Compare and contrast the forages that ruminants can digest versus humans. -Diagram the various uses for beef cattle using flip charts. **SECTION 2:** -Classify the various breeds of beef cattle and their role in the industry. -Identify geographic hotspots for beef cattle production using a map. **SECTION 3:** -Determine where beef retail cuts come from on the animal using an illustrative diagram. -Analyze the uses of certain muscle groups in the beef carcass. -Determine the difference in grades of beef cuts. -Determine the geographic slaughter methods from around the world. **SECTION 4:** -Differentiate between housing styles and beef production systems. -Identify the role beef plays in KY agriculture on a Nationwide scale. **SECTION 5:** -Identify the major reproductive hormones involved in the reproductive cycle of a beef cow and explain their function. -Distinguish between the various levels of the reproductive cycle and explain their role. -Connect reproductive technologies to their given use within the industry. -Calculate EPD's given a beef operation scenario. **SECTION 6:** -Identify career options in animal science. -Compare specific jobs in with required educational levels in the animal science field.	**1. Diagram Charts/Poster Charts** Students will create diagram charts that highlight the external portions of the beef cow, as well as the internal organs. These will be done in layers as students can flip from skeletal, to internal, to exterior. **2. Breed Notebook/ Collective PowerPoint Slides** Virtually, students will be broken into groups after being assigned a breed, each group will research their breed and create a PowerPoint slide on google slides to work alongside their classmates, the groups will come back together to present. **3. Rotation Stations Handout** Students will rotate around the room to various stations broken up into sections of the beef carcass, they will then identify either with physical cuts, or pictures, what the cut is, and where it comes from. **4. Operation Presentations** Students will be split into groups of cow calf, stocker, and order buyer operations, they will be given a herd count and location, it is their job to determine their breed raised, markets they choose to sell to, etc, they will then make a small PowerPoint or presentation to highlight their operation, and try to market their product to the class. **5. The Maury Show** Students will calculate an EPD given a scenario, and in addition, choose from a catalog of sires to select their chosen donor. They will then defend their choice explaining why they chose the bull they did, backing up with facts like calf weight, weaning weight, etc. **6. Interview a Producer/ Industry Project** Students will be given a group of producers (I will have reached out to them prior) that they can contact using email to gather information about their operation or career field. They will present these in a paper or presentation form.	• Speaking • Note-taking • Listening • Communication • Problem Solving • Asking questions • Organizational skills • Creativity • Connectivity • Comparing and Contrasting • Researching evidence • Writing • Supporting facts • Listening • Defending • Evaluating • Measuring • Calculating

(Goodlett, T.L., 2019)

Table 3. Example of a plant science unit plan for an introduction to agriculture course

UNIT 3: Plant Science		Length: 23 Days
Essential Question(s):	What are the four major parts of a plant? What are the functions of each of the four plant parts? What are the parts of a flower? What is a complete flower? How does water affect the germination rate of seeds? How does the sun play a role in the life of a plant? How do plants propagate? What is floral culture? What goes into a landscape? What is hydroponics? What is the difference of hydroponic systems vs a	
Key Terms: *(not limited to the following)*	Anther, Calyx, Chlorophyll, Chloroplast, Complete flower, Corolla, Embryo, Fertilization, Filament, Flower, Fruit, Germination, Germination Rate, Grafting, Imperfect Flower, Incomplete flower, leaf, Nutrient, Ovary, Ovule, Perfect flower, Petal, Photosynthesis, Pistil, Pollen, Pollination, Reproduction, Respiration, Root, Seed, Sepal, Sexual Reproduction, Stamen, Staminate, Stem, Stigma, Style, Vegetable	
Connection to State Standards:	Horticulture OA Develop and implement a crop management plan for a given production goal that accounts for environmental factors OA 6 Determine the optimal air, temperature, and water conditions for plant growth OA 7 Describe plant responses to light color, intensity, and duration OB Apply the principles of classification, plant anatomy and plant physiology to plant production and management OB 1 Explain systems used to classify plants OB 2 Describe the components, types, and functions of plant roots OB 3 Describe the components, types, and functions of plant stems OB 4 Describe the components, types, and functions of plant leaves OB 5 Describe the components, types, and functions of plant flowers (including seeds and fruit) OB 6 Differentiate sexual and asexual reproduction in plants OB 7 Analyze the life cycle of plant growth/development from seed to seed (e.g., annual, biannual, perennial) OB 8 Explain requirements necessary for photosynthesis to occur and identify the products and byproducts of photosynthesis	

continued on following page

Table 3. Continued

Subjects/Daily Lessons	Objectives	Assessment	Skills Acquired
Section 1: Introduction to Plant Science **(2 Days)** - Terminology - Why Plants are important **Section 2:** Plant Structures **(3 Days)** -Parts of a plant -Parts of the flower -Functions of each **Section 3:** Photosynthesis **(2 Days)** -Process behind it -What's happening through the process **Section 4:** Cellular Respiration **(2 Days)** -Anaerobic Respiration -Aerobic Respiration -Air Quality **Section 5:** Sunlight **(2 Days)** - Light Intensity - Light Quantity - Light Color **Section 6:** Plant Production **(5 Days)** - Hydroponics - Aquaponics - Relationship of fish and Plants - Greenhouse - Greenhouse systems **Section 7:** Jobs in the industry **(7 Days)** - Students will be working in two sectors of the industry. Students will then research a third. - Floriculture - Floral Arrangements - Landscape Design - Create their own landscape - Independent Research	Upon completion of this unit, students will be able to: **Section 1** - Describe how important plants are for our everyday life as humans by completing a *Plants Around Us* activity sheet. ***Section 2*** - Locate and label different parts of the plant and flower by dissecting a flower and labeling the terms and function to their location **Section 3** - Explain the process of photosynthesis by completing a Photosynthesis activity. **Section 4** - Explain the difference between Anaerobic and Aerobic Respiration by completing a respiration lab. **Section 5** - Create different growing conditions based on light intensity and light color by completing a *Growing with Light* activity **Section 6** - Describe different growing systems by creating an Information guide for different Horticulture areas learned in class. **Section 7** - Share three different jobs in the plant science industry by completing an ad for a job in that field	- Daily bell ringer questions related to what was learned the day before and what will happen today. This will be turned in every Friday. **Section 1** - Students will hand in a *Plants Around Us* activity sheet that has research on how we use plants around the world and in our personal lives. - Students create a study guide for terminology words that will be used throughout the unit. **Section 2** - Students will dissect and label parts of a plant and the flower. Students will then create a large poster that shows the location of each part as well as where it is located and its function. **Section 3** - Students will act out the photosynthesis process in groups as well as completing a lab. **Section 4** - Students will complete a Respiration lab and create a lab report of their findings. **Section 5** - Students will complete a *Growing with Light* activity. -Students will learn how different colored light affects growth as well as light duration and light intensity. Students will submit their findings in a lab report. **Section 6** - Students will create an information guide that will become available for other students and staff to utilize and learn more about different plant systems. **Section 7** - Students will create floral arrangements and submit their best one. - Students will design a plan for a landscape and submit their drawings and plans. - Students will research different jobs in the plant science industry and create a job add in that field.	-Group Work -Floral arranging -Writing - Employability Skills - Scientific Method

(Jaworski, J.M., 2019)

INTEGRATE CLASSROOM, LABORATORY, FFA, AND SUPERVISED AGRICULTURAL EXPERIENCES (SAE)

Describe the Relationship Between Classroom/Laboratory, FFA, and SAE

The relationship between Classroom/Laboratory, FFA, and SAE are best described as a three circle Venn Diagram. Most Venn Diagrams are used to compare items with certain areas of overlap. Each of the three areas should all work towards the center where overlap is at its highest portion. The Classroom/Laboratory is one area where the teacher is able to reach all students. The FFA Organization may not be able to always reach all students depending on the Career Development (CDE) or Leadership Development (LDE) Event rules and restraints in number of participants. The SAE for All is a very individualized process where students determine the best type of SAE for them whether it be career exploration, ag literacy, service learning, entrepreneurship, research, or placement in a job or career. As discussed in Chapter 2, it is important to have all three portions of the three-circle model for the student to have the necessary opportunities and foundation for success. Educators must be purposeful in their planning of how each portion of the three-circle model will work together.

Figure 1. The Three Circle Model
(The National FFA, 2022)

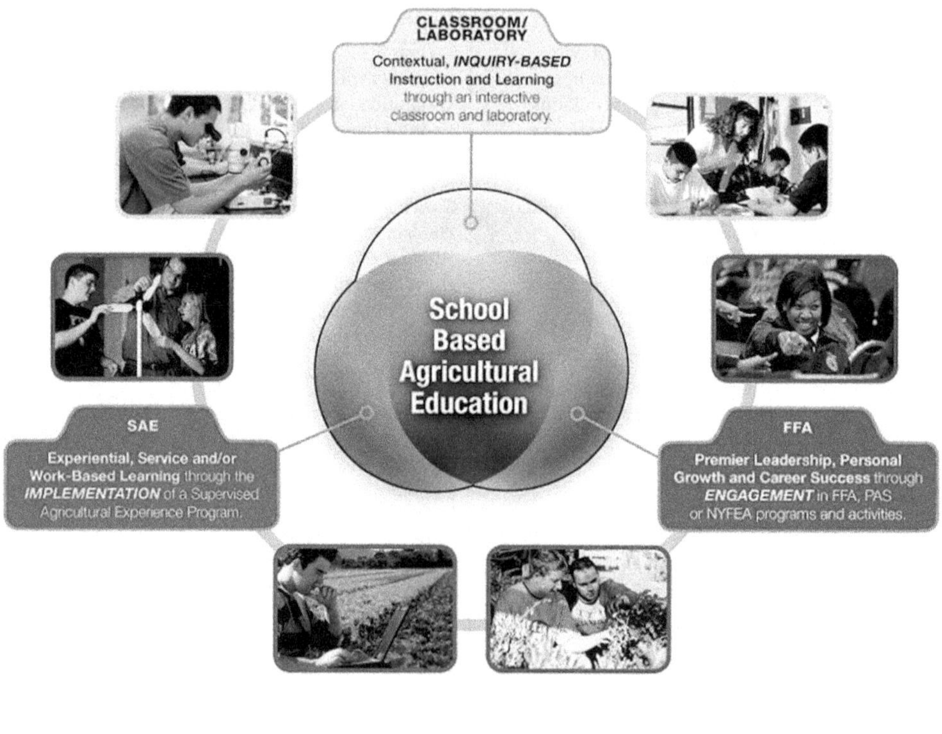

Utilize FFA as a Teaching Tool

School Based Agricultural Education (SBAE) Programs should be dependent on local chapters—local programs, local resources, local economy, and local challenges to drive the FFA activities. FFA should be viewed as truly an intra-curricular organization to develop student leaders. Students can be identified through class as being the highest scoring individual in plant identification, best at giving a speech or proposal based on in class presentations, quickly understanding soil colors and profiles from using in class laboratories. Those students who are identified can then be asked to participate and be further developed as a team for soil judging, nursery landscape, or ag marketing.

For example, one such community located in the Southeastern United States was home to 11 timber/logging companies and four sawmills. This community had two high schools where forestry, natural resources, and wildlife curriculum were needed to supply the local workforce. These high schools taught the curriculum to support the community. Students who had an interest in a possible future career in forestry, natural resources, and wildlife were encouraged to further explore options through contests in the FFA.

Students in an SBAE Program can utilize the FFA to earn recognition in a series of contests known as Career Development Events (CDEs) and Leadership Development Events (LDEs). Students can also participate individually, or with various teams, depending on their interest and the interest or need of the local community. The National FFA Organization (https://www.ffa.org) has a myriad of resources available to assist the educator with further developing students for participation in CDEs and LDEs.

Utilize SAE as a Teaching Tool

Experiential education is a teaching approach or theory in which educators plan purposeful experiences for their students, and guided reflection, to increase knowledge and skills. Agricultural educators are encouraged to utilize the Supervised Agricultural Experiences (SAE) as a purposeful experience for their students. SAE is experiential education and the opportunity for students to utilize the knowledge gained in the classroom and apply it to real-life settings. Each student begins with a foundational SAE that allows them to experience career planning and exploration, identify employability skills, understand personal financial management, safety in the workplace, and agricultural literacy as a consumer. Depending on the age, and other factors some students can advance quickly through foundational SAEs from merely awareness to more advanced understanding. Once a foundation has been set a series of types of experiences including internships/placement, ownership/entrepreneurship, research, school-based enterprise, and service learning. The SAE for All: Evolving the Essential Website (https://saeforall.org) which was developed by The Agricultural Education Council in conjunction with the National FFA Foundation is a great place to learn more specifics about each type of SAE.

It is essential that the SAE is supervised by the educator. Most SAEs consist of a yearlong or yearlong plus program taking place and not just a project. Each student's SAE must be carefully planned with clear, concise, and measurable goals. Parent support and approval is essential for an SAE project or program.

A valuable skill for everyone is the use and management of keeping records. This helps better manage a project or program for making decisions regarding growth, sales, and investing. These are skills that can easily transfer into adulthood when people make decisions regarding financing for a car or home, taxes, savings, and investments. Most recently a national record keeping system, known as The AET, has made student record keeping easier for students with SAE projects. This computer-based system has even served as a basis for local and state record keeping contests.

Identify the Role of Classroom/Laboratory Instruction

The third and final circle in the three-circle model is the classroom/laboratory. This is any instructional setting where teaching and learning occurs. Traditionally it occurs within the confines of a classroom. However, it can occur in a traditional science type lab, an outdoor classroom, field, forest, river, stream, or agribusiness. This portion of the three-circle model allows content to be seen, evaluated, and experienced in the community. Students are able to develop skills in science, math, management, leadership, technology, and communication through the content of agriculture.

Inquiry-Based Methods

Use of inquiry-based methods serve to promote engagement, curiosity, and experimentation within the classroom. According to Pedaste, et al (2014); there are five phases of true inquiry-based instruction. These include:

a. Orientation
   This is the phase where curiosity regarding a subject is stimulated through describing an issue or problem. Generally, a problem statement is generated during the orientation phase.
b. Conceptualization
   This phase includes the process of learning more and understanding the problem. Students will question the topic and begin generating research questions. At the same time, a hypothesis is generated where a research thesis or problem statement can be tested.
c. Investigation
   This third phase is where exploration and experimentation occur. Students can take the hypothesis and begin testing it during this phase generating data to be used in analysis.
d. Conclusion
   This stage occurs when students utilize the data generated to make informed decisions based on the investigation phase.
e. Discussion
   The final stage may be the most important as it where the findings from the previous stages are shared with others in the form of presentations creating discussion about the entire process and findings.

Problem-Solving Method

The Problem-Solving Method is another teaching strategy which should be utilized in the curriculum. This method allows students to learn by working on problems which utilizes new knowledge and function at a higher level of thinking. Students are expected to observe, analyze, interpret, and find solutions to a problem which occurs in a real-life scenario. This theory to practice process comes from basic common knowledge and then builds upon experience in new knowledge gained. The overarching goal is for students to be aware there is more than one way to approach and solve a problem, approach problems in a systematic way, implement solutions, and analyze data and fact by creating a valid experience.

The problem-solving method can be divided into three distinct phases—planning, execution, and evaluation. The teacher will work to assist the students in defining the problem. A drawing, diagram, and known concepts are utilized to help further define the problem. The execution phase includes allowing students to determine which problem, or portion of the problem, needs to be considered first. Students will then generate a list of various solutions. The teacher will guide students on which solution(s) can be evaluated before students implement the solution. The final phase is the evaluation phase where students are again led by the teacher in determining which solution(s) hold the most merit and are realistically feasible for the problem. Students are then usually tasked with writing up the results as a form of assessment. The written report typically includes the problem, the various solutions, the planning phase of determining which solution is best and how the problem is solved (Maheshwari, 2017).

EVALUATE INSTRUCTIONAL MATERIALS

Identify the Types of Instructional Material

It is important for educators to determine the most appropriate instructional materials to utilize during planning and teaching. Young, less experienced teachers will often rely on these instructional materials when beginning their careers. Older, more experienced teachers may have various experiences in the classroom with what works and may be more apt to utilize lessons they have developed, changed, or improved upon previously. If any one particular content area is a favorite or more comfortable area, teachers will tend to utilize their own knowledge and lesson planning skills in that area.

Instructional materials are the teaching and learning materials utilized in the classroom to convey knowledge. It can be already developed lesson plans, PowerPoints, case studies, laboratory manuals, test banks, worksheets, and quizzes. There are numerous sources available for educators, but not all are created equal. D'Angelo et. al. (2018) emphasized the importance of making sure the instructional material supports the objectives for the lesson. They also caution the educator to make decisions regarding instructional material based on the learners "age, emotional development, skill level, and preferred learning styles" (p.1). D'Angelo et al. (2018) and his colleagues identified six different categories where most instructional material can be placed. They are as follows:

1. Information based resources
   This is easily the most common type of resource as it is only content based. These resources do not usually include lesson plans but are considered viable sources of information where educators can attain content information from experts regarding a specific topic. Examples include state extension websites, the United Stated Department of Agriculture, The United States Food and Drug Administration, and a myriad of other government funded state sites.
2. File sharing resources
   This type of resource almost always includes a lesson plan. It is typically written, tested, and shared by educators. Many educators find that it can be easily modified to meet the needs of the students if an educator is willing to adapt and put their own style of teaching into the basic lesson plan. These are sometimes free resources including those from National and State Associations. A recent increase has been seen in educators sharing resources with other educators for pay like Teachers Pay Teachers.
3. Lesson Plans provided by Organizations
   These lesson plans and resources are typically free and are located on commodity group websites or offered through professional development from various commodity groups. Educators must

be cautious and evaluate the information as it does promote the agenda of the organization which may not be in line with the objectives of the lesson. Examples include National Ag in the Classroom, The Beef Board, American Farm Bureau, National Pork Producers Council, and other state and local commodity groups.

4. Fee-Based Instructional Materials
   These resources are created and marketed by curriculum companies. The resources are usually of high quality and include everything an educator may need as far as content, lessons, worksheets, and other supporting material. The largest drawback to these resources is sometimes the upfront cost and unknowing what objectives align with state and local objectives for the classroom.
5. Web-based Inquiry Resources
   This type of curriculum is one that contains modules, simulations, demonstrations and even gaming for students. Technology and access to the internet are required by the student in order for all activities and modules to sometimes work correctly.

Identify Sources of Instructional Materials

The author of this chapter has worked to create a list of various resources available to educators via the internet. Some of the resources are open educational resources available to anyone while others require an educational email address for access. While many are free, some do require a fee to access as well (Egbert, 2017). This complete list can be found in Appendix 1.

Create Criteria to Use in Evaluating Instructional Material

When an educator is utilizing curriculum, either self-designed or designed by others, it is important to evaluate the material. A series of questions that the educator may ask includes: How well is the curriculum/instructional material working? Does the instructional material meet the objectives and standards of the curriculum?

Does it allow for higher levels of learning on Bloom's Taxonomy or remain mainly in the lower levels? What are the strengths of the instructional material? What are the weaknesses of the instructional material (Egbert, 2017)?

There are six criteria used to develop instructional materials for the educational setting and classroom. Those are listed below.

1. Content
   The content in the instructional material is an opportunity for students to apply and develop their own ideas. It should encourage critical thinking skills of students and evaluation of content for future use in real-world applications. The instructional content should be relatable to experiences of the students and apply equally to students without regard to gender, cultural, religious, or SES background. The instructional material needs to reflect diversity through inclusive language, illustrations, photographs, and activities.
2. Organization and Structure
   The instructional material should provide cohesion that builds on previous understanding and knowledge of the students. The material resources need to provide various strategies for

investigations of social/political/local issues as well as be easily adapted for learning styles of various students.

3. Student Experiences
Students learn through active engagement. The resource/materials should provide knowledge and use of inquiry-based, open-ended learning, and problem-solving strategies as guided by the educator. It is highly recommended that technology be incorporated due to virtual and distance learning. However, the most important goal of the instructional material is that it should be relevant to previous student experiences and serve to add knowledge.
4. Support Materials
These are the materials used to help the educator in meeting the objectives of the lesson and serve to reinforce the student experiences. These should include adequate background content information for the educator to communicate to the students. It should also further knowledge and exploration into the community or with parents by engaging the students through various learning strategies. The materials used for support such as worksheets, labs, quizzes and each should have easy adaptations for different developmental levels of students.
5. Student Assessment Materials
Assessing the learning of students is important. When utilizing instructional materials, the educator must decide if the assessment is free from bias no matter the regard to gender, cultural, religious, or SES background. It is recommended the assessment is oriented towards real-world application and experiences and there are multiple forms of assessment throughout the material.
6. Program Development and Implementation
Various schools and districts utilize entire curricula that has been implemented for identified learning areas. These programs are research based, flexible for any size and scope of program and usually offers training, technical assistance, or follow-up for educators.

Select Instructional Material and Educational Technology

Many programs and districts look for curriculum to be utilized throughout many of its content areas. The selected material should be research based and include the six factors listed above. When implementing or adopting new curriculum, the rubric shown in Table 4 can assist in making decisions and selection of curriculum and instructional materials.

Summary

Planning for teaching is one of the most important aspects in education. Benjamin Franklin is known for the famous quote "If you fail to plan, you are planning to fail." This was in reference to personal finances but can be said of the same for educators and planning. Planning takes skill which can be developed through practice of writing objectives, daily lesson plans, unit plans, and previously discussed curriculum maps for an entire course. There are steps which can easily be followed to create a unit and lesson plan to ensure learning occurs for everyone. The use of the three-circle model in agricultural education works to compare learning to occur inside and outside the classroom with guided real-world application. Educators also have access to numerous educational resources and instructional material. It is important to evaluate the material when determining which is best for use in the classroom through identifying the types of resources/instructional material as well as the support necessary for its use.

Table 4. Rubric of criteria for development of instructional materials

	0--Unsatisfactory	1-- Average	2 – Good	3-Outstanding
Content	• No understanding of content and its connection to curriculum, students, or delivery methods.	• Basic understanding of content but is not yet developed. • Some opportunity for students to apply their own ideas • Few opportunities for critical thinking processes • Unable to make connections and application to real world • Obvious biases exist	• Content allows for students to apply some of their own ideas but not consistently • Encourages lower end critical thinking processes • Can be applied to the real world with guidance by instructor • Some bias present and does not reflect complete diversity and inclusion	• Content allows for students to apply own ideas • Encourages critical thinking skills • Real-world application • No bias for gender, culture, religion, or SES status • Reflects diversity and inclusion throughout
Cohesion and Structure	• No regard to previous content learned • No inquiry-based investigations • Must be taught "as is" with no adaptability	• Basic understandings built on previous knowledge • New knowledge with little regard to what has been previously gained • Little to no inquiry-based investigations present • Leads to little adaptability	• Some, but not all, cohesion built on previous knowledge • Provides for some, but not many, inquiry-based investigations • Can be adapted with effort	• Cohesion is built upon from previous knowledge • Provides for inquiry-based investigations • Easily adaptable for various learning styles of students
Student Experience	• No opportunity for active engagement of students • No knowledge or use of inquiry-based learning strategies • No technology support • No connection made to previous student experiences	• Minimal opportunity for students to learn through active engagement • Technology is not supported or incorporated • Little to no connection made to previous student experiences.	• Some opportunity for students to earn through active engagement • Technology not easily incorporated • Somewhat relevant to student experiences	• Many opportunities for students to learn through active engagement • Materials provide knowledge and use of inquiry-based, open-ended, and problem-solving strategies • Incorporation of technology is easily adaptable • Relevant to student experiences and adds knowledge
Support Materials	• No support materials given • No quizzes, labs, worksheets, or resources given	• Little support material that moves the lesson objective forward • Little background content available • Little variety in learning strategies supported • Very few support materials available	• Support material somewhat assists the educator in meeting lesson objectives • Some background content information • Some varied learning strategies supported • Some support material available such as worksheet, labs, and quizzes	• Support material assists the educator in meeting lesson objectives • Adequate background content information • Varied learning strategies supported • Rigorous support material available such as worksheet, labs, and quizzes
Student Assessment Materials	• Little to no materials present • Bias exists throughout • No use of real-life application	• Little variety or rigor in available materials • Bias in regard to gender, culture, religion and SES present • Little to no use of real-life application	• Some variety and rigor in some assessment materials • Not consistent in amount and type present • Materials are free from some bias • Some use of real-life application	• Various and rigorous assessment materials present • Materials free from bias • Use of real-life applications
Program Development and Implementation	• No research-based evidence of curriculum • Little to no flexibility in size and scope • No follow-up in training or assistance	• Little of the curriculum is research based • Curriculum has little flexibility in size and scope • Little to no follow-up in training or assistance	• Some of the curriculum is research based • Curriculum is somewhat flexible in size and scope • Some training is available, but minimum • Some follow-up or assistance is available	• Curriculum available is research based • Curriculum is flexible in size and scope • Training in use of curriculum is available • Follow-up training or assistance is available if needed

(Epps, 2021)

REVIEW QUESTIONS

1. What is the difference between a unit plan and a daily lesson plan?
2. Describe the steps in creating a unit plan.
3. How are FFA and the Supervised Agricultural Experience (SAE) related to the classroom and laboratory?
4. What are the stages in inquiry-based learning? Why are these useful?
5. How are inquiry-based learning and the problem-solving method similar? Dissimilar?
6. What types of instructional material/resources are available to educators?
7. What criteria should be used to evaluate instructional material/resources?

REFERENCES

D'Angelo, T., Barry, D., Bunch, J. C., & Thoron, A. (2018). *Selecting educational resources.* AEC 640. http://edis.ifas.ufl.edu

Egbert, J. (2017). *Methods of education technology: Principles, practice, and tools.* Pressbooks. https://opentext.wsu.edu/tchlrn445/front-matter/introduction/

Epps, R.B. (2021). *Evaluating curriculum material.* AED 583 Course Material.

Goodlett, T. L. (2019). Assignment for AED 583: Curriculum Writing in Career and Technical Education. Community and Leadership Development Department, University of Kentucky.

Jaworski, J. M. (2019). Assignment for AED 583: Curriculum Writing in Career and Technical Education. Community and Leadership Development Department, University of Kentucky.

Maheshwari, V. K. (2017). The problem-solving method in education. *Philosophical commentary on issues of today.* http://www.vkmaheshwari.com/WP/?p=2375

National F. F. A. Organization. (2022). *The three-component model.* Agricultural Education About FFA. https://www.ffa.org/agricultural-education/

Pedaste, M., Maeots, M., Siiman, L. A., deJong, T., van Riesen, S. A. N., Kamp, E. T., Manoli, C. C., Zacharia, Z. C., & Tsourlidaki, E. (2015). Phases of inquiry-based learning: Definitions and the inquiry cycle. *Educational Research Review, 14*, 47–61. doi:10.1016/j.edurev.2015.02.003

KEY TERMS AND DEFINITIONS

Experiential Education: Use of real-world application and experiences in order to teach students.

Inquiry-Based Learning: Teaching strategy used to pique student curiosity and developing a method to answer student questions or interests.

Lesson Plans: Road map created by the teacher which clearly details how learning in the time given is to occur along with assessment.

Planning: Essentially a road map that teachers develop for their students to ensure student learning occurs and how it will occur built upon student's previous knowledge and new knowledge incorporate.

Problem-Solving Method: Teaching strategy used to focus on student interests and engagement where the scientific method is used to solve a problem.

Standards: Local, state, or national level of attainment set forth for student achievement based upon a norm.

Unit Plans: Long term, but detailed, road map to ensure a small content is taught by the teacher, assessed, and mastered by the students before moving forward.

APPENDIX: RESOURCE LIST

Instructional Technologies

Mind Mapping- https://www.mindmeister.com/
Youtube for Teachers- https://youtu.be/NLNuv7jAIhg
Schooltube- https://www.schooltube.com/
Edpuzzle- https://edpuzzle.com/home
Coggle- https://coggle.it/
Labquests
Class Kick- https://classkick.com/
Wordle
QR Code Generator- https://www.qr-code-generator.com/
Podcasts
Google Earth Projects- Google Earth Projects
Screencastify- https://www.screencastify.com/education/resources
Online Bulletin Board- https://padlet.com/
Flipgrid- https://info.flipgrid.com/
Video Messages- https://info.flipgrid.com/
Scavenger Hunts- https://www.goosechase.com/edu/
Collaboration- https://web.seesaw.me/
Socrative- https://www.socrative.com/
Organizational Software- https://www.mystudylife.com/
Edmodo- https://new.edmodo.com/
Quiver App- https://quivervision.com/

Learning Management Systems

Google Classroom- https://classroom.google.com/u/0/h
Schoology- https://www.schoology.com/
Canvas- https://www.instructure.com/canvas/

Assessment

Quizlet- https://quizlet.com/
Zzish- zzish.com
Quizalize- https://www.quizalize.com/
Online Jeopardy- https://jeopardylabs.com/
Poll Everywhere- https://www.polleverywhere.com/
Plickers- https://get.plickers.com/
Kahoot- https://kahoot.com/

Create Fillable PDFs

Kami App- https://www.kamiapp.com/

Distance Learning

Skype
Zoom
Google Hangout
Microsoft Teams

Website Creation

Google Sites- https://sites.google.com/new
Wix Website Creator- wix.com

Presentation Software

Emaze- https://www.emaze.com/
Powtoon- https://www.powtoon.com/
Nearpod- nearpod.com
Storyboard that- https://www.storyboardthat.com/

Note Taking

Evernote- https://evernote.com/
One Note- onenote.com

Blogs

Dairy Carrie- https://www.dairycarrie.com/category/humans-of-agriculture/

Agricultural Education Content

Online Textbooks- http://www.freetextbooklist.com/life-sciences/agriculture/
AgEdNet- https://www.agednet.com/
NAAE Communities of Practice- https://communities.naae.org/welcome
NAAE Online/Distance Learning Resources- https://communities.naae.org/welcome
iCEV- https://www.icevonline.com/curriculum/agricultural-science
National Agriculture in the Classroom- https://www.agclassroom.org/
FFA Ag Explorer- https://agexplorer.com/
Georgia Ag Ed Lesson Plans- http://www.gaaged.org/curriculum/
Glen Rose FFA- https://www.glenrosearkansasffa.com/lesson%20plans.htm

Louisiana Ag Ed Lesson Plans- http://www.la-ffa.org/lesson-plans
Minnesota Ag Ed Lesson Plans- https://www.mnffa.org/teacher-resources
Arkansas Ag Ed Lesson Plans- https://www.arkansasffa.org/inservice
Ag Education Discussion Lab Facebook Group

Movies with Agricultural Connection

Founder
Temple Grandin
Ivy League Farmer
Food Inc.
Dirty Jobs
Farmland
Deep in the Heart

Youtube Channels

The Hoof GP
Peterson Farm Brothers
WV Department of Agriculture
Why I Farm
Discover Agriculture
USDA
University of Kentucky College of Agriculture, Food, and Environment

CONTENT SPECIFIC RESOURCES

Animal Science

"The Herd" App- https://apps.apple.com/us/app/the-herd/id1272532256
Beef. It's What's for Dinner- https://www.beefitswhatsfordinner.com/cuts
Nose to Tail App- https://nosetotailapp.com/
Livestock Judging- https://www.livestockjudging.com/
Kentucky Agricultural Facts-Kentucky Commodities
Virtual Field Trips- https://www.farmfood360.ca/

Plant Science

Websoil Survey- https://websoilsurvey.sc.egov.usda.gov/App/HomePage.htm
Plant Images App- https://apps.apple.com/us/app/picturethis-plant-identifier/id1252497129
Plant Snap- https://www.plantsnap.com/
Pasture Map- https://pasturemap.com/
Ag PhD- http://www.agphd.com/

Agricultural Mechanics

Miller Welding Calculator- https://www.millerwelds.com/resources/weld-setting-calculators
Weldучation- https://apps.apple.com/us/app/weleducation-basic/id1281494916
Welding- https://www.toolingu.com/
Metalworking Certifications- https://www.nims-skills.org/
Autodesk- https://www.autodesk.com/education/home
CAD/3D Printing- https://www.onshape.com/en/
Woodworking Curriculum- Woodworking Curriculum
Harbor Freight Online Resources- Harbor Freight Online

Environmental Natural Resources (ENR)

Journey 2050 App- https://www.journey2050.com/
Forest Tree Identification- https://apps.apple.com/us/app/forest-tree-identification/id1293993638
Bayer Agrisciences- Bayer Agriscience
USDA- USDA Natural Resources

Food Science

Food Safety/Nutrition- https://www.teachfoodscience.org/resources.asp
IFT- https://www.ift.org/community/educators/k12-teaching-resources
NAAE- https://communities.naae.org/community/instruction/agriscience/foodscience/content
Georgia Extension- Georgia Extension
FDA- https://www.fda.gov/food/students-teachers/science-and-our-food-supply

Agricultural Business

Ag Careers- https://www.agcareers.com/
Farm Futures- https://www.farmprogress.com/farm-futures
Cattle Fax- https://www.cattlefax.com/
Ag Web- https://www.agweb.com/
Farm Journal TV- https://www.agweb.com/agday
Real Ag- https://www.realagriculture.com/

Chapter 7
Planning Daily Lessons

Eric D. Rubenstein
University of Georgia, USA

Scott W. Smalley
Iowa State University, USA

ABSTRACT

The planning of daily lessons involves a variety of components to deliver a successful lesson. This chapter will look at the integration of a daily lesson plan with a focus on how to write the objectives, essential questions, the integration of Bloom's taxonomy in lesson objectives and connecting objectives and essential questions to an assessment. An additional focus is placed on the components of the daily lesson plan looking specifically at the alignment of standards, accommodations for students' IEP and 504 plans, integrating interest approaches and lesson activities.

INTRODUCTION

A lesson plan is the teacher's road map of what students will be learning and how it will be completed effectively to meet the learning outcomes. As the facilitator of learning, you serve an important role in the learner being able to reach their goals. A variety of components go into making a successful lesson plan. This road map is a core component in teaching as it provides a broader series of organization and management of the classroom. A single lesson plan could be a couple of day lessons or be designed for a week-long lesson.

In designing a lesson plan, a variety of elements need to be taken into consideration such as the learner, strategies for facilitating the learning, how the learner should be grouped, materials needed, what objectives you will accomplish, what assessments might be utilized and how the lesson connects to future lessons.

In developing a lesson, it is important to ask yourself questions of how the lesson will meet the needs of your learner. Think collectively about your learner and their individual learning needs. During this planning process, you must be mindful of special needs and barriers which may require additional as-

DOI: 10.4018/978-1-6684-3420-8.ch007

sistance. Additionally, you must be cognizant of the materials needed to make all learners successful when implementing an activity or lesson.

In the development of your lesson plan, it is critical to ensure you have a clearly written plan that is easy to follow and understand. By developing a clear, thought-out plan, this road map will be easy for anyone to use to successfully implement the lesson plan you have developed. In this chapter learners will be able to: 1.) construct lesson objectives/essential questions based on the given academic standard, 2.) create a daily lesson plan with each of the required components, 3.) explain the purpose of summative and formative assessments, and 4.) select summative and formative assessments that measure desired learner outcomes (Phipps et al., 2008).

STRUCTURE OF A DAILY LESSON PLAN

There are numerous structures of a daily lesson plan. Lesson plans may be decided upon by the school district or the state, or preference may be decided upon by the individual teacher. Regardless, a daily lesson plan is an effective tool to keep the instructor on-track and prepared. The most common components across daily lesson plans are the following components: identification of the course, grade level, unit of instruction, daily lesson title, student learning objectives and/or essential questions, state and/or national standards, content of the lesson (concepts to be taught or investigated), identification of teaching strategies/activities, lesson summary/closure, reflection, and assessment of student learning (Phipps, et al. 2008).

Objectives/Essential Questions

All lessons that are taught must accomplish a purpose that improves the knowledge, skills, or competencies of the learners. To make sure that learners are comprehending and grasping the new content being taught, teachers use objectives and essential questions to ensure that students can move forward to the next lesson.

Therefore, teachers must ensure that their objects are written in a way that clearly explains the expectations of the lesson and the content that will be addressed. The use of both objectives and essential questions provide structure to any lesson and classroom setting. The use of either objectives or essential questions in your planning will vary based upon the district or school that you are employed in during your teaching tenure. Therefore, a need exists for all teachers to be able to properly construct and use both objectives and essential questions.

How to Write Lesson Objectives

Objectives provide the teacher and the learner with pertinent information that help to determine what the learner is expected to do by the end of the lesson, how they will be assessed, and what threshold needs to be met for the class to move on to the next topic in the lesson progression. Objectives can be written in a variety of different ways. In this chapter will be based on the use of three components to objective writing. Those three components are the: performance, condition, and criterion (PCC) (Mager, 1984).

The performance in an objective is the behavior, knowledge, or competency that learners are to develop by the end of the lesson. For example: describe the four types of injections that are used in animal agriculture. The performance should always start with an action verb that aligns directly with the standard that is being taught throughout the lesson and the assessment that is being used to measure learners' comprehension of the content. Therefore, the assessment should be created prior to the objective being written. When writing the performance component of the objective, the secondary school teacher must ensure that it is a tangible action or behavior that the learner can perform or engage in during the lesson. If the secondary school teacher asks a learner to describe or list, then the assessment can be a paper-and-pencil test. However, if the secondary school teacher asks the learner to demonstrate a skill during the lesson, then the assessment needs to be some form of a psychomotor assessment. Below in the example objective the performance is bolded to help exemplify what this could look like in an agricultural education classroom.

Example Objective: The student will be able to **demonstrate the 4 types of injections used in animal agriculture**, on an orange with no notes, with 100% accuracy.

The condition of the objective provides details to the learner about how this behavior, competency, or skill will be assessed. This can be a reference to the actual assessment, or it can provide details about how the assessment will be administered. For example, on some assessments learners may be able to use their notes or outside references. The learner needs to know how they will be assessed and what the assessment will look like so they can effectively prepare and master the material. In some cases, the condition can also include a time limit or any other factors that will be imposed on the assessment. Below in the example objective the condition is bolded to help exemplify what this could look like in an agricultural education classroom.

Example Objective: The student will be able to demonstrate the 4 types of injections used in animal agriculture, **on an orange with no notes**, with 100% accuracy.

The criterion is the level at which the learner and the class must achieve to move on to the next topic within the unit. The level that is used in each objective should be measured and achieved by the class before the teacher moves to the next lesson within a given unit. Furthermore, the level should be attainable by the class and should ensure that no learner's comprehension of the content will be adversely affected as the unit progresses. Therefore, if a skill, knowledge, or competency is vital for their success within the unit, then the criterion should be set at a higher level. This is one way that formative assessments can be effectively used within our classrooms. The criterion should also be measurable, so be careful not to use a criterion that is not able to be measured or calculated. While you may know what you are expecting, your administrator and students will be left confused, and this could cause some additional comprehension issues later in the unit. Below in the example objective the criterion is bolded to help exemplify what this could look like in an agricultural education classroom.

Example Objective: The student will be able to demonstrate the 4 types of injections used in animal agriculture, on an orange with no notes, **with 100% accuracy**.

When writing objectives, there are two forms of objectives used in classrooms to help learners succeed. The two forms are terminal and enabling objectives. Terminal objectives include the performance, condition, and criterion in each objective and are typically only seen by the teacher or administrator. An enabling objective includes only the performance component that will help learners achieve the terminal objective. The enabling objectives are listed or presented for the students in various formats (on the chalk/white board, in a PowerPoint, on a bulletin board, etc.). Depending on your administrator's expectations, your use of terminal and enabling objectives may vary. Examples of a terminal and enabling objectives are listed below.

Terminal Objective:

1. The student will be able to differentiate between game and non-game species, on a questionnaire individually during class, with 90% accuracy.

Enabling Objectives:

1. Explain why wildlife identification is important.
2. Identify game and non-game species.
3. Describe the characteristics of game/non-game species.

How to Write Lesson Essential Questions

Essential questions are another effective way to help direct learning within a classroom in a more informal manner. While they may not have a condition or criterion component when writing them, they are an effective way to ensure that learning is taking place within any classroom. Furthermore, essential questions can encourage students to develop their own questions about the content and further their comprehension or engagement in the course curriculum. When creating essential questions, there are many different approaches that can be used; however, teachers should always remember that the questions posed to their learners should always be answerable by the end of the lesson. Therefore, the questions should pertain to the most important concepts that were taught during the lesson. These essential questions can be used in a variety of ways throughout the class session. Each question could be used to help end a specific learning activity within the lesson, or a teacher could use the questions during a summary activity to check for comprehension before ending a class session. More specifically, essential questions should always be open-ended and require students to provide more than a one- or two-word response (McTighe & Wiggins, 2013; Wiggins & McTighe, 2005).

Similar to objectives, essential questions should be written after completing the summative assessment for a specific unit. The essential questions should be developed at the same time and split into appropriate lessons so that specific learning activities can be created and implemented with learners to create a cohesive lesson. The essential questions should link directly with the assessment technique and should provide learners with an opportunity to practice what will be expected of them on the summative assessment. Therefore, if the assessment is expecting learners to engage in a physical hands-on application of content, then the essential question should provide the learner with the ability to practice a similar application during the lesson activities or lesson summary. Each essential question that is asked during a lesson should be given context, so the learner has a clear set of expectations when they are responding.

When writing an essential question, the teacher should think about the specific outcome that they are hoping to achieve and how they can further a learner's engagement or question development within the lesson topic. When developing essential questions, the teacher should think about the format of their questions. Most essential questions will start with who, what, when, where, why, and how. The number of essential questions in a lesson will vary depending on the content being taught. Below are some sample essential questions:

1. How do embryos sustain life once conception occurs?
2. Why is the slope an important concept in landscape design?
3. What are the major components considered when selecting trees within a timber set for harvest?

4. What are the negative and positive aspects of using fossil fuels for energy including impacts on the environment?

Connection of Lesson Objectives/Essential Questions to Standards

Regardless of whether you are using objectives or essential questions in your lesson plans, there must be a connection to the specific state or national learning standards. Each state has different requirements when it comes to the learning standards that are associated with agricultural education courses. When planning a lesson, you must look at a variety of things to ensure that your lessons and objectives meet the requirements for the specific course you are teaching. For example, the state standard will have an action verb included in the text, so you should work to ensure that your learners are able to perform at the given level of the standard on Bloom's Taxonomy. Furthermore, the secondary school teacher must ensure that the state or national standards are effectively and thoroughly included in the instruction of our courses. In the age of accountability, most states require that program completers be assessed using some form of pathway assessment. These assessments vary by state but aim to ensure that the standards of the course are being met in all programs and that your students are ready to step into either a career or college classroom at the end of their time in your agricultural education program.

Standards typically have two parts to them, a main overarching standard and then sub-standards or elements that provide specificity for the teacher as to the exact content that should be taught during the course. When planning for instruction, teachers must examine both the overarching standard and the sub-standards to ensure that their lesson meets the expectations of the course. Teachers must examine the action verb in the standard and sub-standard to match language in the objective or essential question. Below is an example from the Natural Resources Systems Career Pathway standards:

- NRS.01.06. Apply ecological concepts and principles to living organisms in natural resource systems.
    - NRS.01.06.01.a. Differentiate between population ecology, population density and population dispersion and describe the importance of these concepts to natural resource systems.
    - NRS.01.06.01.b. Analyze the factors that influence population density and population dispersion in natural resource systems.
    - NRS.01.06.01.c. Create a management plan for a population of a species in an ecosystem given its population ecology, population density and population dispersion in natural resource systems.
    - NRS.01.06.02.a. Research and summarize examples of invasive species.
    - NRS.01.06.02.b. Analyze factors that influence the establishment and spread of invasive species and determine the appropriate steps to prevent or minimize the impact of invasive species.
    - NRS.01.06.02.c. Evaluate the presence and impact of invasive species on natural re-sources in a given area and devise a plan to prevent, control or eliminate invasive species from that habitat (National Council for Agricultural Education, 2022, pg. 6).

When examining the example standards, there are several different action verbs that are used for each of the sub-standards. The action verb for the standard asks learners to do something with the content, not just regurgitate the information on an exam. Therefore, the objectives and the learning activities of a lesson must coincide to ensure that learners are able to obtain and utilize the content in the appropri-

Figure 1. Bloom's taxonomy

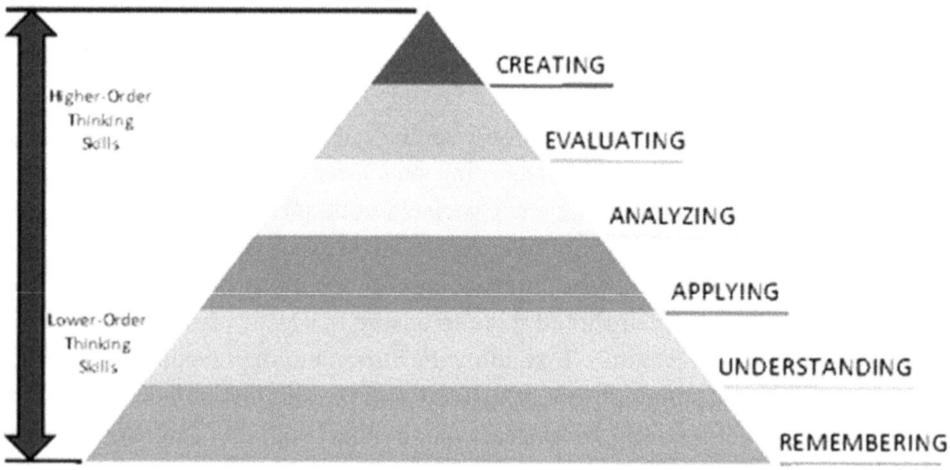

ate format. To accomplish this, teachers must utilize the six levels of Bloom's Taxonomy when they are developing objectives and essential questions.

Bloom's Taxonomy in Lesson Objectives

When planning for any form of instruction, teachers must take into consideration the six levels of Bloom's Taxonomy. The six levels consist of: remember, understand, apply, analyze, evaluate, and create (Blooms et al., 1956). For further consideration, examine the image below to see how each level is built on one another.

Bloom proposed a five-level taxonomy that examined student learning with each level advancing the students' ability to conceptualize and utilize the content in a more advanced manner. In 2001, a revised taxonomy was released with an additional level. While each level plays an integral role in a student's learning, teachers should strive to help advance a learner's comprehension into one of the higher levels of Bloom's Taxonomy. The secondary school teacher will examine each level, and the associated action verbs, to ensure that they can develop lessons that have learning activities that adequately help learners acquire the content and use it at an appropriate level.

To remember content is to have the ability to memorize and immediately recall the information. Most times this level of learning is associated with the term rote memorization. This level of knowledge acquisition typically includes teaching approaches such as: lecture, defining words/vocabulary, watching videos and answering questions, and other individualized applications. When writing an objective to this level of Bloom's Taxonomy teachers should use action verbs such as: define, label, recite, and state. Teachers should aim to evaluate student learning at the remember level with paper examinations or by having a student recite a memorized address or speech.

Teaching to the understand level of Bloom's Taxonomy allows a learner to be able to explain content in their own words and language. This requires learners to have a solid comprehension of the content to be able to explain an abstract idea or concept in their own words. Teachers who aim to have their students acquire this level of Bloom's Taxonomy should utilize action verbs such as: compare, describe, discuss, paraphrase, or summarize. When teaching to the understand level of Bloom's Taxonomy teachers can utilize instructional strategies such as lecture with questioning and demonstration. When assessing students at the understand level, teachers can utilize pencil paper exams with open ended questions or oral examinations/discussions.

The apply level of Bloom's Taxonomy provides students with the opportunity to take the knowledge, skills, or competencies they have learned and transfer that knowledge to a real-world situation that they are facing. Learners must have a firm comprehension of the content and be able to draw conclusions as to how this content can be applied in their personal experiences and the real world. Apply, interpret, model, produce, and discover are all action verbs that can be used in objectives written to help learners develop their knowledge, skills, and competencies at the apply level of Bloom's Taxonomy. Teachers can use questioning techniques and case studies as instructional approaches when teaching at the apply level. When assessing students, teachers can ask students to answer questions based on a given scenario or have them demonstrate a skill in a mock situation.

In many high school classrooms, the highest-level learners attain on Bloom's Taxonomy is the analyze level. At this level, learners are expected to take content and build cognitive connections or schema between their past experiences, knowledge, skills, and competencies and the new content being taught. Many times, this can be a time when learners determine that there was mislearning or incorrect instruction in a previous classroom or experience. Teachers need to be ready to help learners work through the relearning process and correct any misconceptions that they may have about the new content. Teachers can use demonstrations, individualized applications, group exercises, inquiry-based instruction lessons, and case studies to help learners learn at the analyze level. This should be reflected in the lesson objectives by using action verbs such as: compare, contrast, analyze, classify, distinguish, subdivide, connect, and differentiate. Therefore, this type of instruction should be assessed using similar methods when learners are asked to critically examine a given situation and provide solutions to a problem, or when learners compare and contrast two differing methods to solve a given problem while giving the strengths and weaknesses of both methods.

In advanced high school or early college courses, learners will begin to advance to the evaluate level of Bloom's Taxonomy. At the synthesis level, learners should be able to take information to create a stance or belief on a given topic and substantiate that stance or belief with facts. Many teachers will typically use classroom debates and inquiry-based lessons where the students answer a personally developed question. Appraise, critique, criticize, recommend, and assess are all action verbs that can be used when teaching learners at the evaluate level. Assessment of learners at the evaluation level could include having learners write a formal paper with citations or having students engage in a debate related to the topic.

Finally, learners can progress to the create level of Bloom's Taxonomy. Learners at the create level are working to advance knowledge in the topic area by leading research or data analysis to produce new knowledge. In many cases, this level is not truly achieved until graduate school or late in a student's undergraduate education. This is an advanced level for many students to achieve, as most of our educational experiences look at the comprehension of already established knowledge about a topic. When teaching to this level of Bloom's Taxonomy, teachers should utilize inquiry-based instruction lessons that are completely learner derived and created. This provides learners with an opportunity to explore topics that have personal relevance or interest. When writing lesson objectives for a create level lesson, teachers could use action verbs such as: develop, produce, create, design, and construct. Therefore, when assessing this level of learning teachers should create an environment for students to present their individualized work in a meaningful and educational manner either through individual or group presentations.

Connecting Objectives and Essential Questions to Assessments

Regardless of your use of objectives or essential questions, a direct connection to the assessment methods used to measure a learner's comprehension is needed. When planning a unit, the summative assessment should be created prior to any individual lesson development. Following the assessment creation, a teacher should begin planning the major content areas for each lesson. Once those content areas are developed, each component of the assessment should be associated with one of the specific content areas identified. Then, based on what is expected of the learner on the assessment, an action verb should be selected that best meets the expectations of the learner on the summative assessment. Therefore, if the learner is asked to answer a multiple-choice question on the assessment, then the secondary school teacher needs to make sure the action verb is associated with the remember level of Bloom's Taxonomy. Every objective in a lesson plan should follow a similar pattern to ensure that our assessments and objectives/essential questions directly inform the development of lesson activities to ensure that every component of a lesson and unit are connected.

Assessment

Assessment is a significant part of a learner's educational experience and shapes how they learn material provided to them within the learning environment. The word assessment refers to a wide variety of methods used by educators including how you evaluate or document the learning process, and the educational needs of students in your classroom. An assessment should evaluate the learner's growth and progress, provide data about learning and attainment of objectives, assist with allocating resources, and provide evidence of learning (Wiggins & McTighe, 2005).

Throughout the learning process, a student has the ability to evaluate their own performance and approve of it. An authentic assessment method allows a learner to develop their skills through formative and summative assessments. Assessments can be conducted in a variety of settings from the classroom to laboratory experiences. Each assessment plan should be connected to the learning objectives of the lesson (Phipps, et al. 2008).

Within agricultural education, an instructor needs to be mindful of the environment the assessment is taking place. For example, the assessment may take place in the classroom or a laboratory setting. This could include a mechanics, greenhouse, shop, livestock, or land laboratory. No matter the setting, an assessment of learning can occur either in a formative or summative fashion.

Summative Overview

A summative assessment will occur to evaluate the student learner at the end of an instructional period. This could be at the end of a unit, semester, or project. This assessment should be used to determine to what degree the learner has learned the material that has been taught to them. The assessments are used to determine the learning progress and evaluate the progress towards accomplishing the learning objectives. The result of the assessment would typically be recorded as a grade.

Some examples of summative assessment could include an end of unit or chapter test, final project, identification portfolio, or an achievement on a standardized test. Additional summative assessments at the end of a unit could include a presentation, a written report, or a self-evaluation. A summative assessment offers an opportunity for the learner to reflect on what was previously learned and allows the overall learning objectives of the course to be reinforced. Summative assessments are beneficial for a teacher because they can use the results in guiding their decisions for future curriculum development.

Formative Overview

Formative assessments take place throughout the learning process. Feedback can be gathered throughout the learning process to see where the learner is doing well and where they are struggling. This allows an instructor to adjust their teaching to meet the needs of the learner. If a learner is identified as having a problem or struggling, the problem can be addressed. A formative assessment can occur early on during the process to ensure students are able to learn the material.

Some ways formative assessment information might be collected could include through observations during in-class activities, through daily assessments in class or on homework, through a written journal, in question-and-answer sessions, or through observing learning experiences. Additionally, assessments could include a simple entry ticket, exit ticket, or a poll of the students. Any of these assessments can provide feedback to a teacher if the individual lessons meet the short-term objectives.

Formative assessments can have a significant impact on the learning process. In addition, they can increase class participation because the learner will become more involved in a variety of tasks as the course occurs. This type of assessment can help a teacher recognize weak areas and assist learners who are struggling with the content to receive additional assistance. A formative assessment can provide guidance to the teacher whether to move forward or stay longer on a particular concept.

Integration into Daily Lesson Techniques

Being able to continuously utilize assessments throughout the learning process will assist in the development of self-awareness of your learners. The main goal with integrating assessments should be to receive feedback as the learner and instructor regarding how one is doing as the learner and/or instructor. Assessment is an ongoing process throughout a lesson to ensure the material being presented is at a level that learners can comprehend. It is important to make sure the assessment being utilized is authentic. As you are planning what to teach, it is just as important to plan how you will assess the learning. Educators must think of assessment as a means of checks and balances of the teaching and learning process within the classroom so that they can adequately change the course of the lesson/unit if needed.

COMPONENTS OF DAILY LESSON PLAN

Alignment of Standards

The process to align curriculum to standards is particularly important to ensure the curriculum is consistent and matches the expectations that are outlined within the standards. A standard indicates what learners should know and be able to do within the specific content area. Standards are written with a clear description and focus on what students should be able to do at a particular age/grade level. When aligned correctly, the course components will work together in achieving the learning objectives of the lesson. When you align a standard, you are looking to ensure the goals, objectives, and standard components of the lesson are guiding the direction of the lesson plan. As a result, you are working to make sure they are all connected to the same concept. When all the elements are connected, the lesson will flow, and your learner will be more successful in mastering the objectives of the lesson.

Alignment of Accommodations for IEP/504

When an Individualized Education Program (IEP) and 504 plan is aligned correctly for the learner, several benefits can follow. An IEP will lay out the special education instruction and services a student will need to thrive in the learning environment. This written document is developed by a team of educators and is reviewed annually. A 504 plan is designed to ensure a learner with a disability receives the accommodations needed to make them successful in the learning environment. The main difference between the IEP and 504 plan is that the 504 plan does not provide individualized services to a learner (Frank, et al., 2020).

The process used to develop an IEP/504 outlines specific accommodations which can be considered when designing a classroom lesson. Alignment attempts to ensure learners are achieving the goals of the general education classroom with the provided supports and accommodations needed.

As the educator, you can seek assistance from special education teachers if you are not familiar with how to meet the needs of a student with an IEP/504. A learner with disabilities will have more opportunities to meet expectations with the support of accommodations provided. Additionally, special education teachers will likely have additional suggestions and resources on how to modify assignments or aid in completing these efforts (Talbert, et al., 2013).

Interest Approach

An interest approach engages students with the material in a way that prompts them to prepare to learn. This is a critical step in designing a lesson as it can help students focus on the learning that will take place during the lesson. An interest approach will aid in organizing the lesson, and it will help structure the lesson to enhance the overall process (Talbert, et al., 2013). Maximizing a learner's engagement at the start of a lesson will assist in maximizing the attentiveness towards a specific topic. The approach should be connected to the students' previously learned knowledge in the course and what will be learned moving ahead.

A variety of techniques can be used when creating an interest approach. Some approaches to an interest approach might include a story, demonstration, short video clip, pictures, a scenario, or article. No matter the approach, as the teacher you are working to increase the engagement of your learner in the topic. This increased engagement will be helpful in keeping the learners eager to learn within your classroom or laboratory setting.

Lesson Activities

Many factors go into a lesson activity, and it all begins with the ability to gain the attention of your learner. The interest approach at the beginning of the lesson ensures that you have momentum moving into your lesson. Keeping in mind your learners will have a variety of learning style preferences, the lesson might infuse a variety of technology or collaboration. Some practices to help keep your student's attention during a lesson might include:

1. For every ten minutes of instruction, allow learners to process the information learned by asking a question or discussing the information learned with a classmate.
2. Provide movement within your lesson by allowing a learner to complete a demonstration, diagram, or illustration of the material learned.
3. Allow your learners to process the information being learned by pausing from time to time throughout the lesson.
4. At the completion of a lesson, allow learners to summarize material learned by using an exit ticket to identity areas of success or struggles.

Adjusting involvement within a lesson allows the pace of the lesson to be changed. This can be accomplished by providing a variety of activities for your learners to complete and taking breaks to process information rather than going from one activity to the next. By adjusting the pace within a lesson, learners will be able to absorb the information being taught in the lesson. "Activities" should not be misinterpreted as simply "doing things;" activities evolve from instructional strategies. For example, within on class period students may be involved in group discussion, observing a PowerPoint presentation, and writing a summary of the lesson. Variety and change of pace are accomplished without students leaving their seats.

Increasing the retention rate within a lesson is critical for a learner to be able to utilize the materials as they move to future lessons. A learner's retention of information can be viewed within the learning pyramid. A learner will retain less information from lecture, reading or audio visual. A student learner will retain more information by being actively engaged in a demonstration, group discussion, practicing or by teaching others.

As the facilitator of the learning, it is important to focus on the type of learners you have and how to best deliver the material as the instructor. By diversifying your method, you will be able to aid your learners in the retention of additional materials.

Summary of Lesson

When summarizing a lesson, it allows both the learner and teacher to monitor the comprehension of material. Being able to summarize what was covered and learned during the lesson allows the learner to reflect on the content. Within the reflection, the learner can check for understanding of the material to determine what additional learning needs to occur.

A few examples of how to summarize the lesson could include:

- Utilizing an exit ticket provides a quick way to assess the learner. This exit ticket can include open-ended questions, a true/false question, or multiple-choice questions. This method can be beneficial to provide information about the learners' strengths and areas for growth.
- Sentence wrap-up: can provide an opportunity for the learner to write a quick couple of sentences to summarize what was learned. This summary can be kept within a notebook or in a student journal.

- Concept Map: can be utilized to assess the learning by having the learner write the key vocabulary during the lesson and provide a definition. This provides an opportunity to view what the learners understood.

No matter the approach in summarizing the lesson, it is essential to bring closure to the lesson for the day and address what additional support might be needed for the learners moving ahead in future lessons.

Examples of a Daily Lesson Plans

In example one of a daily lesson plan, information on the daily topic, grade, estimated time to teach the lesson, and grade level are identified. This example focuses on all the basis components described above with a particular focus on active learning. Depending on the amount of time to teach the lesson there may be more then one learning activity utilized during instructional time. Not all lessons must have three activities. As stated above it is good practice for teachers to plan for transitions during instruction and have student participation and engagement. This example also allows for planning for a reflective/ summary to draw lesson closure. At the end of the template/example the teacher will need to address both forms of assessment. Thinking about assessment ahead of the instruction delivery of the concepts will assist the teacher in measuring the objectives outline in the daily lesson plan (Phipps, et al. 2008).

In this second example of a daily lesson plan, similar components are utilized as in example one. Both examples list instructional needs (materials, equipment, resources), objectives, and interest approach/ essential questions. The focus on this example places the instructor to focus on safety and the learning situation. Further, the focus is less on an activity basis and encompasses an application portion to the lesson. This lesson does not identify a lesson closure component which should be contained within the instructor directions.

SUMMARY

This chapter has highlighted the planning of a daily lesson plan, which can serve as a road map for the facilitator. Within the learning process, the success of a lesson begins with the facilitator. This chapter has emphasized that the planning process of a lesson begins with the end in mind. Developing strong objectives for the lesson will provide the facilitator of learning and the learner with pertinent information that will clarify what is expected of them by the end of the lesson. From the objectives, the learner will be able to identify the specific behavior, knowledge, or competency that they are expected to learn by the end of the lesson.

The development of the essential questions is another way to effectively help direct the learning within a classroom in an informal manner. With the guidance of an essential question, learners can develop their own questions about the content being shared. The essential questions should be open-ended and allow learners to provide more than a one- or two-word response. No matter if you are using objectives or essential questions to guide your lesson, they need to be connected to the specific state of national learning standards.

The objectives and essential questions should be connected to an assessment. When planning a lesson, the use of summative and formative assessments is usually included within the lesson. The summative assessment for each unit should be created prior to the individual lesson development. A summative

Table 1. Daily plan example 1

<table>
<tr><td colspan="3">Daily Plan – Example 1</td><td colspan="2">Instructor:</td></tr>
<tr><td>Course:</td><td colspan="3">Grade Level:</td><td>Estimated Time to Teach:</td></tr>
<tr><td colspan="3">Unit:</td><td colspan="2">Lesson:</td></tr>
<tr><td colspan="5">Materials, Supplies, Equipment, Reference Materials, Additional Resources</td></tr>
<tr><td colspan="5"></td></tr>
<tr><td colspan="5">Standards (Sate and/or AFNR)</td></tr>
<tr><td colspan="5">•
•</td></tr>
<tr><td colspan="5">Essential Questions/Objectives</td></tr>
<tr><td colspan="5">1.
2.
3.</td></tr>
<tr><td colspan="5">Accommodations</td></tr>
<tr><td colspan="5"></td></tr>
<tr><td colspan="3">Interest Approach</td><td colspan="2">Estimated Time:</td></tr>
<tr><td colspan="5"></td></tr>
<tr><td colspan="3">Learning Activity 1</td><td colspan="2">Estimated Time:</td></tr>
<tr><td colspan="2">Instructor Directions / Materials / Teaching Procedure</td><td colspan="3">Content Outline</td></tr>
<tr><td colspan="2"></td><td colspan="3"></td></tr>
<tr><td colspan="3">Learning Activity 2</td><td colspan="2">Estimated Time:</td></tr>
<tr><td colspan="2">Instructor Directions / Materials / Teaching Procedure</td><td colspan="3">Content Outline</td></tr>
<tr><td colspan="2"></td><td colspan="3"></td></tr>
<tr><td colspan="3">Learning Activity 3</td><td colspan="2">Estimated Time:</td></tr>
<tr><td colspan="2">Instructor Directions / Materials / Teaching Procedure</td><td colspan="3">Content Outline</td></tr>
<tr><td colspan="2"></td><td colspan="3"></td></tr>
<tr><td colspan="3">Summary (Reflection)</td><td colspan="2">Estimated Time:</td></tr>
<tr><td colspan="5"></td></tr>
<tr><td colspan="5">Assessment</td></tr>
<tr><td colspan="5">Formative:</td></tr>
<tr><td colspan="5">Summative:</td></tr>
</table>

assessment will assist with the learning progress and evaluate the progress a student is making towards accomplishing the learning objectives. The use of formative assessments can occur throughout the learning process. Based on what is being assessed in the formative assessment process, a facilitator of the learning can adjust the lesson to meet the needs of the learners.

A variety of components have been outlined within the chapter to highlight the elements of a daily lesson. The interest approach assists in getting your learners prepared to learn and connected to the material for the lesson. Maximizing the learners' engagement will assist in maximizing the attentive-

Table 2. Daily plan example 2

<table>
<tr><th colspan="2">Daily Plan – Example 2</th><th>Instructor:</th></tr>
<tr><td colspan="2">Course:</td><td>Estimated Time to Teach:</td></tr>
<tr><td>Unit:</td><td colspan="2">Lesson Title:</td></tr>
<tr><td colspan="3">Objectives</td></tr>
<tr><td colspan="3">1.
2.
3.</td></tr>
<tr><td colspan="3">Resources & References</td></tr>
<tr><td colspan="3"></td></tr>
<tr><td colspan="3">Supplies & Equipment</td></tr>
<tr><td colspan="3"></td></tr>
<tr><td colspan="3">Safety (this section presents a list of safety considerations in teaching the lesson)</td></tr>
<tr><td colspan="3"></td></tr>
<tr><td colspan="3">Situation</td></tr>
<tr><td colspan="3"></td></tr>
<tr><td colspan="3">Interest Approach</td></tr>
<tr><td colspan="3"></td></tr>
<tr><td>Instructor Directions</td><td colspan="2">Content Outline and/or Procedures</td></tr>
<tr><td></td><td colspan="2"></td></tr>
<tr><td>Instructor Directions</td><td colspan="2">Application</td></tr>
<tr><td></td><td colspan="2"></td></tr>
<tr><td>Instructor Directions</td><td colspan="2">Evaluation</td></tr>
<tr><td></td><td colspan="2"></td></tr>
</table>

ness towards the lesson. The interest approach should be connected to the lesson activities that will be completed with your learners. The momentum and interest gathered with your learners is important to keep as you head into the activities planned for your lesson as they will assist in the engagement of the lesson. As you have learned, the planning of a lesson includes a variety of components which need to be well thought out and connected to help successfully deliver the road map plan to meet the needs of your learners.

REVIEW QUESTIONS

1. What are the three major components of an objective? Please describe each of the three components.
2. What is the difference between a terminal and enabling objective? How are each used within the classroom setting and lesson plan?
3. Compare and contrast objectives and essential questions. Describe when it is best to utilize an objective or essential question within a lesson.
4. How do you connect the state or federal standards to an objective or essential question?

5. What ways can you connect higher-ordered thinking and Bloom's Taxonomy to a lesson?
6. Compare and contrast formative and summative assessments. How are they utilized within a daily lesson plan?
7. In what ways would you assess your learners to measure their comprehension?
8. How do you support your IEP or 504 plan learners within this lesson?
9. Describe how your interest approach and lesson activities should connect throughout a lesson.

REFERENCES

Bloom, B., Englehart, M., Furst, E., Hill, W., & Krathwohl, D. (1956). *Taxonomy of educational Objectives: The classification of educational goals. In Handbook I: Cognitive domain.* Longmans, Green.

Frank, C. L., & Richards, S. B. (2020). *Essentials of Special Education: What Educators Need to Know.* Routledge. doi:10.4324/9780367815530

Mager, R. F. (1984). *Preparing instructional objectives* (2nd ed.). Lake Publishing.

McTighe, J., & Wiggins, G. (2013). *Essential questions: Opening doors to student understanding.* Ascd.

National Council for Agricultural Education. (2022). *National AFNR Content Standards.* Retrieved from https://ffa.app.box.com/s/n6jfkamfof0spttqjvhddzolyevpo3qn/file/294153554237

Phipps, L. J., Osborne, E. W., Dyer, J. E., & Ball, A. L. (2008). Handbook on Agricultural Education in Public Schools. *Thomson Delmar Learning.*

Talbert, B., Allen, S., Vaughn, R., Croom, B., & Lee, J. S. (2013). *Foundations of Agricultural Education. Pearson Education.*

Wiggins, G., & McTighe, J. (2005). *Understanding by Design.* Pearson Merrill Prentice Hall.

KEY TERMS AND DEFINITIONS

504 Plan: A plan developed at the school to customize a student's learning environment to meet their specific needs.

Action Verbs: An action that is used to describe a verb, for example: describe, create, or illustrate. These are used at the beginning of every objective to describe the action a student will demonstrate at the end of a lesson.

Assessment: A method used to determine what content, skills, or knowledge a student has retained from a series of learning activities and lessons.

Bloom's Taxonomy: A hierarchical system that classifies the order of learning levels from the simplest (bottoms of the triangle) to the most complex forms (top of the triangle).

Essential Questions: An open-ended question used to probe for deeper meaning and understanding of a concept or issue during a lesson.

Individualized Education Plan (IEP): An individualized education plan that lays out the special education instruction, supports, and services a learner needs to make progress.

Interest Approach: A learning activity that is used to focus a student's attention on the content that will be taught throughout a lesson and create engagement in the lesson activities.

Learning Objective: A statement written by a teacher that explains what a student will be able to do at the completion of a given lesson, unit, or course.

Lesson Activities: An organized set of activities which are designed to present a manageable piece of information to your learners.

Standards: A learning measure that is written by a governing body to give teachers direction on what students should learn and be able to do at the completion of a course.

Chapter 8
Developing Instructional Methods

Joey Blackburn
https://orcid.org/0000-0002-9483-0288
Louisiana State University, USA

Kristin Stair
Louisiana State University, USA

ABSTRACT

Determining which methods of teaching to use is an important step in developing content for your classroom. Teacher-focused methods involve more direct instruction to students and include methods such as lecture, questioning, demonstration, resource people, field trips, and checking for understanding. Learner-focused methods of teaching include more student input and collaboration and include discussion, debate, problem solving approach, contextualized teaching and learning, flipped classes/team-based learning, case study, inquiry-based learning, experiential learning, problem-based learning, project-based learning, cooperative learning, and role playing. Each method of teaching has different strengths and weaknesses and allows teachers to achieve their learning objectives in different ways and increase their variability.

INTRODUCTION

Deciding *how* to teach is just as important as deciding *what* to teach. Instructional methods allow educators to teach content in a variety of ways and engage with distinct types of learners. One instructional method is not necessarily more effective than others, instead, the objectives of the class should guide which methods are selected. For example, if a teacher wants to show students how to do something, using a demonstration allows them to show the actual skill while clarifying steps, addressing safety concerns, and ensuring all students are ready to complete the skill on their own.

The objectives of this chapter are to:

DOI: 10.4018/978-1-6684-3420-8.ch008

Table 1. Continuum of instructional methods

Methods						
Lecture	Resource People	Demonstration	Discussion	Cooperative Learning	Project Based Learning	Independent Study
Teacher focused	⟵				⟶	Student focused

- Describe the differences between teacher based instructional methods and student based instructional methods
- Identify primary teacher focused instructional methods including lecture, demonstration, resource people, and field trips
- Identify student based instructional methods of teaching including discussion, debate, cooperative learning, role playing, inquiry-based learning, experiential learning, case study, project-based learning, flipped classrooms/Team-based learning problem, problem-solving approach, and problem-based learning
- Define independent instruction and identify methods in which individual instruction can be used.
- Describe instructional methods which allow teachers to incorporate core content into their agricultural classrooms including contextualized teaching and learning and literacy strategies.

TEACHER FOCUSED VERSUS STUDENT FOCUSED METHODS

Instructional methods can be teacher-focused or student-focused and therefore are appropriate in different ways depending on what content is being taught (Borich, 2007). In general, passive methods such as lecture are very teacher focused while methods that allow student input and student collaboration are more student focused.

The first step in selecting which instructional methods to use is to develop high quality learning objectives. Learning objectives help develop content and determine what students should be able to know, do, and understand by the end of the lesson. Objectives also serve as indicators for which methods might be most effective.

For example, if the objective is:

Given 10 breeds of cattle, the learner will be able to identify three characteristics of each breed with 100% accuracy

A teacher will likely begin by presenting content on breeds of cattle and their characteristics using lecture or another teacher-based strategy, then incorporate practice in the form of individual or group work to help students be able to identify those characteristics on their own.

For an objective such as:

After being assigned a role, the learner will develop a 5-minute statement supporting whether they are for, or against, Genetically Modified Organisms (GMO's).

Lecture might be utilized to teach students about GMO's and address the controversial points of the technology, however, a teacher would likely either use discussion, debate, cooperative learning, problem solving approach, or project-based learning for students to learn from others and organize their thoughts into a five-minute presentation.

Similarly, hands-on skills such as welding, working with plants, handling animals, or building structures, are incredibly difficult to teach using lecture alone. While lecture might be a part of these lessons, methods such as demonstration could also be used to enable students to learn the steps and then repeat those steps on their own.

Each instructional method lends itself to addressing diverse needs. The combination of using multiple methods allows teachers to address various aspects of their objectives as well as reach different types of learners. A combination of strategies helps teachers increase their variability as well as making the classroom more engaging.

TEACHER FOCUSED INSTRUCTION

Teacher focused instructional methods are methods teachers use to share information directly with their students. These methods are typically focused on communication that moves from the teacher to the students with the intent of sharing specific information, therefore, they are typically considered to be *teacher focused* instructional methods. These methods are particularly useful when there is essential information that needs to be conveyed to all students regarding specific topics.

Four of the most common teacher-focused instructional methods include:

- Lecture
- Demonstration
- Guest Speakers and Resource People and
- Field Trips

Traditional Lecture and Modified Lecture

Throughout the history of education, lecture is consistently mentioned as a method used to provide instruction directly to learners. **Lecture** is typically defined as a method of teaching where information is conveyed from a teacher or expert to a group and is most clearly described by how information flows from the presenter to the audience (Phipps et al., 2008; Talbert et al., 2014). Because of this flow, lecture is particularly good at sharing information with others, especially for larger groups that all need the same content at the same time. It is often used for introducing new concepts and factual information. However, because it is such a teacher focused method, it can often result in a lack of student engagement or disinterest since they are not actively participating in the content (Newcomb et al., 2004).

In order to develop engaging lectures, there are some essential components of instructional planning that should be included:

- Lecture should begin with an interest approach. Not only should interest approaches serve to develop student interest, but they should introduce content and be used to connect content with prior knowledge while helping to establish why the topic is important. Interest approaches can take many forms, but generally include stories or scenarios, realia, questions, pictures, or problems to solve

- Lectures should tie into educational objectives and those objectives should be shared and reviewed with students
- Content should be delivered in a way that includes visual elements to help students visualize important content or concepts
- Lectures need to include frequent opportunities to check for understanding by the incorporating questioning, activities, or student engagement.
- Lectures should include a strong closure that summarizes the main points and reviews key concepts based on the lesson's educational objectives. Because lecture tends to limit communication from the audience to the speaker, care should be taken to include other opportunities and teaching methods which allow for increased communication and engagement.

Modified Lecture

Lecture in its most traditional form relies on providing information directly to the students for most of the lesson, however, when teachers do not routinely engage with students, students often become disassociated from the content and their attention may wane. To increase student engagement, many teachers prefer to use a *modified lecture*, which uses smaller segments of lecture that last approximately 10-15 minutes, broken up by the inclusion of other activities or teaching methods.

Table 2. Overview of lecture as an instructional method

Lecture: Used to introduce new content to a group Overview of steps in a lecture: • Introduce content and develop interest • Review educational objectives • Deliver content • Interact with students by including stories, asking questions, and using visuals to engage learners • Review main points based on instructional objectives and supply closure Good for: • Sharing information to a class as a whole • Ensuring that all students are receiving the same content at the same time • Identifying key concepts Disadvantages: • Can be less engaging than other methods • Can discourage student interaction

Demonstration

In agriculture, we often spend time teaching students how to perform skills. For skill-based activities, demonstrations allow teachers to directly teach the elements, steps, or components of a skill. **Demonstration** involves the intentional teaching of the steps necessary to complete a skill (Phipps et al., 2008) and includes the following components:

- An overview of the skill including the importance of the skills and the need for understanding the skill
- The inclusion of a list of steps written out and provided to students
- The teacher demonstrating the steps one at a time, being sure to identify critical areas and safety issues
- Demonstrating the steps a second time
- Allowing a student to come up and follow the steps in front of the class
- Asking for any clarifying questions
- Allowing all students to attempt the skill as a class.

Having the steps written out in advance allows students to learn the steps and follow along. Similarly, by reviewing the steps twice as a teacher, students are provided with many opportunities to see the steps in action. It is also suggested that before a demonstration begins, the class be notified that one of them will be called to perform the demonstration and follow the steps with assistance in front of the class. By calling on a student, rather than asking for volunteers, it helps to ensure that all students are watching and paying attention. Additionally, this step allows students one more chance to see the process and builds confidence as they see a peer able to perform the skill. Lastly, asking for additional questions gives the teacher the opportunity to clarify any remaining uncertainty before allowing students to perform the skill on their own.

Table 3. Overview of demonstration as an instructional method

Demonstration Used to teach students how to perform a skill Overview of steps in a demonstration: • Break down the skill into small, manageable steps • Introduce content • Provide a list of steps to the students • Inform students that the material will be reviewed twice and then a student will demonstrate the skill • Conduct the demonstration, reviewing each step individually, including safety information and important tips • Review the process a second time, still highlighting key information • Randomly call on a student to come forward to demonstrate the skill, allowing the class to help review steps or safety information as needed • Ask clarifying questions • Allow all students to complete the skill individually Good for: • Ensuring that all students are made aware of the key steps of an activity • Making sure that all students are aware of safety procedures related to agricultural skills • Reinforcing correct procedures in a controlled environment before having all students attempt the skill Disadvantages: • Steps might not be fully understood even after the demonstration • Demonstration may inadvertently teach students incorrect steps • Demonstrations work best with basic processes that have simple steps such as ear notching, root cuttings, changing the oil in a small gas engine, etc. More complex processes such a raising a pig, planning a vegetable garden, and building an engine cannot be taught using this method.

Guest Speakers and Resource People

Teachers are considered the experts in their area, but sometimes bringing in speakers can be more dynamic and may allow students to connect with content in new ways (Phipps et al., 2008; Talbert et al., 2014). **Guest Speakers/Resource people** can help break down biases that students may have while helping them build connections to agricultural professionals. In order to use resource people effectively, it is important to choose speakers carefully to ensure that they will be able to connect with the class and meet the needs of students (Newcomb et al., 2004). Some speakers may be knowledgeable but may be unable to connect to students or share information at the appropriate level of understanding. Making sure that speakers are aware of what knowledge students already have on a topic can be helpful to ensure that they speak to the correct level of the learners. Resource people will often save part of their time at the end of their presentation for questions. To help questions run smoothly, the teacher can use the class period before to introduce the speaker and have students develop a list of questions they have about the topic or the speaker. This ensures that students will have questions to ask when the speaker comes to class.

Table 4. Overview of guest speakers and resources people as an instructional method

Guest Speakers and Resource People Used to allow professionals to share their agricultural knowledge with students Overview of steps in using resource people/guest speakers: • Identify a speaker that is knowledgeable on the content area and can convey that information to students • Follow-up with the speaker before the presentation to confirm time, content, and share directions for checking into the central office and coming to the classroom • The class period before the speaker, review the information and share information about the speaker with the students. Have students develop a list of questions about the topic or the speaker • Introduce the speaker to the class • Allow students to ask questions from their list at the end of the presentation. • Review main content from the presentation if needed • Send a thank you note (or have the class send a thank you note) to the speaker Good for: • Connecting students to real-world professionals Disadvantages: • Some professionals are excellent at what they do but may not be able to connect to younger students • The speaker may not adequately describe their content area or meet the class objectives

Field Trips

Field trips are opportunities to travel outside of the classroom to help students see agriculture in action. Taking students on field trips allows them to be an active participant in an educational experience while they learn relevant content within the curriculum. Therefore, field trips can be a wonderful opportunity to help students see the real-world connections to content (Newcomb et al., 2004).

In today's classroom, field trips can cause many logistical issues as teachers have to arrange travel, manage costs, find drivers, receive permission from the school (and sometimes the school district or school board), and obtain parent permission. Some schools limit student travel to maintain time in core subject instruction which can be a barrier to organizing trips. It is important to know the rules and restrictions of the school district before planning trips and it is encouraged to plan early since some approvals may require school board approval which may require advanced notice on a school board agenda before the trip can be approved.

With all these factors in mind, there has been an increase in virtual field trips which are tours of facilities available through video and virtual avenues. Many agricultural producers have started offering virtual field trips or interactive experiences to allow the general public to see what production operations actually look like. With permission, teachers can also visit sites, record the facilities, and talk with professionals to share with the class at a later time.

Similar to utilizing resource people, developing questions in advance of going on a field trip or participating in a virtual field trip is a good strategy to help make sure students can actively participate. For virtual trips, students could ask questions and then those questions could be submitted to the production operation or sent to an expert to help generate answers.

Table 5. Overview of field trips as an instructional method

Field Trips Used to allow students to see the agricultural industry in a real-world context Overview of steps in using field trips: • Identify a location that will provide students with an engaging experience related to the content. • Contact administrators to clarify the procedure for field trips. It is important to begin this process at least six months in advance to make sure there is enough time to complete the approval process for the school system. • Arrange transportation and logistics such as permissions slips, lunches, and notification of other teachers. • Review the requirements for participation and the expectations for the field trip with students in advance of the trip • Review the objectives of the field trip and ask students to generate questions related to the content of the field trip. • After the trip, review the main learning objectives from the trip. • Send thank you notes (or have the class create thank you notes) to trip organizers and presenters Good for: • Helping student see aspects of agriculture that are not easily conveyed in a classroom • Helping students make connections to industry professionals • Providing agricultural content in an immersive way Disadvantages: • Classroom management can be difficult to manage in off-campus settings • Not all aspects of a field trip may be engaging or meaningful for students • Not all speakers may be engaging or be able to connect to students in a meaningful way • There are challenges in organizing and facilitating field trips • The speaker may not adequately describe their content area or meet the class objectives

STUDENT FOCUSED INSTRUCTION

Student focused instruction utilize methods that focus primarily on the student or on student interaction. In teacher focused methods, the teacher serves as the primary source of content and knowledge, while in student focused instruction, students learn from each other or interact with each other to meet common goals. However, it is important to mention that even if students are learning from each other, these methods still require careful planning and development, as well as facilitation and structuring on the part of the teacher. Well-developed activities, frequent checks for understanding and effective classroom management are essential for student focused instruction to be successful.

A few examples of student-focused methods that will be discussed in this chapter are:

- Discussion
- Debate
- Cooperative Learning
- Role Playing
- Inquiry Based Learning
- Experiential learning
- Case Study
- Project Based Learning
- Flipped Classroom
- Problem Solving Approach
- Problem Based Learning

Discussion

As opposed to lecture, discussion allows the flow of communication to move between students and between the students and the teacher. **Discussion** allows for students to explore in-depth problems or issues and reach a solution as a group. Discussion is especially useful because it allows students to learn from each other.

Discussions can easily move off task if not structured well at the beginning. When using discussion, it is critically important to have specific questions identified in advance and provide those questions to the students. The list of leading or engaging questions should be provided and then a time limit is established for students to answer the questions (Newcomb et. al., 2004). Often with discussion, a presenter and a scribe are chosen to represent the group, however this can cause disengagement for those

Table 6. Overview of discussion as an instructional method

Discussion Allows students to explore ideas and learn from each other Overview of steps in a discussion: • Introduce content • Share discussion questions and the time limit with students on the board or on a handout. • Place students into groups • Allow students to discuss each question with each other and produce shared answers • Have all students document their questions and group answers • Call on one student from each group to share the groups answers • Review main points as a class Good for: • Allowing students to teach and learn from each other • Encouraging creative answers to questions • Providing new perspectives on content Disadvantages: • Poorly planned activities can result in limited interaction and discussion • Classroom management must be carefully monitored to ensure groups are on task and working toward their group goal. • Frequent monitoring needs to be used to make sure groups are not getting off track or have questions that hinder the discussion process

who do not have a role. Instead, another way to approach discussion is to have each student record the answers that are discussed in the group and then, if answers are to be shared with the class, a presenter is randomly chosen by the teacher to present their group's responses which encourages all students to be responsible for their group content.

Groups for discussion can be chosen in advance or students can choose their own groups depending on the goals of the lesson. When using discussion regularly, groups should be changed often so that students experience different ideas and opinions.

Debate

Debate can often be incorporated as part of a discussion activity, or it can serve as an independent method. A **debate** allows students to explore issues that have multiple perspectives by researching and supporting those perspectives (Phipps et al., 2008). Debates can be wonderful opportunities for students to explore multiple sides of a complex issue and can be structured in a variety of ways, for example:

- Students can be assigned one side of an issue to explore and defend
- Students can choose one side of an issue to explore and defend
- Students can be placed into teams to address one side of the issue against another team of students who address the opposing side of the issue.

After developing a statement and main points that support their side of an issue, a debate is typically structured so that one side presents their main points within a given time frame and then the opposing side is allowed to provide their main points. Often, debates can also allow a small amount of time for opposing individuals or groups to ask questions.

Table 7. Overview of debate as an instructional method

Debate Used to explore different sides of a complex issue Overview of steps in developing a debate: • Identify a complex topic that has multiple perspectives • Determine the format for how the debate will be • Present the topic to the class along with essential facts around the issue • Set ground rules for the debate including expectations and expected format • Assign topics and groups or allow students to choose their side of the issue • Set times for how long students or groups must review their content and construct their debate • Allow each side of the debate to present their content within the given time limit • Allow the opposing side to ask questions or ask for clarifying information • Give each team time to respond to the questions and then allow the other team or side to present • Review main content from both sides of the debate as a class Good for: • Exploring multiple sides of an issue • Providing opportunities to research complex agricultural issues • Teaching students how to synthesize research and present that research to an audience Disadvantages: • Debates without good ground rules can be chaotic • Students might become emotional if they believe strongly in their side of a debate or feel that their points are being attacked • Care should be taken to help ensure that all students are participating in the debate and are respectful to the views of others

Debates can become intense, especially when dealing with controversial issues in agriculture. It is important to set ground rules in advance for how statements should be made and make it clear what the expectations are for interacting with peers. Sometimes, assigning roles can be an effective way to help diffuse personal reactions to an issue and can possibly help students who already have strong feelings about an issue explore the other side and better understand the thoughts of others.

Cooperative Learning

Cooperative learning is a method of teaching that has been used since the 1920's but has seen a resurgence in recent years (Kagan, 1994). Cooperative learning utilizes small groups to work together toward a common goal. Unlike regular group work, cooperative learning contains specific components that go beyond using groups to answer questions or generate discussion. Five elements for cooperative learning have been identified (Johnson et al., 2000):

1. Positive interdependence
   Allows students to take an active role in learning, work together, and feel responsible for their success as well as the success of the group. Team members should rely on each other to achieve the group's goals.
2. Group processing
   Cooperative learning activities include the opportunity for students to set goals as a group and determine their process toward achieving those goals.
3. Face-to-face promotive interaction
   Students work together in a way that allows them to connect, discuss and share ideas. While some aspects of the projects are completed individually, students should work together on common project elements.
4. Individual accountability
   Each student has their own role in the cooperative learning process and each individual is necessary for success.
5. The development of collaborative skills
   Cooperative learning should include opportunities for students to interact with each other in the development of social skills and allow them to learn skills which help them work as a team. Skills such as leadership, decision making, and conflict resolution are all important components of the cooperative learning process.

A few examples of cooperative learning include:

- Think-Pair-Share
  Each student answers a question individually and then answers are compared to formulate an answer as a team or small group.
- Jigsaw
  Students are placed in initial groups that learn specific content or explore a specific focus area, then once the students have mastered their piece of the content, they are grouped with others with different content to share information, find a solution or develop a final product.

Table 8. Overview of cooperative learning as an instructional method

Cooperative Learning Involves intentionally grouping students to allow them to accomplish a common goal Overview of steps in developing a cooperative learning activity: • Design an activity that allows for multiple roles. • Introduce content to the class. • Review expectations and the specific roles that will used within groups • Assign students to groups. • Ensure that students are participating in groups in such a way that there are opportunities for both individual accountability and group accountability. • Allow students to share, present or compete as a group. • Review key points and critical content. Good for: • Allowing students to work in teams toward a shared goal • Teaching students how to work in teams • Reinforcing individual effort while contributing to a team goal Disadvantages: • If individual students do not complete their components, it will cause the group to not be able to complete the full assignment or activity • Classroom management is critical to ensure all students are on task and contributing • Teams may not work well together or individual members may not be compatible

- Numbered Heads Together
  Students are placed into groups, and each student is assigned a number. Each group helps its members become familiar with the content, then the teacher calls on specific students from each group according to their number to answer questions about that content.

Role Playing

Role playing involves any activity that places students into the role of other people, systems, or processes (Talbert et al., 2014). Role playing is often considered to be a strategy used with younger students; however, it can be highly effective for students in secondary education. Role playing allows students to practice different views, careers, or the interaction between systems by putting themselves into a role within that activity (Newcomb et al., 2004). In order to successfully integrate role-playing, a teacher must give clear directions on what the activity will consist of and what each role will involve. For example, role playing might be used to allow students to participate in mock interviews where they take turns with a partner to play the role of the person being interviewed and then the person doing the interviewing. An activity such as this allows students to explore both sides of the interview process, develop questions that might be asked, learn how to respond to interview questions and evaluate each other and their success in both roles. For this activity to be successful, the teacher must first set ground rules about the activity and the students' behavior within those roles, have students develop their behaviors or questions, and set the format and time limits.

Role playing could also be used to:

- Allow students to explore different jobs or roles within an agricultural industry
- Have students explore the various parts of a system such as digestion, water cycle, or photosynthesis, by assigning them a part of the system and acting out what happens in that stage of the process.
- Allow students to role play a career and what may be involved in that career.
- Have students interact with each other in roles such as that of a customer and a business manager

Role playing can be used in an unlimited number of ways if clear guidelines and rules are in place. Care should also be taken to design activities with a variety of roles to allow students to participate in a variety of ways. Some students may be more outgoing and would be likely to take roles that require more talking or action, while others may be more comfortable in roles that do not require as much acting, especially when acting out agricultural or environmental processes.

Table 9. Overview of role playing as an instructional method

Role Playing Used to allow students to immerse themselves in a specific role related to classroom content Overview of steps in developing a role playing activity: • Identify specific roles that students will need to explore. • Develop roles and critical components or information for each role. • Teach essential content. • Review expectations and the specific roles that will be used. • Assign students to roles. • Allow students to demonstrate roles within groups or to the class as a whole. Good for: • Helping students see the connection between individual components in a complex process • Helping students practice specific roles or careers • Making abstract processes more relatable Disadvantages: • Some students may too shy or be unwilling to participate • Learning or participating in the role incorrectly can reinforce false information

Inquiry-Based Learning

Inquiry-based learning can trace its roots to the works of John Dewey (1910/1997) and has historically been used in science classrooms when guiding students through the scientific method. In its simplest form, inquiry-based learning is designed to teach through curiosity and student led exploration. In this instructional method, students design questions, topics to explore, or areas they want to research and design a process to find answers. Inquiry-based learning can be used in a very student-focused manner where students create their own questions and design their own investigation, or it can be utilized in a very structured way where students are given a problem to investigate. There are four levels of inquiry-based learning which can be described as (Banchi & Bell, 2008):

- Confirmatory Inquiry - students are provided with a question to investigate as well as the procedure to be utilized in the investigation. Results of the investigation are already known; students are confirming the results.
- Structured Inquiry - students are provided with a question to investigate as well as the procedure to be utilized in the investigation. Results are not known; students are guided through making conclusions
- Guided Inquiry - students are provided with a question to investigate. Students design and implement the investigation and draw conclusions.
- Open inquiry - students create their own questions to investigate, design and implements an investigation, and draw conclusions

When using inquiry-based learning, a topic or questions is proposed to the class, then students will generate their own questions and methods for solving the problem. Rather than sharing information directly with the students, they are encouraged to explore and find answers on their own. One of the most difficult parts of the inquiry-based process for many teachers and students, is resisting the urge for teachers to answer student questions during the process. With inquiry-based learning, students are encouraged to generate their own questions and find their own answers with the teacher serving as a facilitator and a guide to keep groups moving forward.

Table 10. Overview of inquiry-based learning as an instructional method

Inquiry-Based Learning Used to encourage students to inquire into questions to stimulate higher level thinking Overview of steps in developing an inquiry-based learning activity: • Determine what level of inquiry is most appropriate for the class and topic • Teach only the essential content that students need to begin researching on their own • Provide questions for the students to investigate or teach students how to create their own questions for the investigation • Provide students with a method to answer the questions to investigate or they are guided on how to create a proper research design • Closely observe groups as they are working to correct incorrect information and help facilitate groups if needed • Guide students through analysis of results and drawing of conclusions • Have students present what they have learned to the class • Reflect on the process with students to help them identify roadblock to solving the problem and how they can overcome those roadblocks • Review key points with the class as a whole Good for: • Promoting higher order thinking skills by allowing students to think and reason through the learning process • Exploring Questions/topics that do not have single correct answers • Deepening student understanding of material • Creating student ownership of their learning Disadvantages: • Teachers must learn to give up control of learning to the students • Students may feel uncomfortable being charged with their own learning, especially if this method is being used for the first time in the classroom • Students may have trouble recognizing their success • Students may have trouble working in groups • Successful inquiry-based learning my require multiple attempts to set a culture of inquiry

Problem-Based Learning

Problem-based learning is similar to inquiry-based learning in that an overall goal of problem-based learning is to allow for students to work collaboratively toward and end goal (Bransford et al., 2000). The main difference in the two is the focus on authentic, real-world problems in problem-based learning. Like inquiry-based learning, supporters have described this method as a way to foster critical thinking and problem-solving skills and improve student motivation in a way that is enjoyable for students (Burris, 2008). Problem-based learning tends to utilize ill-structured problems that will be vaguely defined in the beginning, have multiple correct solutions, and require the learner to seek additional information prior to formulating a solution. The role of the teacher in problem-based learning will be to guide, facilitate and empower students as they make meaning of the knowledge gained in the process. Emphasis is placed on the discovery process used when using problem-based learning rather than focusing on a single correct answer.

Table 11. Overview of Problem Based Learning as an Instructional Method

Problem-Based Learning Problem-Based Learning is a student focused approach focused on students working collaboratively to solve authentic, open-ended problems. Overview of steps in developing problems-based learning: • Design instruction based on ill-structured, open-ended problems/questions. • Determine the amount of instructional time dedicated to the problem/question • Examine and define the problem. Depending on the students, this may require teacher assistance • Students must determine needed information/tools/resources to solve the problem • Possible solutions are tested and evaluated • Solution to the problem is reported and presented\ Good for: • Fostering critical thinking and problem-solving skills • Reaching higher order thinking levels • Longer term knowledge retention • Engaging students in the learning process Disadvantages: • Students may have trouble formulating questions and/or potential solutions • Requires a lot of front-loading preparation by the teacher • Can be viewed as disorganized and chaotic

Experiential Learning

Agricultural education has often been described as *hands-on/mind-on* in terms of the way it is taught. A common instructional strategy associated with this description of agricultural education is *experiential learning*. Experiential learning is often associated with Supervised Agricultural Experience (SAE) programs; however, many scholars and practitioners have described experiential learning as a part of the total agricultural education program which involves classroom/laboratory instruction, SAE, and FFA.

Experiential Learning Theory (ELT) is often discussed when applying experiential learning into the agriculture classroom (Baker et al., 2012). Learning, as viewed through an ELT lens, is a process rather than a product. Learners begin the ELT cycle through a concrete experience as the learner experiences and participates in a new situation. Agricultural education has historically excelled at providing concrete experiences for learners, ranging from first-year students engaging in public speaking for the first time in the classroom, to students learning how to safely weld in the agricultural mechanics laboratory (Baker et al. 2012).

Reflective observation occurs after a concrete experience and is a point in time where the teacher needs to assist the learner in reflecting on the experience (Kolb, 1984; Kolb & Kolb, 2005). This time of reflection can allow students to further process concepts and allows time for correction if needed. In the welding example, the teacher would assist the student in breaking down how they performed the weld and reinforce behaviors and concepts that need to be learned.

Abstract conceptualization can be thought of as further exploring concepts and theories to help students see how the theory interacts with the skill. In this step, based on the observations and reflections, a student forms theories about their experience (Kolb, 1984; Kolb & Kolb, 2005). An example could be after participating in welding, students hypothesize how the process works followed by a lesson from the teacher on the scientific processes involved in welding.

During active experimentation, the learner may be allowed to experiment with new concepts and strategies to reinforce learning (Kolb, 1984; Kolb & Kolb, 2005). While the first instinct of many teachers may be to avoid active experimentation in a lesson, it is a vital process for the learner to become proficient. An example of active experimentation in our welding example could be allowing students to

Figure 1. Kolb's Model of Experiential Learning; adapted from Kolb, D.A. (1984). Experiential learning: experience as the source of learning and development. Englewood Cliffs, NJ: Prentice Hall.

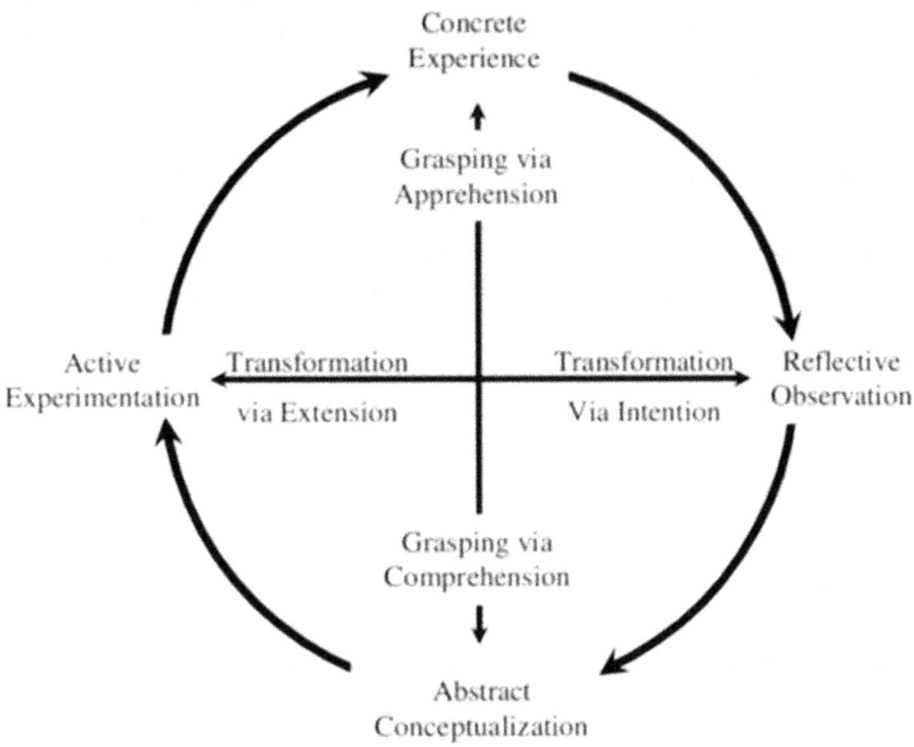

Table 12. Overview of experiential learning as an instructional method

Experiential learning
Experiential learning is as much a classroom culture as it is an instructional method, it can promote higher order thinking, deepen understanding, and bolster creativity Overview of steps in developing an inquiry-based learning activity: • Design – determine whether the students will be provided a concrete experience or abstract conceptualization first. This will depend on the content, students, and teacher's preference. • Develop a concrete experience – typically, this experience is a hand-on, minds-on experience that may occur in the laboratory, field trip, or even an FFA activity • Provide an opportunity for reflective observation where students are guided to reflect both in action and on action to gain deeper understanding of their experience. • Allow for abstract conceptualization where students are presented with and/or guided through the conceptualization of theory related to the experience • Allow students to practice active experimentation and are given opportunities to experiment with their experience and to confirm or disconfirm prior assumptions Good for: • Deepening student understanding of content • Reinforcing the importance of experiences • Connecting content to real world problems Disadvantages: • Requires a lot of front-loading preparation by the teacher • Teacher must give up some classroom control • Students may be uncomfortable through the ELT process because it is primarily student led

try various welding patterns that were not directly taught (e.g., circular vs. weave) to become comfortable in the process.

It is important to note that experiential learning is a cycle and an ongoing process for students and that the model may also be modified as need for individual students to move through the cycle successfully.

Case Study

Typically, **case studies** utilize written scenarios which requires the learner to apply knowledge to solve problems or answer questions. High quality case studies are grounded in authentic situations - real or hypothetical - that are written in a manner that places the student within the situation. Teachers should select or prepare case studies in advance that will require students to identify key questions or problems. Case studies may have one correct answer or may allow for many different responses. Case study pairs well with other methods of teaching where students are encouraged to explore problems or look for real-world connections.

Project-Based Learning

Students learning by completing projects is a concept that is not new to agricultural education. However, **project-based learning** can include more than just allowing students to complete a content focused project. It has been suggested that while learning activities such as designing posters, dioramas and PowerPoints are common instructional activities, they may not allow for deeper processing that can be developed through a project-based instructional method (Miller & Krajcik, 2019; Larmer & Mergendoller, 2010). While there are many ways to implement project-based learning as a teaching method, in

Table 13. Overview of case study as an instructional method

Case Studies
Allow teachers to utilize authentic problems in a controlled setting.

Overview of steps in developing a case study activity:
- Determine where the case study best fits within the lesson. Case studies can be successfully utilized to introduce a lesson, as the core lesson content, or as a closure activity.
- Prepare the case studies in advance. Good case studies take time to develop or modify. Case studies, ideally, are based in authentic situation in which the students could reasonably find themselves. Best written in third person.
- Determine the overall goal for the case study. The best case studies have complicated answers and require students to justify their thoughts.
- Review essential content or generated answers with the class as a whole

Good for:
- Introducing content where it is not feasible to have the real items/examples in class
- Fostering critical thinking skills in a safe environment

Disadvantages:
- Typically requires students to have excellent reading comprehension skills
- Can intimidate students who are poor readers
- Case studies that are developed too broadly or without enough detail may be difficult for students to solve.

Table 14. Overview of project-based learning as an instructional method

Project-Based Learning
Project based learning is designed to provide students the opportunity to learn and develop skills by engaging in projects set in authentic, real-world contexts.

Overview of steps in developing a project-based learning activity:
- Begin by generating interest and presenting students with an engaging situation or problem to solve.
- Present introductory content and information which students need to know to complete the project. This may include lecture, discussions, independent research, or laboratory activities.
- Place students into group and allow the groups to determine how they will solve the problem or address the situation.
- Allow students to work together to determine what type of project they will create to solve the problem or address the issue.
- Have students develop a "need to know" list for their group to help guide their project development and to determine what they will need to research on their own before creating their project.
- Provide time in class for students to work together and provide guidance as needed to keep students on task, help them resolve conflicts, and correct information that may not be factual.
- Provide opportunities for groups to evaluate each other and give feedback before presenting their final product.
- Allow students to present their project.
- Review key findings or results to the class as a whole and clarify any missing or incorrect information

Good for:
- Building collaborative skills among students
- Honing critical thinking and problem-solving skills
- Building authentic skills by engaging students in real-world style problems

Disadvantages:
- Requires careful planning by the teacher
- Can be unstructured and chaotic if not carefully designed
- Come students may not contribute as much as others, so setting expectations among groups is essential when developing projects
- Students may have difficulty determining the direction of the project, especially if this is the first time this method is being used.

general they will include projects that are purposely designed, are grounded in relevant content, include social interaction, and include strategies to increase cognitive engagement.

Research related to project-based learning has identified seven essential components that projects should include (Larmer & Mergendoller, 2010):

- Significant content that ties into state standards and is important for students to know
- The opportunity to develop a "need to know" list and help students see why the project is important
- A driving question which allows students to focus their efforts and is complex enough to encourage them to solve the problem and develop their interest.
- The opportunity for students to choose components of the projects or choose how to solve the problem. While students may not choose the entire project, they should have some choice in how elements of the project are completed.
- The chance to develop social skills and 21st century competencies such as collaboration, communication, and teamwork.
- The opportunity to ask questions and develop deep inquiry into the problem.
- The ability to critique and review each other's work and provide feedback
- The opportunity to present to a public audience

Projects should begin with identifying a critical question or problem to solve that is relevant to agriculture. To help structure projects in a way that allows students to be successful, guidelines should be carefully designed to help assist students in understanding what elements should be developed and assist the groups in developing their components. Also, since this method of teaching should be authentic in nature, rubrics should be used to help provide project guidance, allow students to evaluate themselves and provide guidance on how to complete projects successfully (Borich, 2007; Larmer & Mergendoller, 2010).

Flipped Classrooms

The **flipped classroom** approach to teaching began in the late 1990s to early 2000s when the popularity and availability of educational technologies like learning management systems and high-speed internet saw rapid diffusion (Michaelsen et al., 2004). Simply defined, flipped classrooms utilize modern educational technology to make content available online for students to access outside of class time. With a flipped classroom, most of the content such as readings, presentations, and videos which are typically provided in class, are shared with students to complete out of class as homework or independent work. After students have completed work on their own, time in class is used to ensure students understood the content clearly and focuses on hand-on activities to enhance learning. Flipped classrooms require students to be highly self-regulated and disciplined to access the content prior to class time, which is the major challenge of the method. For many teachers who use this method, the out of class activities are assigned as homework, and then an entry test or check-in activity is used at the beginning of the next class period to ensure that students understood the content. Students who did not complete the content must do that during class before being able to participate in in-class activities.

Table 15. Overview of flipped classroom as an instructional method

Flipped Classrooms Allows more class time to be dedicated to application activities Overview of steps in developing flipped classroom resources: • High quality online experiences are designed in advance for successful implementation. Ideally, content such as short videos and lectures are used rather than large reading assignments or chapters of reading material. • Design accountability measures for student when they return to class to ensure that content was reviewed. • If using TBL, quality assessments (IRATS and TRATS) must be developed, and procedures put in place for immediate feedback • Review is conducted to ensure that students have grasped main points • Instructional activities, labs or hand-on content is delivered during class time Good for: • Classes with a high focus on hand-on activities • Fostering accountability of learning • Promoting higher level thinking skills Disadvantages: • Students are likely to not navigate through online content if accountability is not put in place • Requires an elevated level of advanced preparation by the teacher to ensure success • Can be difficult for students who are not familiar with learning content on their own.

Team-Based Learning

Team-based learning (TBL) is a modification to the traditional flipped classroom approach that aims to address the common challenge of self-regulation among learners while also impacting critical thinking and problem-solving skills (Michaelsen et al. 2004). Like the flipped classroom approach, much of the content to be taught is made available online for students to access outside of class time. Accountability is a major theme in TBL and to accomplish this, students are placed in small (3-6 members) teams and assessed individually and as a team, based on the content. During class time, teachers begin with an Individual Readiness Assurance Test (IRAT) that is the individual accountability portion of the approach. After the IRAT, the students are then allowed to join their team members and complete a Team Readiness Assurance Test (TRAT), which is the same assessment as the IRAT, but they are allowed to complete as a team. Following the IRAT and TRAT, the teacher can clarify content and address knowledge gaps. The remainder of the class is then used for activities and application of knowledge.

Problem-Solving Approach

The **problem-solving approach** to teaching is arguably one of the most heavily researched instructional approaches in agricultural education. This approach is rooted in the pragmatic nature of vocational agriculture when the goal of agricultural education programs was to prepare young men to return to the farm (Crunkilton & Krebs, 1982). Teachers could easily incorporate authentic farm problems into daily instruction and help the young men learn to solve them. While agricultural education has transformed from focusing on agriculture as a vocation to a model of agricultural literacy and leadership development, the need for students to solve problems remains an important part of the classroom (Parr & Edwards, 2004).

Broadly, a **problem** can be defined as "a situation in which you are trying to reach some goal and must find a means for getting there" (Chi & Glaser, 1985, p. 229). People solve problems on a daily basis, and while some problems are simple, others are more complex. As teachers, helping students to become better problem solvers and critical thinkers should be an overarching goal (Blackburn & Robinson, 2016). This begins with teacher being able to identify and define problems in a more pragmatic, classroom friendly manner. Some problems are easily identified and present themselves plainly, such as a small engine not starting when cranked, or a livestock animal exhibiting signs of illness, while others may not be as obvious to the casual observer. Other potential problems that can be especially useful in the classroom can include the following: uncertainty of beliefs; something needing an explanation; there is ambiguity to be resolved; there is a question to answer; or there is a decision to be made.

Dewey (1910/1997) began to outline a mental process to begin to solve some of these ambiguous problems. He termed his process as a **Chain of Reasoning** where the thinker: (a) experiences a felt difficulty; (b) creates a suggestion for a possible solution (c) defines the problem; (d) develops possible solutions; (e) employs deductive reasoning; (f) tests the hypothesis, and (g) resolves the situation (Dewey, 1910/1997). As time passes, this process was re-envisioned into the **scientific method** most students are familiar with from early science classes. When employing the scientific method, the learner will: (a) experience a thought-provoking situation; (b) define the problem; (c) collect data and information; (d) formulate possible solutions; (e) test proposed solutions, and (f) evaluate results. Similarly, implementing the problem-solving approach in the classroom involves: (a) identifying the problem/need; (b) defining the problem; (c) generating alternative solutions (i.e., hypotheses); (d) determining criteria to evaluate alternatives; (e) evaluating alternatives; (f) choosing a preferred solution; (g) implementing the solution; and (h) evaluating the results (Phipps, Osborne, Dyer, & Ball, 2008)

A benefit of the problem-solving approach as an instructional method is the flexibility the teacher has when employing it. This approach can be incorporated as a small component of a larger lesson, or it can be the basis of the lesson or unit itself. For example, a teacher could utilize the problem-solving approach as a short activity on vaccinating cattle or an entire unit on troubleshooting small gasoline engines. Crunkilton and Krebs (1982) outlined the basic instructional process for the problem-solving approach as:

1. Interest Approach
2. Objective to be Achieved
3. Problem to be solved/questions to answer
4. Formulation of problem solutions
5. Testing of solutions
6. Evaluation of Solutions
7. Closure of the lesson

Types of Problem-Solving Techniques

Problems come in many variations. Some problems are simple, some are complex, some are well-structured, and some can be ill structured. As a teacher, presenting a structured format in which students can solve problems can be challenging. In general, there are five types of problems teachers may desire to utilize in the agricultural education classroom: (a) forked road; (b) possibilities-factors; (c) effect-cause; (d) steps and key points; and (e) situation to be improved (Newcomb, McCracken, Warmbrod, & Whittington, 2004; Phipps et al., 2008). Determining which problem-solving technique is most appropriate can aid in the development of student problem solving skills.

Forked Road Problem Solving Technique

This problem-solving technique is regarded as one of the more simplistic techniques. In the forked road technique, the teacher presents the students with a problem that only has two possible solutions - hence, forked road. An example of a forked road type problem could be having the students role play a soybean producer faced with the decision to plant traditional or herbicide resistant (e.g., GMO) soybeans. In this scenario the teacher can either guide discussion of the factors a producer would consider in making this decision or they could charge the students with determining the key factors.

Table 16. Forked road problem solving-technique example (Phipps et al., 2008)

Define the Problem / Decision to be Made: **Should I plant herbicide-resistant or traditional soybeans.**		
	Choice One	**Choice Two**
Factors to Consider	Glyphosate Resistant	Traditional
Initial Seed Cost	High	Lower
Input Cost – herbicide	Lower	Higher
Input Cost – Tillage	Minimal	High
Yield	Higher	Lower
Marketability	Restricted	Open
Chemical Weed Control	Yes – All weeds	Yes, Monocot weeds only
Organic Option	No	Yes

Possibilities-Factors Problem Solving-Technique (Phipps et al., 2008)

The Possibilities-Factors problem-solving technique is similar to the forked road problem solving technique except that the students will face more than two possible solutions to the problem. A possibilities-

Table 17. Possibilities-factors problem-solving technique example (Phipps et al., 2008)

Define the Problem / Decision to be Made: **What covering should a new greenhouse have?**				
	Possibilities (Solutions)			
Factors to Consider	Single Pane Glass	Double Pane Glass	Polycarbonate Panels (6mm)	6 mil polyethylene
R Value – Effectiveness at thermal insulation	.95	2.00	1.54	.87
Cost	High	High	High	Low
Durability	Breaks Easy	Breaks Easy	High	Low, tears
Repairable	No	No	Sometimes	Usually
Weight	Heavy	Heavy	Medium	Low
Clarity	High	High	High	Low
Life of Covering	Many Years	Many Years	15 years	5 years

factors problem solving technique example could be the agriculture teacher challenging students in a horticulture class to pick the best covering for a greenhouse. The teacher could challenge the students to determine types of greenhouse coverings and their advantages and disadvantages or narrow down choices of coverings for them to research.

Effect-Cause Problem-Solving Technique (Phipps et al., 2008)

In the Effect-Cause Problem Solving Technique, the students are presented a problem and are charged with determining the cause. This problem-solving technique is particularly useful when guiding students through the troubleshooting process. For example, the teacher can present students with a small gasoline engine exhibiting fault symptoms and guide the students through the process of thinking through possible causes of the fault in order to repair the engine.

Table 18. Effect-cause problem-solving technique example (Phipps et al., 2008)

Define the Problem: **Small gasoline engine hunting/surging. Drastic change in RPM when speed is set**		
Possible Causes	**Related Facts**	**Accept / Reject**
Misadjusted Carburetor	Fuel Mix May be too lean (Air/fuel ratio) • Start engine, let warm up • Adjust throttle to half open • Adjust idle screw until engine runs 2800 – 3200 rpm • Manually hold throttle against idle stop screw • If hunting/surging ceases, problem is in governor system • If hunting/surging continues, problem is in fuel system o Dirt/Debris in carburetor (fuel idle circuit) o Remove and clean carburetor	
Damaged Governor Spring	Check spring for damage: bent, broken	
Wrong Governor Spring	Determine top no-load speed (application) desired and replace with correct spring	
Binding in Governor linkage	Remove governor spring, manually adjust governor linkage to "feel" for binding • Paint (new engines) • Corrosion (old engines)	
Decision / Recommendation:		

Steps and Key Points Problem-Solving Technique

In the steps and key points problem-solving technique the teacher presents a problem to the students that requires multiple steps to complete, often in a specific order. For example, a quite common lesson of administering intramuscular injections to cattle could be taught through the steps and key points problem-solving approach.

Table 19. Steps and key points problem-solving technique example (Phipps et al., 2008)

Define the Problem: **How do I administer intramuscular injections in cattle?**	
STEPS (What to Do)	**KEY POINTS (How to Do It)**
1. Gather Needed Materials	1. Needles, syringes, medication, latex gloves 2. Check to be sure needles are clean and sterile 3. Ensure medication/vaccine is not out of date
2. Determine proper injection route	1. Check label - Intramuscular - Subcutaneous
3. Determine Correct Dosage	1. Check Label - Dosage by weight - Consistent Dosage
4. Determine injection site	- Check label for maximum dosage per injection site - Do not exceed 10 mL per site
5. Administer injection	1. Quickly insert needle into injection site 2. Depress plunger 3. Remove needle quickly

Situation to be Improved Problem-Solving Technique (Phipps et al., 2008)

In the situation to be improved problem-solving technique, the teacher will present a given situation to the students and determine actions to undertake to improve the situation to a desired or ideal state. For example, the teacher could present a current garden soil situation and have the students work through factors to propose solutions to improve the garden's productivity.

Table 20. Situation to be improved problem-solving technique example (Phipps et al., 2008)

Define the Problem: How do I improve the productivity of my garden? Adapted from Phipps, Osborne, Dyer, & Ball (2008)			
Factors to Consider	Current Situation	Ideal Situation (What Should be Done?)	Recommendations
Soil type	Clay Loam	Loamy Sand	Add Sand
pH	6.0	6.2 – 6.7	Add Lime
Size	10' X 15'	20' X 30'	Increase Size
Water Access	Hand Watered	Drip Irrigation	Install Irrigation
Slope	Considerable–10%	Mild or no slope	Build Terraces
Organic Matter	Low	Moderate to High	Add O.M. – Manure, clippings
Fertility	Unknown	High	Perform Soil Test, fertilize as needed

Table 21. Overview of problem-solving approach as an instructional method

Problem-Solving Approach The problem-solving approach promotes problem solving and critical thinking skills by empowering students to seek answers to authentic problems. Overview of steps in developing a problem-solving approach lesson: • Determine which problem type is appropriate for the given problem that you want to present • Introduce the lesson topic with an interest approach • Communicate the objectives of the lesson • Identify the problem to be solved/questions to answer • Either guide the students through the formulation of solutions or allow freedom for students to formulate solutions on their own • Provide opportunities for students to test and evaluate solutions Good for: • Authentic problems where a correct solution is desired • Empowering students to think critically Disadvantages: • Students may have trouble formulating questions and/or potential solutions • Requires a lot of initial preparation by the teacher

INDEPENDENT INSTRUCTION

Independent instruction is any form of instruction where students are leading their own learning. As compared to other methods, independent instruction requires teachers to facilitate the learning environment while students take the lead on what and how they learn. Independent instruction can be highly effective at meeting the needs of each student, but the experience should still be planned, implemented, and managed by the teacher.

Independent Study

Many agricultural programs offer advanced courses or content that allows students to work on individual projects or coursework (Talbert et al., 2014). **Independent study** is an excellent way to help students work through content that is uniquely relevant to their needs. For independent learning to be successful, some of the important components of teacher directed lessons must still be considered to help ensure students reach their goals. This can often be accomplished using **learning contracts**. These contracts outline the project or area to be studied and students help to identify their individual goals. At each session, the teacher can review objectives with the student and help students set their goals for that session which will help them reach their larger goals. It is important that at the end of the learning session, the teacher and student check-in together to see if goals were accomplished and how the goals might need to be changed for the next learning session.

Learning Stations

Learning stations are more teacher focused than independent study and involve the development of educational learning stations by the teacher. Students are then allowed to rotate through stations, or choose which stations they participate in. As with independent study, student expectations and goals should be identified at each station and the teacher should have learners check in at the end of the session to determine if the goals from each station were achieved and to clarify any areas that might still be unclear after participating in a station.

INCORPORATING GENERAL EDUCATION CONTENT INTO AGRICULTURAL EDUCATION

Science, Technology, Engineering, and Mathematics (STEM) Integration

Science, technology, engineering, and mathematics (STEM) has been a hot button topic in education since Congress created a STEM Caucus in 2005 and by 2008, STEM was a well know acronym across the U.S. (Loews, 2015). Within agricultural education, STEM concepts have always been present, regardless of the use of the acronym as demonstrated by early writings in agricultural education which included topics such as agricultural chemistry, agricultural physics, and agricultural mathematics (Stimson, 1919; Stimson & Lathrop, 1954; True, 1929). However, the result of the debates between Dewey and Snedden/Prosser ended with subjects like agriculture being viewed and taught as only vocational in nature. The world of agriculture and education changed over the course of the 20th century and eventually the idea of being strictly vocational in nature gave way to the idea of teaching agricultural literacy that opened the door to thinking of agriculture as a STEM discipline.

A common argument heard over the years is that teachers *"went to school to be an Ag teacher, not a math or English teacher!"* Teachers should view STEM integration as *highlighting* concepts already present in the agriculture curriculum, not forcing concepts where they are not naturally present (Stone III et al., 2008). Because agriculture is already rich in concepts such as science, math, and technology, making connections to these areas can be extremely beneficial to help students reinforce the content in core subjects as well as developing a deeper understanding of how those concepts relate to agriculture

Contextualized Teaching and Learning

Contextualized teaching and learning (CT&L) is an instructional strategy used to purposefully highlight existing Science, Technology, Engineering and Mathematics (STEM) content present in career and technical education fields (Stone III et al., 2008). A seven-element approach to Contextualized Teaching and Learning been designed in a way that allows students to transfer contextual learning to other contexts or situations. The first step is for agricultural educators to recognize the currently existing STEM concepts already present in the curriculum. The teacher must then determine how to teach the agriculture content while highlighting the desired STEM concepts.

So how can CT&L actually be implemented in the agricultural education classroom? The **Seven Elements of CT&L** (Stone III et al., 2008) are:

1. Introduce the agricultural content
2. Assess the students' awareness of the STEM concept
3. Teach the STEM concept in the context of the lesson
4. Teach the STEM concept in a related context
5. Teach the STEM concept in an unrelated context
6. Return to the original CTE context and have students demonstrate understanding

7. Assess student learning

Here is an example of highlighting square footage calculation in an agricultural structures course (Blackburn, 2018):

1. Present an interest approach/motivation for a lesson on building a portable building
2. Assess students' current knowledge of square footage by determining if students know how to calculate area of simple shapes such as squares
3. Calculate square footage of a portable building, emphasizing that is the same procedure utilized in the example
4. Calculate square footage calculations of other agricultural structures such as hay sheds, milking parlors, or broiler houses. Can also teach importance of proper square footage for human and livestock safety
5. Work through square footage calculations of non-agricultural structures such as homes to emphasize the procedure of calculating square footage remains the same regardless of content.
6. Have students demonstrate knowledge and increase rigor by having students calculate square footage independently and utilizing shapes/buildings of varying shapes/difficulties, such as triangles, pentagons, parallelograms, etc.
7. Assess knowledge of the agricultural context and STEM concept. Depending on instructional objectives, this assessment could be a traditional assessment (i.e., paper based), or authentic in nature where students are tasked to build a portable building to a certain square footage.

Working through these steps allows student to not only have a deeper understanding of agricultural content, but it can help them make connections in other subjects. Students may not have an interest in math, for example, but may better learn the content when it is introduced in agricultural education and they see the practical applications of the material (Blackburn, 2018; Stone et al., 2008).

A secondary purpose of intentionally including STEM content is that agriculture as an industry is quickly becoming more scientific in nature. Many agricultural careers rely on the use of science and math skills. By helping students see these connections, it enables them to better prepare themselves for the expectations of a future career in agriculture.

Literacy and Reading Strategies into Agriculture

Literacy strategies include language development as approached through multiple components of listening, writing, speaking, and reading. In general, these areas are often expanded by encouraging reading and writing within the content area, providing targeted vocabulary practice, providing opportunities for students to read complex and technical texts, providing writing and reading instruction, and providing literacy interventions (Quigley & Coleman, 2019).

Incorporating **literacy strategies** into our agricultural content is of critical importance. Students in agriculture will often be expected to understand technical reading materials and conduct technical writing, whether it be for agricultural research, or synthesizing news and current events as a supplement to instruction. However, many teachers fail to incorporate literacy instruction into their courses which can be detrimental for students, especially when teaching a diverse group of individuals. Researchers have found that agriculture teachers spent class time engaging in reading, but spend significantly less time helping students learn how to process what they read (Park & Osborne, 2006). Because agriculture has a unique vocabulary, the intentional incorporation of literacy strategies can be extremely beneficial. Focusing on all areas of literacy such as listening, speaking, writing and reading can help students to develop a wide variety of literacy strategies (Wallace et al., 2004).

Listening is used often in formal education since the majority of educational content is routinely provided through lecture. Listening can be further improved by utilizing cooperative learning, group processing activities, and notetaking to encourage students to improve their listening skills in the classroom.

Speaking can be incorporated both formally and informally within the agricultural education curriculum. At the most basic level, allowing students to respond to questions and discuss their thoughts with others can be a method of providing literacy development opportunities. More formally, verbal literacy can be encouraged through presentations, public speaking, and debates.

Many teachers include writing as a formal part of their instruction by having students author research papers. While this is a good strategy for developing writing literacy, it can be supported and expanded within agricultural content in a variety of ways.

One Minute Writing

One minute writing can be used as an opportunity to check for understanding or to identify previous knowledge on a topic. Students are given a prompt and are asked to write for one minute using full sentences and correct grammar. One-minute writings can be a highly effective way to determine how well students understand content that has just been taught. One minute writing can also be used to identify areas where they already have prior experience in a subject or areas that are unclear on after formal instruction.

Breaking Down Large Writing Assignments

Often, papers are assigned in such a way that students complete the entire paper at once and submit it for a grade. Breaking down large writing assignment into smaller components gives students the opportunity to submit multiple drafts of assignment sections and provides additional support and feedback throughout the writing process. Breaking down writing assignments also helps students who may be overwhelmed with large assignments and teaches them to break large projects into smaller steps. Strategies such as this can best be used by providing multiple due dates for sections of a larger project and allowing students to submit certain sections of the assignment at each benchmark with feedback provided at each step.

Including Grammar and Spelling in Assignments

When writing assignments are used in the classroom, teachers can support the literacy development of students by asking them to submit their answers in full sentences and grading them on grammar and spelling. To build additional literacy support, teachers can give students the opportunity to correct mistakes and resubmit graded work.

Teaching reading involves instruction in three specific stages: Pre-reading, during reading, and post-reading Saricoban (2002). This means that there are multiple ways to include reading instruction along with agricultural content.

Multiple strategies can also be used in a variety of ways. For example, graphic organizers such as K-W-L charts can be used prior to, during and after reading. Some examples different activities that can be used at different reading stages can be seen below, however it is important to know that there are many distinct types of reading strategies that can be used and customized to address specific classroom needs.

Pre-Reading Activities

Reading strategies can first be approached through pre-reading and typically allows students to anticipate what a text is about while connecting to their prior knowledge and helping them be prepared to read new content.

Anticipation Guides

Anticipation Guides are used to allow students to analyze their own opinions about a text in advance of reading. Anticipation guides allow the teacher to develop a list of statements about the reading and allows students to indicate whether the statements are true or false or if they agree or disagree with the statement.

Table 22. Anticipation guide example

Before Reading			**After Reading**		**Support from text**
Agree	**Disagree**	**Statement**	**Agree**	**Disagree**	
		Plants produce carbon dioxide			
		The process of plants creating glucose from sunlight is known as photosynthesis			
		The process of respiration provides plants with sugars that are used for growth			
		Photosynthesis occurs during the night, while respiration occurs during the day			

Frayer Models

Frayer Models are a form of graphic organizers which are used to explore vocabulary in more depth and allow students to identify several aspects of vocabulary words. Frayer models are typically designed as four quadrants with a circle in the middle. The vocabulary word is placed in the middle. In each quadrant, a different aspect of the vocabulary word is explored (definition, characteristic/facts/properties, examples, and non-examples).

Table 23. Frayer model example

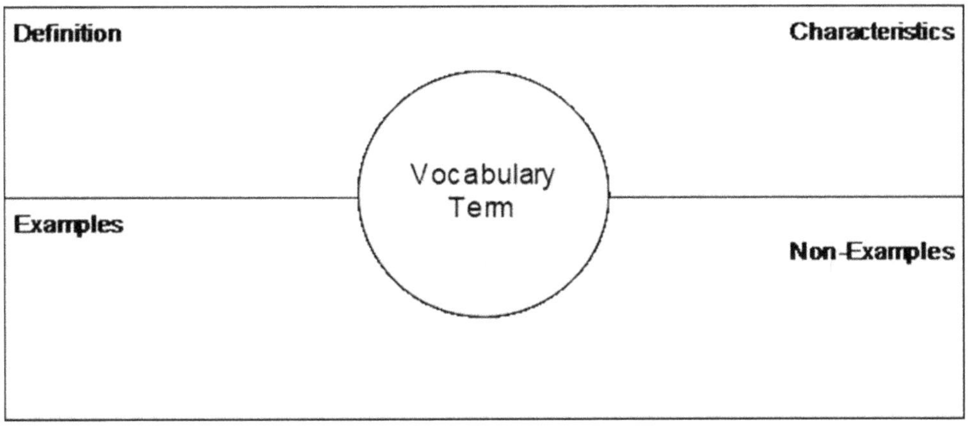

Vocabulary Preview

Vocabulary preview provides students with a list of key vocabulary words and definitions prior to reading so that they can become familiar with terms before they encounter them in the text. Both Frayer models and vocabulary previews can be especially useful in agriculture since there is often unique vocabulary that students may not have already been exposed to.

Table 24. Cornell notes example

Title of Content	
Key ideas and vocabulary	Notes from the reading passages
Summary of content	

During Reading Activities

During reading, students can use multiple types of strategies to help make sense of content as they read it. This allows students to learn to organize content as well as identify valuable information and critical ideas.

Cornell Notes

Cornell Notes is a reading strategy developed at Cornell University and typically consists of identifying keywords, taking notes around those keywords, and then summarizing the passage (Pauk & Owens, 2010). The paper on which the notes are taken is divided into two columns with the left column taking approximately ⅓ of the page which will include key ideas, vocabulary and questions, and the right column taking approximately ⅔ of the page which is used for notes from the reading. A summary of the passage is then included at the end of the page.

Annotated Text

Text passages can be annotated in a variety of ways: Highlighting/underlining, summarizing, adding comments/questions, illustrations, etc.

Some examples include:

- Students can underline critical points in the reading passage which they identify as being important to the topic.
- Students can summarize concepts in the margins of a reading passage, allowing them to return to those concepts later
- They can ask questions that need additional clarification which can be asked later in class or explored on their own
- They can use margins to include symbols relating to the text or illustrate concepts so they are easier to remember.

Post-Reading Strategies

Retelling

Retelling involves students reading a passage and then providing an oral or written summary about the concepts included in the passage. While often used with fiction materials, the process can be implemented with content focused text such as articles or textbook readings as well as technical manuals.

Related News Search

After reading, students can take what they learned and find related news, stories, or articles. This can be a practical way for students to connect agricultural content to real-life current events. This can also allow a teacher to check for understanding and ensure that the key components of the text are being understood and used correctly.

INCORPORATING STUDENT ENGAGEMENTS INTO INSTRUCTIONAL METHODS

Regardless of which instructional methods one chooses to use, there are instructional strategies that should be used throughout all types of lessons. In particular, strategies such as questioning and checking for understanding can be used in a variety of ways within lessons and as part of instructional methods to increase engagement and ensure students are meeting learning objective.

Questioning

One of the most common instructional strategies that can be used within all instructional methods is questioning. Several types of questions can be employed to increase variability within lessons. For example, questions can be:

- Open ended questions with no specific answer
- Closed ended questions with one correct answer
- Directed questions addressed to one student
- Non-directed questions addressed to the class as a whole

There is a variety of ways to ask questions to increase learner engagement, and a mix of these question types can and should be utilized throughout lessons. For example, a teacher could ask a closed ended question to the class as a whole and then ask an open-ended follow-up question to a specific student. Many questions are emergent from the lesson, meaning the events in the class may encourage a teacher to ask related questions. It is also important to include questions and follow-up questions that are planned along with instruction; this is particularly important when there are critical questions that students should consider.

One way to increase engagement when asking directed questions is to ask the question first, and then call on students. This allows all students to hear the question and think through potential responses, which they may not do if another student is called on and then the questions is asked.

Questions utilizing Blooms Taxonomy can be classified as lower order or higher order in nature. No matter which way questions are asked, a variety of question types should be used at a variety of levels. A

Table 25. Possible question stems based on bloom's taxonomy.

Level of Blooms Taxonomy	Possible Question Stems
Creating	Can you design a . . . to . . .? How would you devise your own way to . . .? What would happen if . . .? Can you create a new and different . . . for . . .?
Evaluating	Can you produce a better solution to . . .? How effective is . . .? What are the pros/cons of . . . ? Is . . . a positive or negative . . .
Analyzing	Which events could not have happened? How is . . . similar/different to . . .? What are some other possible outcomes? Why did . . . happen? What is the problem with . . .?
Applying	Can you place these into groups? Can you show me how . . .? Which factors would you change . . .? What questions would you ask . . .? From the information you know, can you develop a set of instructions for . . .?
Understanding	- Can you write that in your own words? - How would you explain . . .? - What could happen next? - Who do you think . . .? - What was the main idea . . .?
Remembering	- What happened after…? - How many…? - What is…? - Who …? - Can you name…? - Which is true or false?

mix of lower order and higher order questions provide variability and allows for students to think about information more deeply. Follow-up questions can also be used to encourage critical thinking by asking a more basic question and then, once a response is received, following up with a question that allows for more advanced thought (Croom & Stair, 2005). Always remember to add three to five seconds of *wait time* after posing questions to students and after a student responds (Gage and Berliner, 1998). Allowing students this moment to process the question and compose a response is a step often overlooked and may limit student responses.

Checking for Understanding

Checking for understanding should be used along with all methods of teaching to see if content is being understood throughout the lesson (Fisher, 2006). Checking for understanding can take many forms, however the most common forms are questioning, peer reflections, or any other activities which provide immediate feedback. Checking for understanding will involve short activities which allow students to provide feedback on critical concepts to the teacher or to each other before moving onto new topics.

Some examples:

- Asking specific questions to a variety of students regarding the content that has just been discussed
- Asking students to pair with a partner and summarize content in their own words
- Asking students to use colors to indicate their understanding such as green being very confident and red being not at all confident.
- Asking students to rate their level of understanding on a scale of 1-5
- Using technological tools for students to answer questions so the responses can be seen immediately by the teacher or the class

SUMMARY

There are many distinct types of instructional methods that can be used in a variety of ways. These methods, paired with other class activities, provides opportunities for students to learn content in multiple ways. Choosing the right method of teaching ultimately comes down to the learning objectives and how the teacher wants students to master content. Variability is a vital component of good teaching, so utilizing multiple methods allows teachers to diversify their classroom and helps to reach more learners than if only one method is primarily used.

REVIEW QUESTIONS

1. What factors should be considered when selecting an instructional method?
2. What is the difference between a teacher focused method and a student focused method?
3. Which method is most effective to share direct knowledge with students simultaneously?
4. Which method(s) would a teacher use if they wanted to teach an agricultural skill?
5. Describe the difference(s) between project-based learning and the problem-solving approach to teaching.

6. What are the five required elements of cooperative learning?
7. How does independent instruction differ from other methods of teaching?
8. Based on Kolb's model of experiential learning, describe the differences between concrete experiences and abstract conceptualization.
9. What are the four levels of inquiry-based learning and how do they differ?
10. Which problem-solving technique is most appropriate for problems in which the student must decide between two possible solutions?
11. Which problem-solving technique is most appropriate for problems in which there are multiple steps for the learner to implement?
12. In what ways does Team-Based Learning differ from the traditional flipped classroom approach?
13. What are the 7 Elements of Contextualized Teaching and Learning (CT&L)?
14. What are the three types of reading strategies?
15. How can checking for understanding be used within the classroom?

REFERENCES

Baker, M. A., Robinson, J. S., & Kolb, D. A. (2012). Aligning Kolb's experiential learning theory with a comprehensive agricultural education model. *Journal of Agricultural Education, 53*(4), 1–16. doi:10.5032/jae.2012.04001

Banchi, H., & Bell, R. (2008). The many levels of inquiry. *Science and Children, 46*(2).

Blackburn, J. J. (2018). Teaching STEM in agricultural mechanics. *The Agricultural Education Magazine, 90*(5), 24–25.

Blackburn, J. J., & Robinson, J. S. (2016). Determining the effects of cognitive style, problem complexity, and hypothesis generation on the problem solving ability of school-based agricultural education students. *Journal of Agricultural Education, 57*(2), 46–59. doi:10.5032/jae.2016.02046

Borich, G. D. (n.d.). *Effective Teaching Methods: Research-Based Practice* (10th ed.). Academic Press.

Bransford, J. D., Brown, A. L., & Cocking, R. R. (2000). *How people learn.* National Academy Press.

Burris, S. (2008). PBL: Less content, more process. *The Agricultural Education Magazine, 80*(6), 6–7, 10.

Chi, M. T. H., & Glaser, R. (1985). Problem-solving ability. In R. J. Sternberg (Ed.), *Human abilities: An information processing approach* (pp. 227–250). Freeman.

Croom, B., & Stair, K. (2005). Getting from q to a: Effective questioning for effective learning. *Agricultural Education Magazine, 78*(1), 12.

Crunkilton, J. R., & Krebs, A. H. (1982). *Teaching agriculture through problem solving.* Interstate Printers & Publishers, Inc.

Dewey, J. (1997). *How we think.* Dover Publications. (Original work published 1910)

Fisher, D. (2006). Checking for understanding: Strategies for ensuring learning. *Thinking Classroom*, *7*(4), 48.

Gage, N. L., & Berliner, D. C. (1998). *Educational psychology* (6th ed.). Houghton Mifflin.

Johnson, D. W., Johnson, R. T., & Stanne, M. E. (2000). *Cooperative Learning Methods: A meta-analysis. University of Minnesota.* Cooperative Learning Center. http://www.cooperation.org/pages/cl-methods.html

Kagan, S. (1994). *Cooperative Learning*. Kagan Publishing.

Kolb, A., & Kolb, D. A. (2005). Learning styles and learning spaces: Enhancing experiential learning in higher education. *Academy of Management Learning & Education*, *4*(2), 193–212. doi:10.5465/amle.2005.17268566

Kolb, D. A. (1984). *Experiential learning: Experience as the source of learning and development.* Prentice Hall.

Larmer, J., & Mergendoller, J. R. (2010). Seven essentials for project-based learning. *Educational Leadership*, *68*(1), 34–37.

Loews, L. H. (2015, April 2). When did science education become STEM? *Education Week*. http://blogs.edweek.org/edweek/curricu-lum/2015/04/when_did_science_ed- ucation_become_STEM.html

Michaelsen, L. K., Knight, A. B., & Fink, L. D. (Eds.). (2004). *Team-based learning: A transformative use of small groups. Sterling*. Stylus Publishing, LLC.

Miller, E. C., & Krajcik, J. S. (2019). Promoting deep learning through project-based learning: A design problem. *Disciplinary and Interdisciplinary Science Education Research*, *1*(1), 1–10. doi:10.118643031-019-0009-6

Newcomb, L. H., McCracken, J. D., Warmbrod, J. R., & Whittington, M. S. (2004). *Methods of teaching agriculture* (3rd ed.). Pearson.

Park, T. D., & Osborne, E. (2006). Content area reading strategies and textbooks in agricultural education. *Journal of Agricultural Education*, *47*(4), 1–14. doi:10.5032/jae.2006.04001

Parr, B., & Edwards, M. C. (2004). Inquiry-based instruction in secondary agricultural education: Problem-solving – An old friend revisited. *Journal of Agricultural Education*, *45*(4), 106–117. doi:10.5032/jae.2004.04106

Pauk, W., & Owens, R. J. Q. (2010). How to Study in College (10th ed.). Boston, MA: Wadsworth.

Phipps, L. J., Osborne, E. W., Dyer, J. E., & Ball, A. (2008). *Handbook on agricultural education in public schools* (6th ed.). Thompson Delmar Learning.

Quigley, A., & Coleman, R. (2019). *Improving Literacy in Secondary Schools. Guidance Report*. Education Endowment Foundation.

Saricoban, A. (2002). Reading strategies of successful readers through the three phase approach. *The Reading Matrix*, *2*(3).

Stimson, R. W. (1919). *Vocational agricultural education by home projects*. The Macmillan Company.

Stimson, R. W., & Lathrop, F. W. (1954). *History of agricultural education of less than college grade in the United States*. Unites States Government Printing Office.

Stone, J. R. III, Alfeld, C., & Pearson, D. (2008). Rigor and relevance: Enhancing high school students' math skills through career and technical education. *American Educational Research Journal, 45*(3), 767–795. doi:10.3102/0002831208317460

Talbert, B. A., Vaughn, R. C., Croom, D. B., & Lee, J. S. (2014). *Foundations of agricultural education*. Pearson Education.

True, A. C. (1929). *A history of agri-cultural education in the United States: 1785-1925*. Washington, DC: United States Government Printing Office.

Wallace, T., Stariha, W. E., & Walberg, H. J. (2004). *Teaching speaking, listening and writing*. International Academy of Education.

KEY TERMS AND DEFINITIONS

Case Study: Method of teaching that utilizes written scenarios which require the learner to apply knowledge to solve problems or answer questions.

Checking for Understanding: Strategies used to determine if content is mastered or to measure understanding.

Contextualized Teaching and Learning: As a method of teaching, it is a purposeful means of incorporating core academic content into the agricultural curriculum utilizing the 7-Element approach.

Cooperative Learning: Method of teaching that uses small groups to work together toward a common goal which includes face to face interactions, group processing, and accountability for both the individual and the group.

Debate: Method that is used to allow students to research sides of an issue and defend their side to others.

Demonstration: Method of teaching that involves the intentional teaching of steps to complete a task or a skill.

Discussion: Method that is used to allow students to discuss within a small group to answer specific questions or solve a problem.

Experiential Learning: Naturalistic learning that utilizes concrete experience, reflective observation, abstract conceptualization, and active experimentation to enhance student learning and critical thinking skills.

Experiential Learning Theory: Developed by Kolb (1984), ELT is a cyclical process that guides students through concrete experience, reflective observation, active experimentation, and abstract conceptualization.

Field Trips: Taking learner off campus to see elements of content in a real-world setting.

Flipped Classroom: Approach that general utilized modern educational technologies to deliver content so that time in class can be utilized for application of knowledge.

Guest Speaker/Resource Person: A speaker that is brought into the classroom to act as an expert and share content knowledge or experiences with students.

Independent Instruction: Method of instruction where students are leading their own learning.

Inquiry-Based Learning: learning is designed to teach through curiosity and student led exploration.

Instructional Method: Methods and strategies used by teachers to ensure educational objectives are achieved within the educational environment.

Learning Contracts: Written contracts used with independent instruction to help students set goals and determine if their goals are met.

Learning Stations: A teaching method where the teachers and students design stations and students are then allowed to rotate through the stations at their own pace, or choose which stations they participate in.

Lecture: A method of teaching where information is conveyed directly from a teacher or expert to a group.

Literacy Strategies: Strategies used to develop student literacy through multiple components of listening, writing, speaking, and reading.

Problem-Based Learning: Learning strategy whereby students to work collaboratively to solve authentic, real-world problems to enhance critical thinking and problem-solving skills.

Problem-Solving Approach: Teaching approach commonly applied in agricultural education based on Dewey's concept of reflective thinking.

Project-Based Learning: The use of project-based assignments that are intentionally designed to be grounded in relevant content, include social interaction, and include strategies to increase cognitive engagement.

Questioning: Method of teaching where the teacher asks questions to determine understanding of content.

Role Playing: A teaching method which puts students into the role of other people, systems, or processes.

Student-Focused Instructional Method: teaching methods that focused primarily on student interaction or revolving around active student participation.

Teacher-Focused Instructional Method: Teaching methods that are primarily focused on providing information from the teacher to the student.

Team-Based Learning: Modified flipped classroom with increased emphasis on accountability by placing students in peer teams and utilizing multiple assessments.

Chapter 9
Managing the Classroom and Laboratory

Michael L. Pate
Utah State University, USA

Robert Bud McKendree
Michigan State University, USA

ABSTRACT

This chapter focuses on management strategies for classroom and laboratory instruction for school-based agricultural education teacher preparation. Specifically, the focus is on teaching methods, techniques, and strategies to proactively sustain positive student engagement in activities to prevent behavioral issues that disrupt the learning process. Characteristics and guidelines for instructional technology are discussed for effective management of student learning. Management practices for both instructional spaces and student behavior are incorporated. Highlights of underlying causes of behavior such as issues within social and psychological disruptions provide context for student engagement efforts. A portion of the chapter differentiates appropriate professional interactions of teachers with students. Behavior scenarios are utilized to provide application of theoretical concepts.

INTRODUCTION

Classroom management starts long before students arrive. Preparation and planning are key components that will contribute to your success in establishing a positive environment for learning. Effective teaching strategies are vital to managing student behavior conducive to learning. This chapter will focus on developing teachers' management strategies through contextualized theories and examples of proactive approaches.

Learning objectives for this chapter are:

1. Evaluate/discern technology for classroom incorporation
2. Employ management strategies within the classroom space

DOI: 10.4018/978-1-6684-3420-8.ch009

3. Employ management strategies within the laboratory space
4. Evaluating strategies for managing student behavior
5. Creating student behavior management plans

Managing an environment conducive to learning focuses on minimizing distractions and maximizes engagement. Education has seen a revolution in the development of online and digital learning environments. Technology has become a fundamental part of the classroom teaching experience. Instructional technology deals with the theory and practice of using technology for education with the ultimate goal of engaging effective learning experiences. The benefits include better access to information, more opportunities for collaboration, and better capabilities for meeting diverse learners' needs. However, using technology requires thoughtful planning and implementation to effectively manage the learning experience.

Instructional Technology is defined as the use of hardware or software to aid in instruction. This technology, or tech for short, usage in the classroom presents both opportunities and challenges for instructors, including the constantly changing tech available, student access to tech, and opportunity cost associated with teaching the tech. While tech presents opportunities and challenges for students and teachers alike, it has arguably never been more accessible. According to the Pew Research Center (2021), 85% of adults own a smartphone, with younger demographics owning smartphones at a higher rate than older U.S. adults (96% of Americans ages 18-29 own smartphones). While owning a smartphone is nothing out of the ordinary, it means 85% of people within the U.S. now have access to more computing power than was available to NASA on the Voyager I and II space probes – this section will identify thought processes behind how we tap into available instructional technologies.

While tech is readily available, the actual tech we have access to is constantly changing. Technology ages quickly. For instance, a five-year-old computer equates to a 93-year-old person (TAZ Networks, 2016) . What does a constantly changing tech landscape mean for educators? When considering tech use in the classroom, it is more practical and effective to focus on an instructional decision-making process than on existing tech at any given point in time. In this section we will explore fundamental thought processes educators should use when deciding on when, where, and how to utilize tech in the classroom. For those who have engaged in the instructional decision-making process and want to incorporate a new tech, professional development and resource-seeking by that teacher are important next steps to increase literacy.

Instructional technology has long been a focus of educators and educational researchers, with some potential barriers to adoption being identified as lack of support, cost of equipment or software, and lack of teacher time to master new tech (Alston et al, 2003; Williams et al., 2014). While these barriers do exist, access to tech – for example, in the form of smartphones – has arguably never been higher, as evidenced by the propensity of U.S. adults to own smartphones. So, how does one make decisions on incorporating tech in their teaching? Ensuring the use of a given tech aligns with the teacher-decided instructional outcomes and the ways students will engage in learning any given content. To guide these decisions, an educator should filter decisions on the use of technology within the classroom through the following questions:

1. How much time will it take students to learn to use the tech/software?
2. How much instructional benefit is there to using the tech/software (i.e., how will the tech/software increase student learning)?
3. Will the tech/software be useable again for other learning within the course?

4. How much does the tech/software cost? Is it a one-time fee?

When used as a whole, these questions can inform our practice as educators. The time it takes students to learn the tech is an opportunity cost – for every minute spent teaching the tech there is one less minute available to teach content within the state-mandated standards. Whether that opportunity cost is justifiable depends on the instructional benefit gained from the new tech/software and how many times we can use that tech/software in the future to augment learning. If a tech requires low student skill development, but offers instructional benefit and has broad applicability within our course, it is likely a good instructional decision to incorporate said tech. However, if it requires students learning new skills and is not easily transferable to other course learnings, or offers little instructional benefit, it would be wise to avoid incorporating said tech. For example, it may be enticing to have students use a timeline application on their smartphone to re-create historical events of the FFA, but what benefit does the use of the new software (i.e., timeline app) provide? Does it justify the opportunity costs of time to download and teaching students how to use the app? Weighing answers to these questions may help us to arrive at the conclusion of, no, that specific tech is not worthy of inclusion for the purpose of learning FFA history.

Tech use within the classroom can augment assessment and the collection of evidence of student learning. Applications on smartphones (e.g., Plickers) as well as other available software (e.g., Kahoot) can leverage existing resources to benefit student learning. Again, a series of questions can help educators make instructional decisions about using tech to augment assessment:

1. How will the tech/software assist the teacher in gathering evidence of student learning?
2. What existing resources does the tech/software leverage?
3. How quickly can the tech/software be used in a classroom setting?
4. How much does the tech/software cost? Is it a one-time fee?

Applications and software that help teachers gather evidence of student learning are abundant, and many are free of cost. Considering the high propensity for students to own a smartphone, leveraging software and applications compatible with those devices is an efficient way for educators to augment assessment with tech. Some software (i.e., Plickers) utilizes the teacher's mobile device and partners it with student-assigned QR codes to quickly collect answers to selected-response questions. Teachers should explore tech options available and make instructional decisions based on the guiding questions provided in this section.

Skills related to effective use of technology are essential for success in a variety of settings outside the classroom. As instructors introducing a new technology to the classroom will require implementation instructions to guide students in developing their skills. Students should demonstrate the ability to use technology to find, evaluate, create, and communicate information. Instructors should evaluate instructional technology to ensure an effective user onboarding experience. Onboarding is the process of improving students' success with a product. Tech can have a variety of onboarding techniques to assist with student onboarding. These include:

- Single sign-on – Simplifies student login.
- Interactive tutorials or Walkthrough – These include in-application tutorials that highlight features to students.

- Checklists – This feature can leverage gamification effect and provide a clear step-by-step procedure to accomplish the learning task.
- Training videos – Provides context-based training videos for students.

Conversely, teachers must also hone their skills to effectively use new technologies for facilitating learning. Four domains of teacher technology use include:

- *Planning and preparation* – In this domain the teacher must demonstrate an ability to inform and guide instructional activities using available technology.
- *Instruction* – Teachers effectively demonstrate use of the technology and exemplify appropriate levels of productivity with the technology.
- *Classroom Environment* – Teachers establish a welcoming and safe environment for students to utilize the technology.
- *Professional Responsibilities* – Teachers demonstrate professional communications through technology and promote collaborative opportunities to share feedback with students and parents.

Technology is one tool that is useful in helping mange an environment conducive to learning while minimizing distractions. Next, teachers should examine common settings used to facilitate student learning such as classrooms and laboratories. Each setting requires similar strategies but may have different peripheral equipment or distractions to accommodate.

CLASSROOM SPACEs

Classroom space has the potential to influence student learning and engagement (Brooks, 2012). The setup of a classroom space can augment learning, or it can hinder learning, depending on the efficiency and welcoming nature of the space. In this section we explore how setup can enhance learning and increase efficiency of instructional time.

Classroom spaces should be setup with an eye toward student learning, addressing questions such as: how easy is the space to navigate for the teacher and students? How will students work collaboratively? How will supplies be disseminated? How will students submit their work? All of these questions can, and should, inform the physical layout of the classroom. In addition, learning is influenced by emotion (Sousa, 2017), a classroom should be a welcoming space that invites students to engage in learning.

Order/Appearance

Maintaining order within the classroom makes it easier to inventory and track learning supplies. Order can be enhanced by working with students on a system for getting out/putting away learning supplies, submitting assignments, and daily cleanup. The key is working with students on the expectations for the physical space – the teacher alone cannot be responsible for maintaining order in the space.

Appearance of the classroom space is influenced by the order maintained as well as the esthetics generated by the teacher. FFA-associated decorations, inspirational posters, and other classroom materials can help create a classroom appearance that is welcoming for students, as well as informative. Other features may include an activities calendar for FFA, an information station for students to get information about upcoming school-related events, or a weekly calendar for classroom activities/learning focuses for each day.

Organization and Storage

Ensuring the space is setup for learning can also be aided by organization and storage systems. Having a plan – and sharing that plan with students – ensures items are accessible when needed. Classroom Organization can encompass the following items:

- Submitted student work
- Learning supplies such as paper, rulers, etc.
- Learning resources/equipment such as printers, computers, scales, etc.
- Information centers such as calendars, FFA updates, task-boards, etc.

Working with students is important for executing the organization and storage plan. Consider labeling organization systems or storage bins, and ensuring the layout is visible by the teacher from multiple locations in the classroom. Once a system is in place, reinforce routines related to organization and storage early in the school year to help students learn the system. Always allow students time to put away their own materials after a learning episode, and then take a moment to quickly check materials were returned/submitted to their proper places. A class of 25 students can return materials to their proper location in a fraction of the time it will take one teacher to return those same materials.

Establishing Routines

Setting up the classroom space is the first step to enhancing learning, but establishing routines is just as important. Routines in the classroom help teachers utilize instructional time efficiently and cuts down on downtime that may lead to student management concerns by developing a system of repetitious procedures. *Routines* in the classroom can be broken down into the following categories: starting instruction, maintaining activity, and stopping instruction.

Starting Instruction

Establishing a routine for beginning instruction helps students know what to expect when entering a learning space. A starting routine should include the following characteristics:

- Provides students a task connected to the day's learning
- Reinforces to students the need to be ready to learn (i.e., materials out and ready)
- Spans an amount of time needed by the teacher to accomplish administrative tasks (e.g., attendance, hall duty, etc.)

Any potential routine to start the day should in some way be connected to the day's learning, this could be a reflection question over previous content or an open-ended critical thinking question to engage students in thinking about a new concept. By connecting it to the learning, we can reinforce prior learning while building a foundation for what will be learned next. This also prevents the perception of "busy work" by students.

Establishing a starting routine with a task connected to the learning also reinforces the need for students to enter the space ready to learn. Consider two scenarios:

1. Teacher A conducts hallway duty while students file into their classroom. As students enter, they are eager to continue conversations they started with friends in the hallway. As more students enter, the noise gets progressively louder, and students become increasingly distracted. The bell rings and the teacher enters the room and begins taking attendance, collecting homework, and passing out graded papers. Five minutes into the 55-minute class period the teacher finally finishes passing out papers, and after an additional 30 seconds they quell the discussions and start the day's learning.
2. Teacher B conducts hallway duty while students file into their classroom. As students enter, they know teacher B is passionate about the content in the course and has high expectations for students. Students know to check their conversations at the door and to start the activity on the board as soon as they enter the room. They also know to get any required materials before going to their seats, and to submit assignments to the submission box. Once seated, they begin the day's activity, which reads: "What was the muddiest point from yesterday? You have until 30 seconds after the bell rings to formulate your response. Once you have written your response, discuss your muddiest point with a neighbor for 2 minutes, clarifying each other's questions if you can. When prompted, end the discussion and revisit your initial muddiest point for a moment – prepare to ask your question if still muddy." The bell rings and the teacher enters the room and begins the usual duties: taking attendance and passing out graded papers. As they distribute graded papers, they prompt the end of the pair-share and refocus students. Three minutes into the class period the teacher begins their lesson by answering student questions about the previous day's learning.

In the two scenarios, which teacher utilized instructional time most efficiently? Teacher B, who, in less time, accomplished the same administrative duties (collecting work, attendance, and distributing graded assignments) *and* facilitated a reflection for students that reinforced learning. Note teacher B, while getting students on task and productive, encouraged on-topic collaboration, whereas teacher A promoted off-topic discussion.

Maintaining Activity

Routines can also be used in maintaining activity throughout learning opportunities. These are routines that aid in the facilitation of instruction and encompass daily tasks, restarting, and continuing learning activities.

Daily tasks and activities

Daily tasks and activities are routines that further aid the facilitation of learning. Examples of these may include procedures for getting learning supplies or materials, submitting work within a class period, or routines for initiating other learning activities. Focus on cultivating these routines with students early in the school year, as they will frequently be used during instruction on a day-to-day basis. Some specific examples of such routines are:

- When working collaboratively, students know to designate one person from their group to retrieve/put away supplies before starting an activity.
- When completing an in-class assignment, students know to put their names in the upper left-hand corner of the paper and place the assignment in the period-specific file folder next to the teacher's desk.
- When beginning a learning activity, students know directions are provided by the teacher before the activity begins – this includes before groups are formed.

Restarts/continuations

How teachers facilitate restarting or continuing a learning experience can greatly impact efficiency of instructional time. Having a set routine, such as designating responsibilities for each group member while in a collaborative project, can save a lot of time. These routines should be as consistent as possible to help students navigate in and out of restarts and continuations. Some specific examples include:

- While working on a two-week collaborative project, students within groups are each assigned a task at the beginning of the project to initiate each day as they enter the room: supplies manager, discussion facilitator, note-taker, and project manager.
- While returning to a science lab initiated the day before, students know to await direction from the teacher after securing their supplies from their designated storage location in the back of the room.
- When ending instruction for the day that includes an unfinished activity, while students are placing their supplies in the appropriate storage area in the back of the room the teacher is within visible range to ensure materials are labeled appropriately and stored in the correct location for easy retrieval the next day.

Stops for Instruction

Frequently during instructional time featuring labs or activities additional direction or instruction is needed. Routine can aid students in these moments using a technique called *anchoring*. Anchoring is a technique in teaching that combines both physical space and movement as well as routine, to help students anticipate when different types of directions or instruction are going to be given. It involves the teacher standing in designated spots within the classroom according to the type of direction or instruction to be given. Some specific examples include:

- Teacher standing by the SMART board may signify teacher-driven instruction is about to begin.
- Teacher standing next to the lab supplies storage location may signify during-activity directions are about to be provided.
- Teacher standing in the middle of the classroom with a hand raised may mean "stop what you are doing and listen."

The location of the anchoring, or the movements/directions associated with them, are less important than the consistency associated with each. A teacher that develops a routine with students increases the ease with which students stop an activity to receive further instruction or direction.

LABORATORY SPACES

Laboratories are unique and offer students interactive experiences that put course content into action. One of the most challenging aspects to maintaining laboratories is the diversity of labs found within agricultural education programs. Teacher may have several different types of labs to maintain and manage from greenhouses to mechanics facilities (Shoulders & Myers, 2012). Organizing and maintaining a laboratory environment for optimum learning present special challenges (Croom & Moore, 2003).

In some instances, there may not be enough equipment for all students in the class to complete the activity (Bear & Hoerner, 1986). These scenarios will require proper planning and organization to keep students on task with limited disruptions. The instructor will need to manage the space for learning as well as utilize resources to facilitate positive student behavior. Ultimately this is to ensure the safety of students and those working around them.

Learning environments that are focus on providing constructive interactions will result in greater success for the teaching and learning process (Banks, 2014). Techniques discussed in the following sections highlight appropriate management strategies to prompt appropriate behavior while minimizing disruptions to the learning process.

Appearance/Clean-Up

The appearance of your laboratory sets the tone and climate of the learning environment. Students will attribute the instructor's personality to the appearance of the laboratory. If the laboratory is disorderly and cluttered, it may send the wrong message about the intentions of the learning activity. Instructors of agricultural education laboratories will have may subject areas and classes to prepare therefore laboratory clean-up must be accounted for as part of each experience. It is important to allow students ample time to return tools and equipment to the proper storage location for the next users. Not only does the appearance of the laboratory speak volumes about your intentions for learning but it is also an important aspect of keeping students safe. Clean-up reduces the opportunity for trips and falls by removing hazards from walkways such as hoses or cords used to operate equipment. Expectations for clean-up should be specifically provided to each student. A "all for one" approach where everyone is expected but not directed usually results in very little productivity when it comes time to clean-up.

Assign specific roles for each student and utilize a rotation schedule. By assigning roles and rotations there is little concern regarding favoritism among students. It is recommended to assign clean-up roles on a weekly basis. This allows students time to learn the task required. Examples are commonly found using a Numerical Rotation System. Students are assigned a number and clean up assignment. These assignments serve as an instructional tool with a focus on service or skill development. Service oriented assignments focus on developing students' management or service such as "safety supervisor" or "safety manager." These assignments offer opportunities to develop students' leadership and soft skills such as interpersonal communication. This in turn has shown to increase student appreciation and respect for the laboratory. When students are participating in self-management, the amount of time instructors have for teaching increases. Skill-oriented assignments focus on clean-up or maintenance tasks rather than management. This assignment is suited well for beginning or intermediate students to develop hands-on experiences such as drill press maintenance/clean-up. These tasks focus on developing competencies such as preventive maintenance of equipment and maintaining safety standards such as removing debris to reduce slip/trip/fall hazards. Advanced course may require different strategies where more elaborate lab projects are occurring with varying completion times.

If your laboratories are smaller, work habits that center around the process theory of "working clean" can allow you to adapt your space and students' daily habits to save time and increase productivity in the laboratory. Working clean is a conceptual strategy used for managing project workspaces. Students must be trained in the mindset of working towards efficiency by adding steps of cleaning and organizing. This essentially gets to a habit of cleaning and putting away tools after the end of a process or task. For this strategy to be effective, an organizational plan for storage is essential.

Organization

A great deal of a teacher's time is spent outside actively teaching in the lab. Estimates show that one-fourth a teacher's time may be spent on material control and lab maintenance. Organization strategies can help reduce time spent on supply and equipment control thereby efficiently utilizing precious time for learning. If an instructor uses poor organization practices thereby inadvertently making student learning in laboratory a function of trial and error rather than formal instruction. Tool and supply organization should focus on five goals.

1. Promote Safety
    a. Placing fuel containers near heat sources is asking for trouble.
2. Limit supply waste/loss to a minimum
    a. When the football coaches help themselves to lumber for a project.
3. Keep tools/instruments in working order
    a. Getting students started on an activity to find out the drill does not work.
4. Teach students responsibility
    a. When students ask, "where do I find a piece of sandpaper or potting soil," you have to stop what you are doing to help them find it.
5. Keep the laboratory neat and orderly
    a. Nothing says "I don't care" like a dirty and cluttered lab.

A few basic organizational strategies include:

1. Fast visual access and inspection for frequently used tools/equipment
2. Locate frequently used tools and equipment close to the assigned work area
3. Assign infrequently used tools and portable equipment to students at the beginning of the laboratory activity.

Storage

When teachers accept a teaching assignment, they also accept the responsibility of the laboratory and its contents. Taxpayer funds used to equip the laboratories should be treated with dignity and responsibility. The reasons for secure storage of tools and equipment include:

1. Pilferage or theft is serious as losses represent waste of public property purchased with public funds.
2. Tight security discourages the borrowing of tools and equipment from the school
3. Missing tools and equipment reduce students' time to work in the laboratory
4. Student morale drops when tools are missing or not working because of abuse from unauthorized users.

Common methods for tool/equipment storage:

- Lockable tool/instrument cabinets
- Individualized kits
- Centralized supply rooms

Maintenance

Specialized equipment and tools will require routine service and calibration. When this is neglected, equipment failures are common resulting in downtime. Students who are idle due to unavailable equipment will result in behavioral issues. Equipment that is not properly maintain also present safety concerns. The process of providing routine maintenance can be done by the teacher, school maintenance, or specialized contract services. The pros and cons of these should be weighted as well as liability or warranty concerns if equipment fails during laboratory activities. Maintenance programs help teachers identify potential hazards as well as prepare by identifying parts or supplies that are needed before the next use of the equipment. The primary goal of teacher management of maintenance is to identify and correct early to prevent downtime during scheduled laboratories. This commonly involves the following tasks by the instructor.

1. Inspect all equipment periodically
2. Perform or schedule services needed
3. Replace consumables
4. Replace or repair equipment periodically
5. Record inspection reports

Checklists or creating a maintenance database are helpful in keeping up with your needs.

Establishing Routines

Before students arrive, it is important to develop a plan to guide your instructional efforts. A laboratory routine will need to address how students begin the activities of the day, maintain student engagement, and clearly outline how to conclude the learning activity. Given limited equipment or workstations, the instructor may involve rotating or scheduling students through various units or laboratory activities to ensure all students remain on task.

Starts for Instruction

Beginning the instruction should establish the expectations and provide clear boundaries for students. Provide an outline of the activity such as a Job Operation Sheet and review procedures. This should include a demonstration of the activity. As students are watching this is an opportunity for mischief, the instructor should provide expectations for students on the purpose of the demonstration. Common elements that should be considered to ensure students stay focus are:

1. Establish the importance of the skill. Students need to know why the skill/activity is important or useful
2. Intrigue their interest. Show an example of a finish product.
3. Have all materials ready.
4. Use effective questioning to check for understanding
5. Emphasis key points such as reinforce safety practices
6. Provide correct information. Instructors should maintain credulity by doing their due diligence about the content.
7. Review the setting or location. Background noise and obstruction to visually seeing the demonstration can create an opportunity for students to be distracted during the demonstration.

Maintaining Activity

With limited equipment, laboratory rotation or stations are utilized to maintain student activity once the session has been started. A variety of methods can be used but generally a schedule of activities should:

- Assign students to areas
- Indicate required activities
- Provide timeline and due dates for specific tasks
- Indicate next rotation

Stops for Instruction

Instructors must account for a scheduled conclusion of the laboratory activity. This should involve signaling students to clean up and put away items prior to leaving the laboratory area. Providing verbal notification as well as having a visible clock help students manage their time to finalize their activities in preparation for the end of class. As machine noise may interfere with verbal instructions to conclude the lab activity, instructors may use a warning light to indicate when students should stop.

STUDENT BEHAVIORS

It is difficult to remain calm under the pressure of feeling disrespected as students misbehave. Student behaviors are learned long before they reach the classroom. Students often perform disruptive actions to communicate. As instructors, it is important to remain calm and not act in a reactive manner. This is goes against the natural response of fight-flight reflex. Emotions can be heightened, and the initial reaction is to deal with the issue immediately, however, instructors need to investigate the underlying causes of the behavior. Maslow's Hierarchy of Needs provides a framework to discuss student motivation or antecedents of student misbehavior.

Underlying Causes of Behavior

PK-5

Children enrolled in PreK through fifth grade often misbehave due to an unmet need. The acronym *HALT* is a good reminder to allow the instructor to assess the situation.

H. = Hungry

Estimates in the United States suggest that nearly 13 million children live with "food insecurity" at home. School may provide most of their meals. A student who has spent a weekend away and returns on Monday may have a lack of focus as well as irritableness.

A. = Angry

Anger and aggression are complex issues. A child may experience frustration when they cannot acquire what is being requested or when they are asked to do something they do not feel like doing. Anger is often expressed when children struggle with situations beyond their perceived control. This can take the form of anxiety or frustration. Often these feelings are associated with abuse or neglect.

L. = Lonely/bored

Children who are lonely or bored may exhibit poor peer relationships. Feelings of boredom may stem from lack of connection or engagement with others. Factors that have been identified with young children expressing loneliness include conflict within the home, moving to a new school, death of a family member or friend. Rejection by peers or bullying may contribute to the feeling of loneliness.

T. = Tired

Estimates indicate sleep problems affect 25-50% of children. Children between the ages of 4- and 12-years old need between 10-12 hours of sleep. Circumstances at home may limit the amount of sleep a child receives resulting in irritability, difficulty concentrating, and depression.

Middle school

The same basic needs noted for Pre-K to Fifth Grade are the same for middle school through high school ages. However, additional complexities arise as children develop into adolescence. The onset of puberty which raise questions about independence and identity. At this stage youth are struggling with their sense of self. Greater importance is placed on acceptance, security, and identity. At this stage students are seeking purpose. Motivation is a key driver. When students have a lack of interest or communications fail in connecting to the importance of the learning activity, behavioral problems can arise.

High school

As students mature, teens cognitive development allows them to think about abstract concepts such as morality and justice. Student often seek opportunities to demonstrate independence and place higher value on receiving appreciation to validate their self-worth. Teens tend to use success at school, sports, or their friendships to compare themselves with others. These students desire to make their own decisions. Feeling of support and validation are often sought by teens among adult leaders but in the form of guidance.

Differentiating Punishment versus Discipline

In their aptly titled book *Empowering Students: Seven Strategies for a Smart Start in School and Life*, Robert Brower and Amy Keller (2006) outline the need for teachers to help students understand a few critical distinctions. Prominent – as much for the teacher as the student – is the distinction between punishment and *discipline*. According to the authors punishment is done *to* a student in retribution for an action, while discipline is done *for* a student and is a teachable moment. The key difference? Helping students understand why they face consequences for their actions and working with them to be productive classroom citizens.

Ensuring discipline does not let students off the hook (more on accountability in a moment), but rather shifts the focus from the teacher/anger and onto the behavior or decisions of the student that have led to a need for consequences. It is critical in this moment the teacher does not act out of anger but rather with a focus on helping the student be a productive citizen. This can be accomplished by a combination of the following:

- Ask the student why they face consequences for their actions.
- Ask what they would do if in the role of teacher and they had to ensure a positive learning environment for all students.
- Ask what a reasonable consequence for their actions should be.
- Ask why they acted in the way they did (especially if a repeat offender).
- Ask how their guardian would respond if they knew of such actions.

Remember, discipline is done for a student and punishment is done to – a teacher cannot accomplish a teachable moment and achieve discipline if they are not willing to explore the justification or reasons behind a student's actions. It may seem counterintuitive to seek input from a student 'in trouble,' but that input me be the key to ensure discipline happens – and a behavior is corrected – instead of punishment.

Accountability

While discipline should be a teacher's goal, it simply cannot be achieved without accountability. The teacher must ensure accountability is present by being equitable in their disciplining of students. A simple, and oft-repeated, phrase sums it up best: what is not corrected is condoned. The best intentions and most thought-out list of classroom rules can unravel quickly if students are not held accountable for their actions. Equitable discipline does not mean there is a standardized response for any given action. In fact, a teacher could quickly burnout if trying to create an air-tight if/then set of rules to strictly follow for any possible student management issue. Rather, students should be held accountable by ensuring discipline for student management issues as they arise – with both a focus on ensuring consequences for the student as well as empathy from the teacher.

What does this look like in practice? A list of classroom expectations and/or rules agreed to by both teacher and students is an excellent starting point. By having a base set of expectations and generating rules agreed to by students, it frees the teacher to use those expectations and rules as a starting point when holding students accountable. A broad set of expectations are a great starting point when discussing class rules with students, for example:

1. Be on time
2. Be prepared to learn
3. Be present/engaged
4. Be respectful of self, classmates, and teacher
5. Be authentic

Expectations

No one wants to guess what is expected. If expectations are not explicitly communicated it leads to assumptions made by students of what the teachers wants from them. This could be disastrous. Teachers who expect students to perform well, behavior as well as academically, often see positive results. However, this is dependent on the consistency from the teacher in reinforcing performance and behavior expectations. Also, positive rapport and professional relationships are key in managing student behavior. Teachers who invest in their students by identifying their likes and dislikes are seen by students as more caring. Respected teachers are less likely to have classroom management problems. Two major types of *expectations* can help in guiding student behavior. These are broad and specific expectations.

Broad – These are goals for general student behavior that teachers should bring to the classroom and are essential elements of behavior that are non-negotiable with students. They serve as guidelines for behavior across settings. Samples include Be safe, Be responsible, and Be Respectful

Specific – Often referred to as rules, these define the appropriate behaviors that instructors want students to observe or demonstrate. These are concrete, observable, and measurable. Variations will depend on the setting such as the classroom or laboratory. A matrix can be used to maintain continuity across settings. Samples include Store and silence cell phones during testing, Keep food and drinks out of the laboratory, or Begin the bell ringer once you are seated.

Levels of Interventions

When a student, or students, are out of compliance with expectations and rules within the classroom, student accountability must be maintained through upholding those expectations and rules. *Interventions* are actions we can take to discipline students, maintain accountability, and help students not repeat the same behavioral mistakes. Below three tiers of interventions are discussed, as well as potential intervention options for each tier.

Positive Behavioral Interventions.

Tier 1: Universal

Supports serve as the foundation for behavior and academics across the school. Generally, schools will provide these to all students, staff, and settings. For most students, these interventions identify how to be successful and to prevent future problems so that instructors can create a positive and reinforcing learning environment. Positive interventions can also be teacher-focused, and present as proactive actions such as:

1. Plan: plan full days of instruction for students. Oftentimes downtime can result in student management challenges.
2. Rapport: know your students and exhibit interest in their interests as well as their learning. While it underscores the value of a teacher's content knowledge, the adage is true: "they don't care how much you know until they know how much you care."
3. Variability: this is not just a characteristic of effective teachers; it is an effective student management tool. Alongside a full hour of instruction, utilize different engagement activities and teaching methods. A good rule of thumb is students should be doing something different every 20-30 minutes.
4. Movement: consistently moving around the room lets students know they are being held accountable with supervision by the teacher.

Tier 2: Targeted

These interventions are strategically placed to assist students who exhibit signs of noncompliance to tier 1 interventions. These interventions are designed to be more intensive to elicit behavior change.

1. Remove temptation: if a student is distracted, calmly remover the distraction (e.g., cell phone).
2. Redirect behavior: ask distracted students a question or to complete a task to help them re-engage.
3. Proximity: use your physical location within the room to redirect or re-engage students.
4. Pit stop: briefly slow down/stop by student to communicate appropriate behavior and keep moving – intentionally not stopping long enough for the student to engage in a rebuttal.
5. Planned ignoring: if minor and the student has stopped, planned ignoring is an effective strategy. This works under the assumption that calling attention to the behavior would cause further distraction by the offending student or students around them.
6. Nonverbal communication: communicating nonverbally to the student to stop the behavior.
7. Verbal communication: directly calling on student to cease the offending behavior.
8. Offer a Choice: provide the student with two choices, (a) act appropriately/get on task or (b) face discipline measures. For example, a disciplinary choice instead of on-task behavior may be the student calling home to explain why they chose not to learn in that class session (this measure can be adjusted based on the student and frequency of infractions).

Tier 3: Intensive individualized

Approximately 5% or less of students who do not respond to tier 1 and tier 2 may need additional intensive resource dependent interventions. Students may need therapy from mental health professionals, special classes, or residential treatment placement for more supervision and structure.

SUMMARY

As discussed in this chapter, management of the classroom and students can be aided by sound decision-making practices by teachers. Those sound decision-making processes encompass technology usage, lab setup and physical layout, classroom setup and physical layout, and expectations and rules for students. Tech use should center on the opportunity cost of the new tech, and whether there is learning value. Layouts of the classroom and laboratory spaces should focus on efficiency and organization and assist the teacher in maintaining an orderly learning space. Finally, expectations and rules are a tool for teachers to hold students accountable and focused on the learning happening during instruction.

REVIEW QUESTIONS

1. What process should teachers follow when determining whether to use a new instructional Technology?
2. How can instructional technology aid with assessment?
3. When organizing a classroom space, what should the teacher consider?
4. What three characteristics should a starting routine include?
5. How can *anchoring* assist with starts and stops during instruction?
6. How does the appearance of a laboratory space impact students?
7. What five goals should tool and organization safety focus on?
8. What maintenance activities are necessary by a teacher?
9. What does HALT stand for, and can it affect student learning?
10. How can teachers hold students accountable for behavior?

REFERENCES

Alston, A. J., Miller, W. W., & Williams, D. L. (2003). The future role of instructional technology in agricultural education in North Carolina and Virginia. *Journal of Agricultural Education*, *44*(2), 38–49. doi:10.5032/jae.2003.02038

Banks, T. (2014). Creating positive learning environments: Antecedent strategies for managing the classroom environment & student behavior. *Creative Education*, *5*(7), 519–524. doi:10.4236/ce.2014.57061

Bear, W. F., & Hoerner, T. A. (1986). *Planning, Organizing, and Teaching Agricultural Mechanics*. Hobar Publications.

Brooks, D. C. (2012). Space and consequences: The impact of different formal learning spaces on instructor and student behavior. *Journal of Learning Spaces*, *1*(2).

Brower, R., & Keller, A. (2006). *Empowering Students: Seven Strategies for a Smart Start in School and Life*. Rowman & Littlefield Education.

Croom, D. B., & Moore, G. E. (2003). Student misbehavior in agricultural education: A comparative study. *Journal of Agricultural Education*, *44*(2), 14–26. doi:10.5032/jae.2003.02014

Pew Research Center. (2021). *Who owns cell phones and smartphones?* https://www.pewresearch.org/internet/fact-sheet/mobile/

Shoulders, C. W., & Meyers, B. E. (2012). Teachers' use of agricultural laboratories in agricultural education. *Journal of Agricultural Education*, *53*(2), 124–138. doi:10.5032/jae.2012.02124

Sousa, D. (2017). *How the Brain Learns* (5th ed.). Corwin Press.

Taz Networks. (2016). *If your computer was human, how old would it be?* https://www.taznetworks.com/computer-human-years/

Williams, M. R., Warner, W. J., Flowers, J. L., & Croom, D. B. (2014). Teaching with technology: North Carolina agriculture teachers' knowledge acquisition, attitudes, and identified barriers. *Journal of Agricultural Education*, *55*(5), 1–15. doi:10.5032/jae.2014.05001

KEY TERMS AND DEFINITIONS

Accountability: How teachers hold students in compliance with rules and expectations, may include actions geared toward discipline.

Anchoring: Instructional technique used in conjunction with routines that places emphasis on the physical location of the teacher while giving different types of verbal instruction.

Discipline: Differs from punishment in that it is focused on helping the student make improvements toward the desired behavior; done *for* a student and not *to*.

Expectations: Broad or specific guidelines for student behavior.

HALT: Acronym connected to hierarchy of needs that may explain student behavior concerns, includes hungry, angry, lonely, and tired.

Instructional Technology: Hardware or software resource used by teachers to aid in the learning process. Can be teacher-focused or student-focused and can encompass many different types of electronic devices.

Intervention: Actions taken by the teacher, administrator, and/or parent to discipline students and maintain accountability.

Organization: Physical layout of a learning space as it pertains to stowage of learning supplies.

Pit Stop: Common intervention used by teachers that involves briefly redirecting a student's behavior.

Routines: System of repetitious actions taken by students and teacher to aid in the efficiency of the learning process, and includes starting instruction, maintaining activity, and stopping instruction.

Chapter 10
Assessing Student Learning

Tyson J. Sorensen
Utah State University, USA

ABSTRACT

The purpose of this chapter is to present assessment concepts and provide guidelines for effective assessment practices within school-based agricultural education. The chapter explains common terms and different types of assessments. It also provides guiding principles for how to effectively construct various types of assessments including both traditional and alternative types of assessments. In addition, this chapter provides guidelines and principles for managing grades in school-based agricultural education programs. This chapter provides practical recommendations and examples within a school-based agricultural education context.

INTRODUCTION

Assessment is sometimes an afterthought and can occasionally be challenging for teachers; but, when done properly, is a very important part of the teaching and learning process. To understand the importance of assessment, the question every educator should ask is: "why do we assess?" Effective assessment is important in education because it helps teachers evaluate the needs and strengths of learners, helps to guide instruction and instructional decisions, and is the way in which student outcomes and progress are measured. Through proper assessment, teachers can determine student learning, assess their own teaching effectiveness, determine grades, and provide class data for administrative purposes. The nature of school-based agricultural education requires that students master performance objectives in cognitive, psychomotor, and affective areas of learning. This chapter discusses assessment considerations and principles for effectively assessing student learning and managing grades in school-based agricultural education settings. Learning objectives include: 1.) describe common terms used in educational assessment, 2.) identify common types of educational assessment, 3.) describe techniques for effectively constructing effective assessments in school-based agricultural education, and 4.) discuss grading practices used in school-based agricultural education.

DOI: 10.4018/978-1-6684-3420-8.ch010

COMMON TERMS USED IN EDUCATIONAL ASSESSMENT

Assessment, Measurement, and Evaluation

Various terms are used to describe processes and tools used to measure student learning. A range of sometimes conflicting definitions in educational assessment can be found across various textbooks and educational resources, and the terms are often used interchangeably by educational professionals. Despite this, it is important to understand the fundamental concepts of educational assessment. Here, some common and fundamental terms will be described.

Assessment is the process (the verb) or tool (the noun) used for collecting information to monitor progress and make educational decisions. For example, the administering (verb) of a test (noun) to a group of students is a form of assessment because it is the process and tool used to collect information about student learning. Other examples include observations, interviews, and behavior monitoring. Assessment is a continuous process that provides teachers and students valuable information in the teaching and learning process. Throughout this chapter, the term assessment and test are often used interchangeable. **Measurement** is the systematic assignment of numbers or names to student learning. It is a systematic way to determine students' learning status. For example, assigning the grade of 75% on a test, or using a rubric to assign the grade of a 3 out of 4 are examples of measurements. **Evaluation** is the procedure used to judge a student's skills or knowledge in relation to predetermined criteria (i.e., learning objective). The term evaluation literally means to determine the *value* of something or someone. Thus, evaluation is used to give meaning to the measurement from the assessment. What does a 75% on a test really mean? Does it mean the student passed? Does it mean the student mastered the objective? Descriptors such as "passing," "approaching mastery," or "adequately met the learning objective" are values derived from measurements that enable educators to make decisions about teaching practices and draw conclusions about student learning.

Validity

Validity should be a primary consideration in both the development of assessments and the interpretation of evaluations. **Validity** refers to the extent in which the assessment measures what it is supposed to measure. This type of validity is called content-validity. When the assessment (e.g., a test) does not produce an accurate result of what students know or can do pertaining to the learning objectives being measured, then the assessment is not invalid. A common example of this would be a teacher giving a test to measure student learning for a specific learning objective with items on the test not relating to that objective.

Validity also refers to the accuracy of the interpretations of the evaluations. For example, teachers sometimes attempt to interpret student learning based on a student's ability to write a well-constructed response to an essay-type question. Teachers sometimes fall into the trap of assessing students on their writing ability versus their knowledge specifically related to the learning objective. In this instance, the test itself might seem to measure what it purports to measure, but the accuracy of the interpretations may be questionable.

Many factors can contribute to variations in validity of assessments. Table 1 highlights the most common issues affecting validity.

Table 1. Common factors contributing to poor validity in assessments

Factor	Description
Overmeasurement	When teachers attempt to assess information or skills not related to the intended learning objectives.
Undermeasurement	When teachers attempt to assess knowledge or skills of students but do not develop a sufficient number of items for measurement. For example, the level of a student's understanding of a complex process such as cellular respiration generally cannot be assessed sufficiently with only one question. An assessment needs to represent all aspects or dimensions of the learning objective.
Mismeasurement	When a certain skillset is required to successfully complete an assessment, but students do not possess those requisite skills. This is common with essay type questions and the required skill of writing. In many cases, teachers inadvertently measure two things (technical knowledge and writing skills) instead of the students' knowledge or skill related to the intended learning objective. To some degree, this cannot be avoided because a certain level of language literacy is required for traditional assessments. Teachers can reduce validity issues by understanding their students' reading, writing, and communication skills. Writing instructions clearly and simply is also helpful.
Answer Signaling	When assessment items provide cues to answers on other items, enabling students with no knowledge of the subject to answer the item(s) correctly. For example, a teacher can provide grammatical cues on previous test items that can indicate a correct answer later in the test.
Test Length	The assessment should be long enough to cover all the knowledge or skills intended to be assessed but not too long to produce test fatigue or burnout.
Time Lapse	The assessment is given within too short a time period or too long a time period that does not allow students to effectively learn or retain assessment concepts.

For tests to be valid, they must be reliable. **Reliability**, a key component of validity, refers to the consistency of results from assessments. The more variation in results means less consistency, and therefore a less reliable (and valid) evaluation. An assessment with high reliability means teachers should be able to administer a test multiple times and get the same scores. For example, an assessment in which a student could take two times, one day apart, should in theory not differ much from day to day. This would indicate good reliability. Reliability and validity are interrelated, because as consistency in responses increase, so does the ability to produce accurate evaluations about students and their learning.

Reliability issues often result from assessments that are too long, too ambiguous, or require too much cognitive focus on the assessment process itself versus a complete focus on the skill or knowledge to be assessed. For example, a student who is required to flip back and forth from page to page to find answers to a matching portion of a written test can be mentally and physically taxing and can take away from the student's ability to answer the questions properly. A test that takes students more than 30 minutes to complete can be too difficult for many young students to focus their attention for that amount of time. Likewise, an assessment with so many intricate details, difficult language, or confusing prompts can cause cognitive overload causing students to focus more on the process rather than the content of the assessment.

Another factor negatively impacting reliability is when an assessment is given to students in an unsuitable environment that causes distractions or other focus away from the assessment. In this case, the assessment itself might be good, but because the student is distracted, the interpretations of the results would possibly not be an accurate reflection of student knowledge or skill in relation to the learning objective.

Subjectivity in scoring is another factor impacting reliability of assessments. When tests are subjective, scoring becomes more ambiguous, bias is often introduced, and scores become more variable. Two teachers should be able to score a student's work and produce nearly identical scores. If the test questions are confusing, the scoring rubric is not clear, the testing environment is bad, or a teacher has some negative bias towards a student, test reliability can be jeopardized. Thus, writing good test items and tests that are easy to navigate, administering the test effectively in a positive testing environment, and consistently scoring accurately and objectively are important aspects of test reliability.

Bias refers to the degree in which elements of an assessment distract attention away or contain offensive elements for specific groups of students. Bias in assessments is a threat to validity because it distorts students' performances. It is possible that students actually can demonstrate mastery of the learning objectives, but because of a biased assessment, student performance may vary. Content underrepresentation is a common problem leading to bias and is the result of teachers not including context that students can relate to. Teachers can avoid bias by contextualizing assessments with the students' backgrounds in mind.

Common Types of Educational Assessment

Assessments can take on many forms. The sections below will discuss the seven major types of educational assessments, including formative and summative assessment, criterion-referenced and norm-referenced tests, traditional and alternative assessments, and diagnostic assessments (see Figure 1).

Formative and Summative Assessment

Formative Assessment refers to assessment that is conducted *during* instruction or a learning activity with the goal of improving learning. Formative assessment is a method for monitoring student progress and providing feedback. Formative assessment serves many purposes including helping students identify their strengths and weaknesses and pathway for improvement; helping to make thinking visible for students and the instructor; helping teachers identify specific student struggles; monitoring learning and giving students immediate feedback before grades are assigned; and helping teachers identify weaknesses with their instruction and concepts or skills that need to be retaught. Formative assessment should be viewed as part of a cyclical process in which teachers instruct and provide learning opportunities® conduct formative assessment® instruct and provide learning opportunities based on formative assessment data (see Figure 2). Therefore, it can be conceptualized as assessment *for* learning. As a result, during formative assessment, grades should not typically be recorded. One key aspect of formative assessment that is critical to its success is regular, prompt, and clear feedback.

There are hundreds of formative assessment strategies a teacher could employ, but the main focus of formative assessment should be to diagnose weaknesses and strengths and determine what students need in order to improve their learning or skills. Teachers can formatively assess students' understanding through verbal means, questioning, writing prompts, traditional assessments, performance assessments, and projects. Some specific examples of formative assessment in an agriculture classroom might include a teacher asking students at the end of class to answer specific questions related to the day's discussion; a teacher providing verbal feedback to a student while learning how to weld; or a teacher asking students to show their level of understanding (self-assessment) before moving on to the next concept by asking them to either show thumbs up, thumbs down, or thumbs sideways. Table 2 lists some general examples of formative assessment strategies that can be employed in an agricultural education classroom.

Summative Assessment refers to assessment that occurs at the end of instruction or a learning activity with the goal of making final evaluations on student performance. Students typically receive

Figure 1.

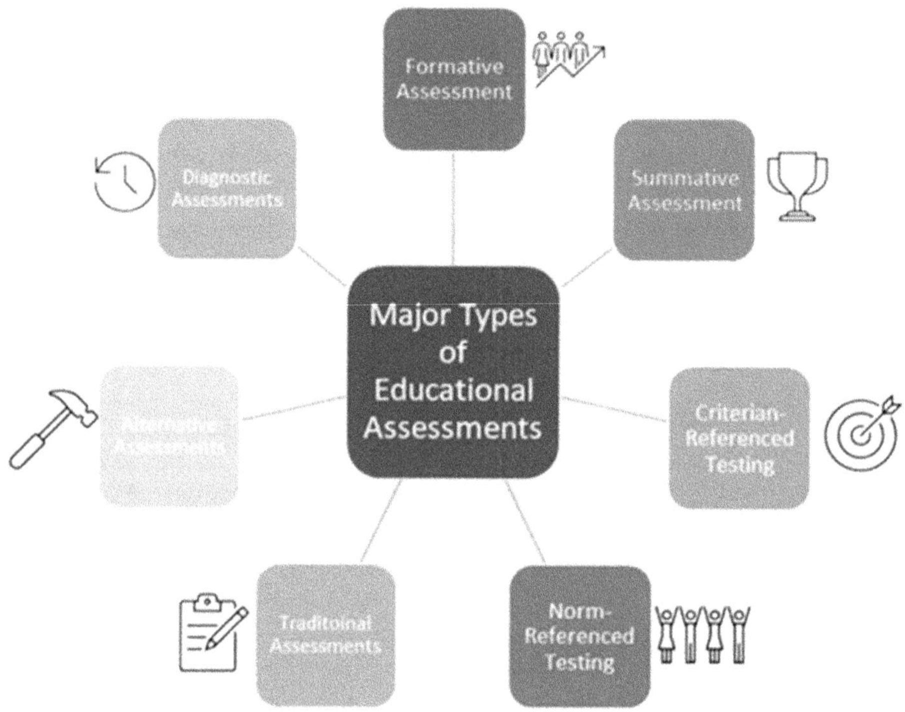

feedback in the form of grades, which are typically recorded. Therefore, it can be conceptualized as assessment *of* learning. Summative assessment therefore requires measurement of a learning objective so evaluation (judgement of student's knowledge or skill in relation to a predetermined learning objective) can occur. Summative assessment usually occurs at the end of a unit of instruction. Common examples might include a computer-based exam with questions measuring student knowledge related to each of the learning objectives for a unit; an assignment requiring students to demonstrate the process of doing specific skills (e.g., suturing, safely operating a machine, etc.) by "passing off" the skills to the instructor; and an assignment requiring students to demonstrate skill proficiency by producing a product (e.g., portfolio of floral arrangements; a weld, a written report). Ultimately, the goal of summative assessment is to determine whether the students have mastered learning objectives. Grades or scores are often associated with summative assessments while qualitative feedback is generally the product of formative assessments. Table 3 lists some general examples of summative assessment strategies that can be employed in agricultural education.

Criterion-Referenced and Norm-Referenced Testing

Criterion-Referenced Testing is the most common type of testing used by schoolteachers and refers to testing to determine a student's level of mastery or performance for a pre-established criterion. Agriculture teachers use state or national standards to formulate the specific learning objectives and guide curriculum and assessment decisions (Refer to Chapters 3 and 4). These standards and specific learning objectives are the pre-established criteria in which students are assessed and evaluated. With Criterion-

Figure 2.

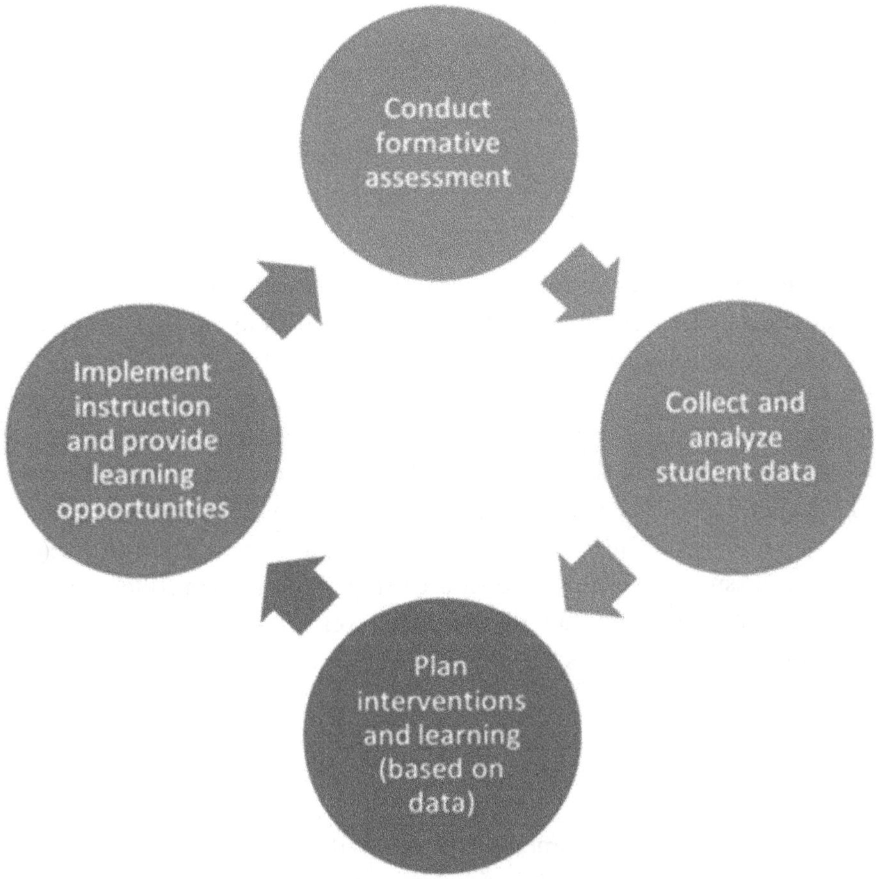

Referenced Testing, the focus is on a student's performance only in relation to the learning criteria. For example, a teacher would ask: "Did the student get the question right or wrong?" and "Did the student show sufficient mastery of the skill being assessed?" In each of the above examples, the student is being assessed in relation to a specific learning criterion.

Table 2. List of formative assessment examples in agricultural education

• Teacher listening to group conversations during in-class activities. • Teacher looking for non-verbal cues (e.g., reading the facial expressions and body language of students). • Teacher making note of student responses during a discussion. • Think-pair-share activity. • Student gives the teacher a short report on their progress. • Short-regular quizzes. • Exit and entry tickets. • Whiteboard slate responses to prompts • Teacher-student one-on-one conversations. • Lab reports or worksheets. • Virtual and online trivia and review games. • Check for understanding with student self-assessment (e.g., fist-to-five/ four finger rating, emoji cards, green-yellow-red cards, etc.) • Teacher checking homework for completion.

Table 3. List of summative assessment examples in agricultural education

• Traditional unit test. • Students develop a product to show their level of mastery of the content or skill (e.g., poster showing cellular processes, podcast describing the pros and cons of GMOs, welded pieces of metal, a floral arrangement, prepared public speech, etc.). • Students pass-off the steps of a skill to a teacher (e.g., set up and take down an oxy-fuel unit, properly restraining an animal, fish filleting and packing on ice, etc.). • Laboratory/field activity (e.g., plant identification, soil survey, timber cruising to calculate board volume, identifying safety hazards, troubleshoot and repair a problem in a compact gas engine, etc.).

In contrast, **Norm-Referenced Testing** refers to assessments that focus on a student's performance in relation to other students in a group (e.g., the norm). Norm-Referenced Testing simply compares a student's performance to other students by asking, "Did the student perform better or worse than the peers?" Norm-Referenced Tests are interpreted in relation to the peer group while Criterion-Referenced Tests are interpreted in relation to the pre-determined learning criterion. Norm-Referenced Testing can be useful for ranking students to make educational decisions. For example, Norm-Referenced Testing might be useful if an agriculture teacher is trying to select the top four performing students for a Career Development Event. Grading on a normal curve is also an example of Norm-Referenced Testing. Outside of the traditional classroom, Norm-Referenced Testing is common in college entrance or placement exams, IQ tests, and most aptitude and standardized achievement tests. They are useful for educational administrators making decisions about student grade readiness, language ability, or need for additional learning support.

Traditional and Alternative Assessments

Traditional assessments refer to assessments that use traditional methods as a means for evaluating student performance. While the term traditional assessment is very ambiguous, it generally refers to tests, quizzes, and homework as assessment tools. These are also often referred to as "paper-pencil" tests. These terms might be a bit outdated, as many of today's traditional assessments are computer-based. Traditional assessments are an efficient and effective method of determining student comprehension of knowledge at the end of the unit or topic. One benefit of traditional assessment is that they are generally easier and take less time for an instructor to design and grade than other alternative types of assessment methods. Another benefit of traditional assessment is the ease in which grading can be objective with minimized biases from a teacher's attitude, opinion, or judgement.

Since traditional assessments easily measure memorized and repetitious types of knowledge, they generally are most effective for assessing lower-level cognitive skills and knowledge. While it is possible, it is often difficult for instructors to design traditional assessments that measure high level thinking, processes (thinking or skills), or authentic skills (Fellenz, 2004). One other challenge that traditional assessments pose is they can often create test anxiety among students, which can influence reliability and validity of test results and evaluations. Strategies for constructing traditional assessments will be discussed later in this chapter.

Alternative Assessments refer to assessments that are different in form than traditional (e.g., quiz) assessments. Evaluation of performance skills within agricultural education usually requires a teacher to observe and evaluate the skill being performed or the product produced. The observation and evaluation of the skill is best suited by alternative assessment methods and tools. Alternative assessments provide

Table 4. Product-based and process-based performance assessment examples

Process-Based Examples	Product-Based Examples
Process of delivering a prepared public speech	A manuscript of a public speech
Process and procedures used to make a flower arrangement	A flower arrangement
Process and procedures used for cutting with an oxy-fuel torch, including set up and take-down	A piece of metal cut from the oxy-fuel torch
Process and procedures used to properly fill a syringe with medicine	A syringe filled with the proper dosage
Techniques of doing side-veneer and splice grafts	Properly grafted scions to stock

many advantages, including the ability to measure real-life skills as well as higher cognitive skills than most traditional assessments can usually do. Additionally, alternative assessments offer the benefit of being able to measure the process of specific skills as well as the end product. As a result, alternative assessments generally produce less test anxiety than traditional assessments and students are generally more motivated by alternative assessments than by traditional assessments. While there are certainly obvious benefits to alternative assessments, the main disadvantage is the time it takes to construct the tasks to be assessed, create the assessment and measurement criteria and tools (e.g., rubrics), and score each student. Alternative assessments also have the potential to be subjective if adequate assessment measures are not available.

Most commonly, there are three types of alternative assessments used in agricultural education which include performance-based assessments, authentic assessments, and portfolio assessments. These terms are not mutually exclusive, meaning that an assessment can be considered both an authentic and performance-based assessment. Each of these alternative assessments are described next. Strategies for constructing alternative assessments (e.g., rubrics) will be discussed later in this chapter.

Performance-based assessments refer to assessments that enable students to demonstrate their knowledge and skills as they perform a specific task. Performance-based assessments usually includes a task for students to perform and a rubric by which their performance on the task will be evaluated. These types of assessments are generally flexible in how they are executed. Furthermore, they generally promote active engagement of students. With performance-based assessments, students are generally evaluated on the quality of their ability to perform specific tasks (process-based) and/or the artifact they create (product-based) in the process. Process-based performance assessments are scored during an observation and measure the students' skills during a task. Evaluations are made of the actual skill being performed in the moment. Product-based performance assessments measure a tangible and scorable product that can be scored at the instructor's leisure. Evaluations are based on the quality of the end-product, not on the process. Table 4 shows examples of product-based and process-based performance assessments that are commonly used in school-based agricultural education.

Authentic Assessments focus on real-world situations as the context for specific skills and tasks to be performed and assessed. Authentic Assessments are generally more complex than performance assessments and require students to perform tasks representative of activities in real-world situations. Examples in school-based agricultural education might include harvesting, processing, packaging, and pricing an entire carcass for sale; organizing, preparing and carrying out floral arrangements for an actual wedding or school dance; planning, constructing, marketing, and selling chicken coops to the community; or planning, planting, growing, harvesting, marketing, and selling produce from a school garden or greenhouse to community members.

Portfolio-based Assessments are assessments that demonstrate student achievement and skill growth over time. It involves a systematic collection of student-developed products, where students assess their own work with teacher input, to highlight individual growth and progress over time and to showcase accomplishments. Teachers generally utilize scoring rubrics and checklists for portfolio-based assessments. Examples of ways in which portfolio assessments could be utilized in school-based agricultural education include: student collection of welds with self-assessments and reflections and showing skill development over time; student collection of floral arrangements highlighting various skills developed over time; and a collection of livestock judging oral or written reasons with self-assessments and reflections and showing skill development over time.

Portfolios are an example of ipsative assessments which enable students to compare earlier work with later work to reflect upon and highlight the improvement. Ipsative assessments, like portfolios, help students learn from their mistakes and motivates them to continue to improve. An advantage of portfolio assessments is the ability to document and evaluate student growth across time. In this regard, portfolio assessments can also be utilized to establish or modify individual student learning goals throughout the course. These assessments encourage student autonomy and ownership which also encourages student self-reflection. Portfolio-based assessments allow teachers to individualize assessments based on student needs and interests, particularly when students begin at different levels of skill and knowledge.

However, portfolio assessments take a great deal of time to execute effectively, and teachers sometimes lack the training or ability to make portfolios meaningful. Since portfolio-based assessments can vary from student to student, it makes this type of assessment difficult to evaluate. Additionally, one key component of portfolio-based assessments is student self-assessment, which can be a difficult task for some students, making the portfolio assessment less meaningful and effective.

Diagnostic Assessments

Diagnostic assessments focus on what students have learned in the past and what knowledge and skill they currently possess. A common type of diagnostic assessment used in agricultural education is the unit pretest. Unit pretests differ slightly from traditional diagnostic assessments because pretests measure knowledge and skills of the upcoming unit whereas traditional diagnostic assessments evaluate knowledge students have learned in the past. Both however are conducted before a unit of instruction. Diagnostic assessments and unit pretests can help teachers plan for instruction because they allow a teacher to determine students' individual strengths, weaknesses, knowledge, skills, etc. prior to beginning instruction. Diagnostic assessments can prevent redundant learning and save time because teachers can identify information or skills that may not need to be taught. One disadvantage is they are time consuming. Additionally, because diagnostic assessments often measure a wide range of skills and knowledge, they are difficult to design in a way that provides a valid evaluation of student knowledge and skill that is useful for teachers in guiding future instruction. Examples of diagnostic assessments before instruction begins can be a short quiz or unit exam, student interviews, classroom discussions, or graphic organizers.

Constructing Effective Assessments

There is a science to constructing tests for valid interpretation of results. Assessments that are constructed properly will reduce measurement error, resulting in more valid evaluations. The ability of a teacher to design valid assessments to measure student knowledge and performance helps teachers make appropriate decisions about instruction. Teachers therefore should strive to create valid assessments that truly do measure what they are intended to measure. The following section provides guidelines for constructing different types of effective assessments.

Traditional Test Assembly

Traditional assessments are common in educational settings, but attention to validity is critical when constructing the assessments. When planning to construct traditional assessments, teachers should consider the following practices to increase the validity of the assessments:

1. Identify the learning objectives for which students will be assessed and ensure that each learning objective is adequately assessed. Evaluate whether undermeasurement, overmeasurement, or mismeasurement is present.
2. Before administering the assessment, it is important to review the unit lesson plans to ensure content was adequately delivered. Sometimes information gets forgotten or left out of the instruction, so adjustments to the assessment may need to be made.
3. It is important to determine how many items to develop for each learning objective that will be assessed. Different learning objectives should generally be addressed by different test items, so it is easier to identify specific weaknesses and strengths of each student in relation to each learning objective. The number of test items for each learning objective is dependent on the relative importance or weight of each objective. How many items to include for a single objective depends on how important it is. For example, if each of the learning objectives are equally important, then the number of questions or the number of scoring points should be relatively equal for each objective measured.

Focusing on the quality of each type of test item can certainly improve assessments. Most often, when teachers develop traditional tests, they choose a variety of test item types (e.g., multiple choice, matching, and essay). How each type of item is assembled as a whole assessment is an important consideration. Assessments should be assembled in a way that that reduce extra cognitive load and maintain the highest level of accuracy possible. Table 5 lists some guidelines for constructing traditional tests.

Constructing Traditional Assessment Items

It is important to know how to construct assessment because if students can get the correct answers without knowledge of the content, the exam will lack validity.

Not giving attention to the details results in exams that may not actually measure what they purport to measure (validity issue). It is also important for teachers to determine the quality of their assessments by finding and fixing errors and poorly written questions. Some common ways for teachers to reduce error and maintain accuracy and consistency in assessment can be to score blindly, prepare a scoring

Table 5. Guidelines for assembling traditional tests

- Group together items of similar format and provide one set of directions per section. This enables students to stay focused on one task-type at a time, reducing cognitive overload. This also simplifies the scoring process.
- Arrange the test from easy to hard if possible so students build confidence from the beginning. This can lower anxiety and produce more accurate and reliable test results.
- Space test items for ease in navigating and reading. Start new sections on new pages whenever possible. Avoid trying to "fit" test items on a page at the expense of students' ability to properly navigate test items comfortably.
- Keep items and options on the same page (e.g., avoid having multiple-choice options span two pages).
- Position illustrations near descriptions and avoid spanning multiple pages whenever possible.
- Check the answer key after constructing the assessment and look for patterns and randomness. Avoid systematic patterns when possible (e.g., all multiple-choice answers are letter D- "none of the above").
- Provide clear directions for each section on how students should complete each task and how they will record their answers. For example, will the students record their responses by writing directly on the test or will they use a separate sheet or a scan sheet?
- Use language and write test items at the appropriate reading level of the students.
- Provide a space on the assessment for the student's name and date.
- Double check the instructions to make sure they are clear. Proofread each item to make sure there is clarity with each item.
- Take the exam or have someone else take it before administering to the class of students. Make edits to test items as needed.
- If applicable, provide details of points per test item in section directions so transparency is open to students throughout the test process.

guide and check for accuracy, and analyze test items that might be confusing or poorly written (e.g., most of the class misses it). Furthermore, beyond looking for confusing or poorly written questions, teachers should ask questions like, "are the items on the test linked to the learning objectives?" "Do the distractors distract?" "Is there more than one correct answer?" when constructing their assessments.

There are two major types of student response items in traditional assessments which include selected response and constructed response. **Selected-response** is a type of test item in which a student selects from a set of response options provided by the teacher. **Constructed-response** is a type of test item in which students generate their own response to a test item.

Selected-Response Guidelines

Selected-response items are generally easy and quick to grade and many learning objectives can be measured in a short amount of time. However, student guessing can be high, which can skew results and valid interpretations of what students actually know and are able to do. Because students usually only need to recognize the correct answer, selected response items generally do not measure high levels of knowledge and skill. Examples of selected-response items include multiple choice, matching, and alternative-choice items. Table 6 lists some guidelines to ensure validity with selected-response items.

Table 6. Guidelines to ensure validity with selected-response items

- Provide clear, full, and simple directions. Provide instructions for how students should respond (circle vs write answer, etc.) to test items.
- Provide clear and simple statements. Do not have ambiguous or overly complex statements. Use simple grammar, syntax, and vocabulary that is easy to understand for students.
- Write stems and answer options to reduce required reading time.
- Avoid clues in the question/statement that would indicate the correct answer. Avoid making the correct answer grammatically different than other options.
- Keep response items of equal length.
- Avoid qualifiers such as "always" or "never."

Multiple-Choice items consist of a question or statement (stem) with one correct answer and several incorrect answer options called distractors. They are generally easy to grade but determining effective distractors can be challenging. Table 7 lists some guidelines for constructing effective and valid multiple-choice items and Table 8 provides some sample multiple-choice items.

Table 7. Guidelines for constructing multiple-choice items

- Provide simple and clear instructions.
- Use longer stems than choices so students do not spend all of their time reading through the choices. Keep it simple, unambiguous, and exclude all extraneous information.
- Use only one correct answer. Distractor items need to be incorrect.
- Use plausible distractors.
- Avoid grammatical clues for the correct answer. For example, if the stem ends in "a" and only one option begins with a consonant, it makes it easy for the student to eliminate distractors solely based on grammatical cues.
- Avoid using negative statements when possible. Avoid double negatives in the stem.
- Questions are usually better than incomplete statements because they easier to cognitively process and there is generally less ambiguity.
- Critical words should be emphasized. If negative statements in the stem are used, emphasize the word by bolding, underlining, or using capital letters.
- Include 4-5 options. Too few increases the likelihood of guessing. Too many becomes too cognitively taxing for students to process.
- Put all options in a logical, chronological, or alphabetical order. This helps students locate answers easier but also helps to avoid the tendency of response preference (e.g., teacher tends to select option "b" as the correct answer most frequently).
- Avoid "All of the Above" options.
- "None of the above" can be used to increase item difficulty when an additional distractor is needed. Avoid using this option only when "none of the above" is the correct response (e.g., use "none of the above" as an incorrect distractor as well).
- Answer options should be of similar length.
- All responses should use the same grammatical form.
- Avoid extreme words (e.g., always, never, all) whenever possible.
- Avoid giving clues to correct answers from other items on the test.

Table 8. Example multiple-choice items

Directions: For each item below, select the correct response by circling the letter that corresponds to the correct answer. For each item there is only one correct answer.

Which of the following swine breeds is red with drooped ears?
A. Berkshire
B. Duroc
C. Hampshire
D. Yorkshire

Where does cellular respiration occur?
A. In the chloroplast and cytoplasm
B. In the chloroplast only
C. In the cytoplasm and mitochondria
D. In the cytoplasm only
E. In the mitochondria and chloroplast

Which of the following is NOT a component of soil?
A. Air
B. Gravel
C. Minerals
D. Organic matter
E. Water

Alternative-Choice, sometimes referred to as binary-choice, consists of test items in which a student selects from only two options such as true or false, yes or no, correct or incorrect, safe or unsafe, or fact or opinion. In contrast to multiple-choice items where three or more options are provided, students only select from two options with alternative-choice items. Another example of alternative-choice items is when a teacher provides a cluster of grouped items, but students still have two options to choose from. An example of this could be when a teacher provides a list of 30 different plants and the student must identify if the plant is a monocot or a dicot by indicating on the test one of the two options for each plant. In general, alternative-choice test items are easy to grade but guessing is high since choosing the correct answer from only two choices is usually quite probably. Table 9 provides a list of guidelines for constructing alternative-choice items while Table 10 provides some examples.

Table 9. Guidelines for constructing alternative-choice items

- Provide simple and clear instructions.
- Ensure that statements have definite answers. Avoid statements of opinion unless it specifically relates to the learning objective to be measured.
- Use short statements and avoid ambiguity in statements.
- Avoid absolute statements such as always, all, never, only.
- Avoid "trick questions." Trick questions reduce the accuracy in measuring what students actually know or can do. Avoid false items based on trivial information.
- Avoid double negatives. Write items positively. Emphasize any negative words such as "not" by underlining or capitalizing it.
- Include only one concept in each statement. Avoid double-barreled items (two concepts in the same sentence usually separated by the word "and" or "or").
- Write approximately equal numbers of each alternative (e.g., avoid having every correct answer be "true").
- Keep answer statements the same approximate length.
- Avoid a systematic order of correct answers (e.g., true, false, true, false).
- Consider having student write the entire word out or provide an option for circling the correct answer. Use symbols or letters for alternatives that are NOT closely related to each other if students will be abbreviating the response. For example, writing a T and F can sometimes be difficult to interpret. Other combinations that could be difficult to interpret are G/O, O/Q, N/M.
- To increase difficulty, teachers can require that students complete a revision of false statements (i.e., underline & revise).

Table 10. Example alternate-choice items

Directions: Some of the statements below are true and some are false. Indicate which statements are true and which ones are false by *underlining* either "TRUE" for the true statements or "FALSE" for the false statements.

TRUE or FALSE Soil is a renewable resource.
TRUE or FALSE Vitamin C is a fat-soluble vitamin.
TRUE or FALSE An individual who is homozygous carries two copies of the same allele for a gene

Directions: Following are alternate-choice type examination items. Indicate whether the statement is a Renewable Resource by writing an "R" in the space provided or a Non-Renewable Resource by writing an "N" in the space provided.

NOTE: Renewable Resource = R Non-Renewable = N

1. _____ Oil
2. _____ Wind energy
3. _____ Coal
4. _____ Gold
5. _____ Soil
6. _____ Uranium
7. _____ Biomass

Matching items consist of matching groups of lists and descriptions of words or phrases in which students identify the matching associations. These types of items are generally easy to construct and score. However, matching type items generally emphasize low-level factual information. Tables 11 and 12 provide a summary of guidelines and an example of constructing matching items.

Table 11. Guidelines for constructing matching items

- Lists should be in the right column and descriptions in the left column. Lists should be shorter than descriptions. Lists should contain short phrases, words, or symbols and should avoid long sentences. Letter the list.
- Descriptions should be in the left column and longer. This saves students time in locating the correct match from the list on the right. Number the descriptors.
- Use a reasonable number of items in the list and descriptions. Between 5-12 is generally appropriate. Too many items become burdensome for students and takes too much time to complete. Too few items and guessing becomes easier.
- Keep the list and descriptions homogeneous. Avoid mixing unrelated content in the same matching blocks (e.g., FFA dates and plant structures in the same list).
- Arrange list in a logical order (e.g., dates in order, alphabetical order, etc.).
- Include more options in the list than descriptors. This ensures students do not automatically get two statements incorrect and helps reduce correct responses from simple elimination and guessing. Instruct students in the directions that there are more options than descriptors. Also consider the possibility that each option in the list may be used once, more than once, or not at all.
- Use plausible distractors for items in the list that will not constitute one of the correct matches.
- Keep lists and descriptions on one page or screen. Avoid having students flip back and forth on two separate pages to complete matching items.
- Use clear directions on what is expected of students, particularly in how they will respond (e.g., place a letter in the box that correctly corresponds to the numbered description [recommended] response, write the correct word next to the description, draw a line to the correct match, etc.).

Table 12. Example matching items

Directions: *Match the Order with the Description by writing the CAPITAL letter from the "right-hand" column in the space provided by the descriptions in the "left-hand" column.*

Note: The listed Orders may be used once, more than once, or not at all.

Descriptions	Orders
1. _____ Wasps and bees.	A. Araneida
2. _____ Moths.	B. Chiloptera
3. _____ Termites.	C. Coleoptera
4. _____ Beetles.	D. Dermaptera
5. _____ Butterflies.	E. Diptera
6. _____ Earwig.	F. Hemiptera
7. _____ Crickets.	G. Homoptera
8. _____ Mosquitoes.	H. Hymenoptera
9. _____Dragonfly.	I. Isoptera
	J. Lepidoptera
	K. Odonata
	L. Orthoptera

Constructed-Response Guidelines

Constructed-response items tend to approximate more authentic, real-world behavior since it is rare in normal life to choose from nicely arranged options to determine a correct solution to a problem. As a result, these types of test response items typically measure a higher level of knowledge and skill than selected response items do. Examples of constructed-response items include short answer, essay, and problem-solving items. One disadvantages of constructed-response items is that scoring can be more challenging and time consuming than selected-response items. Due to the increased complexity in scoring constructed-response items, consistent and accurate scoring is not always achieved. Below are some general guidelines for each type of constructed-response item.

Short Answer items, also referred to as fill-in-the-blank items, consist of a task where students respond to a direct question or an incomplete sentence using a single word, phrase, or sentence. They are generally easy to construct and grade and guessing is low. However, sometimes there is ambiguity where many answers could be correct. These can also create challenges with computer-based scoring if responses are not spelled correctly. Tables 13 and 14 provide guidelines and examples for effectively constructing short answer items.

Table 13. Guidelines for constructing short-answer items

- Write statements or questions to indicate a clear and concise answer. Avoid ambiguity to keep scoring objective.
- A question is usually better than incomplete statements. It is easier to cognitively process a complete question than an incomplete statement.
- Use short-answer items to test for significant and important pieces of information related to learning objectives. If using incomplete statements, omit only key words from statements.
- Use only one or two blanks total. Too many blanks create ambiguity. Be sure the desired responses are clear and concise.
- When using incomplete statements, place the blank near the end of the statement. Incomplete statements are more difficult to understand if the front part of the statement is missing versus the tail end of the statement. Try to present a complete statement or question before asking for the response.
- Avoid grammatical cues to the correct answer.
- Prepare a scoring key that lists all the possible acceptable answers for each response. Placing blanks in a column can make scoring easier.

Table 14. Example short-answer items

Directions: For each question listed below, write the correct response in the blank provided. Please write each response clearly.

1. ______________________ What type of bond joins the bases of complimentary DNA strands?

2. ______________________ Which enzyme separates the double-stranded DNA into single strands during DNA replication?

3. ______________________ In DNA, Cytosine bonds with which nitrogenous base?

Essay items consist of a task where students respond to a direct question or prompt with a written response of a paragraph or more in length. Essays provide the opportunity for teachers to assess complex learning objectives by measuring student's ability to synthesize, evaluate, and compose. It also provides an opportunity for teachers to measure more than one learning objective and enables students to show understanding of possible interrelationships between learning objectives. Guessing is low, but students with good writing skills can sometimes bluff their learning. Essay items are sometimes difficult to score consistently (reliability) across students and teachers sometimes end up measuring writing skills rather than the purported learning objective (validity). Table 15 provides guidelines for constructing effective essay items.

Table 15. Guidelines for constructing essay-type items

• Provide clear and complete instructions for the task. Be clear on the extensiveness of the response desired. • Provide ample time for answering questions. Provide a time limit and/or page limit. • Describe the response format desired such as full sentences or bullet points. • Make decisions regarding how mechanics of writing will be scored before the scoring takes place. Mechanics of writing should generally be scored separately from the technical knowledge based on the learning objective being measured in the essay. • Grade essays using a scoring scheme. Generally, a rubric is most appropriate. Determine in advance what a quality response would be and design a rubric accordingly. • Evaluate answers to essay questions based on the learning objective to be measured.

Problem-Solving items, such as case studies, give students realistic and detailed situations (usually in writing) in which students are asked to analyze the situation or solve a problem based on the situation. Because students are asked to create, evaluate, synthesize, compose, or analyze various situations, problem-solving items provide a great opportunity to assess students in higher-level types of knowledge.

Problem-solving items generally consists of two parts which include the situation or scenario and the student response to questions. The situation is a detailed and realistic scenario that introduces students to a problem or situation that needs to be addressed or solved. It provides background and context for the students. Teachers should provide students with as many relevant details as possible for the students to fully understand the situation. Students are then asked to respond to questions or complete specific tasks about the detailed situation. Problem-solving items can employ a mixture of constructed-response and selected-response items for students to respond to. Table 16 lists guidelines for constructing problem-solving items.

Table 16. Guidelines for constructing problem-solving items

• Describe realistic situations to be solved or addressed. • Provide students with all the necessary information they would need to successfully address the situation or accomplish desired tasks. • Provide clear and complete instructions for completion of the task. • Use appropriate constructed-response or selected-response items and follow guidelines provided in this chapter for writing quality test items.

Computer-Based Assessments

Computer-based testing (see also electronic assessment, digital assessment, online assessment) is becoming more popular as schools have adopted learning management systems and web-based assessment applications have expanded. This technology has made it easier and more efficient for many teachers to conduct assessment. Through computer-based assessment, automated scoring can save immense amounts of time for teachers. Furthermore, the ability to integrate media such as pictures, videos, and games has enhanced teacher's ability to better assess target learning goals. Computer-based assessments also provide teachers with a variety of tools for diagnosing reliability and validity issues with each assessment. However, computer-based assessments can be limited by the features offered by the specific learning management system. Skills-based standards can also be difficult to assess on a computer. Further, teacher training is often required for teachers to take full advantage of the assessment options and tools available within the computer-based system, which can be costly and time-consuming at first. One of the biggest challenges however with computer-based assessment is difficulty controlling various factors of test administration in online settings. For example, controlling the testing conditions to ensure fidelity (e.g., verification of student, cheating) in the process as well as ensuring equal and adequate access for student populations continue to be challenges.

With regard to constructing computer-based assessments, the same general guidelines for effectively constructing traditional assessment items apply. When administering tests in an online format, other considerations should be taken. Table 17 lists some general considerations of online and computer-based assessments (International Test Commission, 2006).

Table 17. Considerations for online and computer-based assessments

- Consider student learning and technology needs, such as those with disabilities or those without internet access.
- If both online and paper-pencil versions of the assessment are given to students, be sure to make them equal regarding student tasks, types of questions, etc.
- Consider equality of access for all students. Do all students have access to reliable internet to be able to complete the assessment? Does the assessment have software requirements that some students may not have access to?
- Consider appropriate control over test conditions, supervision, and possibility of cheating.

Constructing Alternative Assessments

Alternative assessments are generally ideal for skill-centered performance objectives. Alternative assessments generally consist of performance-based, authentic, or portfolio assessments. With each of these, alternate forms of evaluation are generally required. Generally, alternative assessments require two parts, the task and the assessment. *Task*: detailed and clear instructions should be provided for a student task to be completed. Students first need to understand what the expected task is. The task is usually a specified procedure or sequence of steps to be performed by the student. *Assessment*: students are assessed on the specific task(s) using an appropriate assessment tool to measure performance (e.g., end-product or result that meets certain criteria).

When planning to administer skill-based assessments, teachers should first align the student tasks with the intended learning outcome or performance objective. Students should receive clear instructions (e.g., task sheet) explaining the expectations for task completion. Teachers should provide a list of materials, tools, or equipment available for students when completing the tasks. A rubric or checklist should be developed in advance by the teacher and should be shared with the students ahead of time. The rubric or checklist should clearly indicate the observable and measurable actions for students to meet the performance objective. Generally, rubrics and checklists are used as the appropriate measures for alternative assessments. When selecting tasks for alternative assessments, teachers should try to design authentic, real-world tasks whenever possible. Furthermore, teachers should ensure that the task is fair for all students and can realistically be accomplished. When designing the assessment for the student task, teachers should clearly identify the learning outcomes to be measured from the task and ensure that scoring can be done in a consistent and accurate manner.

Constructing Rubrics and Checklists for Alternative Assessments

Rubrics are multi-level rating scales that help teachers assess student performance in objective and reliable ways. Rubrics are also valuable in helping teachers objectively measure open-ended essay questions. Rubrics provide criteria for teachers to consistently assess students, but they also provide students with criteria and expectations for which they will be assessed along with corresponding point values.

Rubrics are shaped like a matrix, consisting of evaluative criteria on one side and levels of mastery on the other side (see Figure 3). *Evaluative criteria* are the characteristics to be rated, such as knowledge, skill, behavior, etc. These are the dimensions of the skill that is being measured. For example, tone, body language, and clarity might be the dimensions of a good public speech. Generally, rubrics can consist of many criteria, but it becomes more challenging to score effectively if too many criteria are assessed in one rubric (Popham, 2020).

Figure 3.

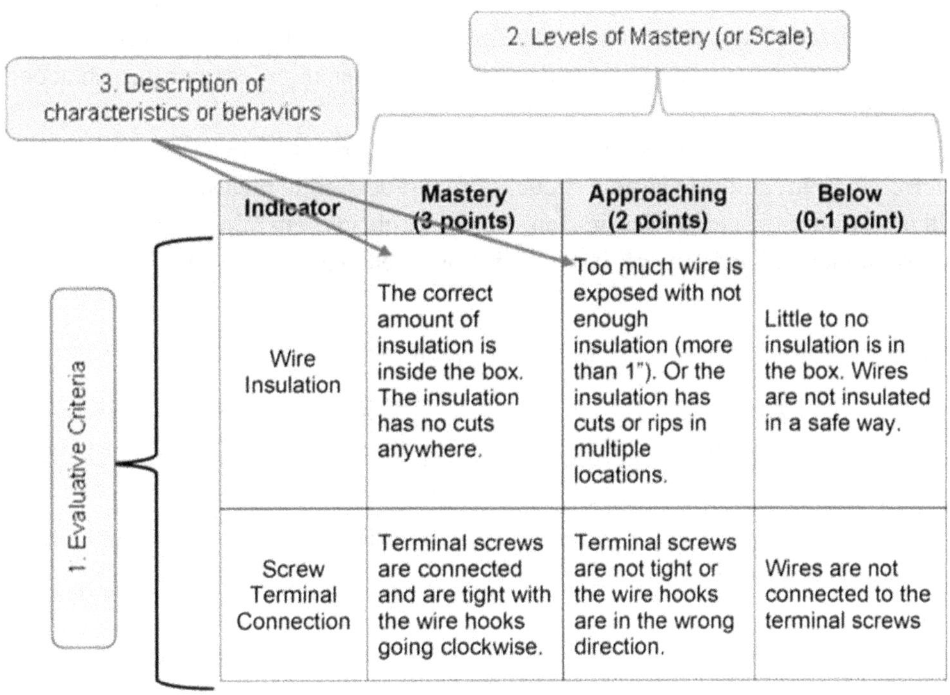

The *levels of mastery*, also referred to as the scale, are labels used to describe the levels of mastery for each criterion. For example, the levels of mastery labels for a specific skill students would be assessed on might include "not meeting, approaching, meeting, exceeding" or "novice, intermediate, high, advanced." The levels of mastery labels should be associated with numbers representative of the qualitative descriptors. For example, 1, 2, 3, and 4. Within each cell of the rubric matrix, a description of each characteristic can be integrated. For clarity of scoring, each description of each characteristic should be mutually exclusive.

When creating a rubric, first identify the learning outcome or skill that needs to be assessed. Second, identify the various characteristics of the learning outcome or skill. This is easiest done by asking, "what are all the essential components of the skills/knowledge/behaviors being assessed?" Once identified, the first column of the rubric should be populated with these characteristics or dimensions. These will ultimately make the rows of the rubric. Third, identify the levels of mastery--the measurement scale. It is wise to limit the number of mastery levels to less than 6 for clarity with scoring. These will be placed on the top row of the rubric, ultimately make the columns of the rubric. The distance between each level of mastery should also be equal. For example, it doesn't make sense to have the levels of mastery be "not meeting standard," "proficient," and "exceeding standard" because those are not equal distances. There should be a descriptor between "not meeting standard" and "proficient," such as "progressing" or "approaching standard." The final step in developing a rubric is to describe each level of mastery for each characteristic. These are generally placed within each cell of the rubric. The easiest way to do this is to begin with the highest level of mastery and describe what that behavior or skill would look like for each criterion. Once the top level of mastery has been described, skip to the lowest level of mastery and describe the observable behaviors or skills at that level. Finally, develop the descriptions for the intermediate levels of mastery. Once each cell has been populated with descriptions, the rubric is complete and ready to be utilized. It is common however for teachers to make minor adjustments to each newly developed rubric after testing it out with students or colleagues in order to ensure an accurate and reliable measurement.

Like everything else in the assessment world, there are various types of rubrics. The one described above is referred to as an **analytic rubric** because of its precision in measurement, providing a description for each level of mastery for each criterion. Analytic rubrics generally take longer to create and take longer to assess students. Furthermore, if a description of each level of mastery for each criterion is not

Table 18. Example of an analytic rubric used in agricultural education

Indicator	Mastery (3 points)	Approaching (2 points)	Below (0-1 point)
Initial Contact	Student identifies themselves with a good first impression. Asks questions to build personal rapport and learn about the customer's needs and wants	Student mostly identifies themselves with a good first impression and asks some questions to attempt to build personal rapport and learn about the customer's needs and wants.	Student poorly identifies themselves with a good first impression and does not ask appropriate questions to build personal rapport and learn about the customer's needs and wants.
Features and benefits	Student successfully applies features and benefits of their product to the customer's needs/wants and allows customer to participate in matching the needs and wants.	Student somewhat applied features and benefits of their product to the customer's needs/wants and somewhat allows customer to participate in matching the needs and wants.	Student poorly applied features and benefits of their product to the customer's needs/ wants and poorly allows customer to participate in matching the needs and wants.
Addresses Objections and Active Listening	Student successfully listens, clarifies customer's objections, and successfully addresses customer objections.	Student mostly listens, clarifies customer's objections, and mostly addresses customer objections.	Student poorly listens, poorly clarifies customer's objections, and poorly addresses customer's objections.
Trial Close and Close	Student successfully identifies buying signals. Student successfully closes or attempts to close the sale.	Student mostly identifies buying signals. Student mostly closes or attempts to close the sale.	Student poorly identifies buying signals. Student poorly closes or attempts to close the sale.

clearly described, reliability in scoring can become an issue. However, because of the detail in analytic rubrics, they can be very useful for teachers in measuring observable behaviors or skills and identifying specific weaknesses and strengths of students. Table 18 provides an example of an analytic rubric used to evaluate students' sales presentations (National FFA Organization, 2021).

Holistic rubrics are general rubrics that consist of only one single level of mastery. Holistic rubrics list and describe the behavior or skill for the highest level of mastery for each criterion. Holistic rubrics can therefore become a "short version" of an analytic rubric. As a result, holistic rubrics lack the precision of analytic rubrics and the ability to provide specific feedback on weaknesses and strengths. The advantage of holistic rubrics is the time that is saved in creating and scoring as well as the simplicity for the both the teacher and student. Table 19 is an example of a holistic rubric. Notice the single mastery criteria is identical to the highest scale rating (mastery criteria) of the analytic rubric in Table 18. Also notice an extra column for "points earned" since cells of the rubric matrix that identify specific characteristic are not present in holistic rubrics.

Table 19. Example of a holistic rubric used in agricultural education

Indicator	Mastery Criteria (1-4 scale)	Points Earned
Initial Contact	Student identifies themselves with a good first impression. Asks questions to build personal rapport and learn about the customer's needs and wants.	
Features and benefits	Student successfully applies features and benefits of their product to the customer's needs/wants and allows customer to participate in matching the needs and wants.	
Addresses Objections and Active Listening	Student successfully listens, clarifies customer's objections, and successfully addresses customer objections.	
Trial Close and Close	Student successfully identifies buying signals. Student successfully closes or attempts to close the sale.	

Checklists are basic rubrics in which only two performance levels are present (e.g., yes/no; pass/fail). Checklists are generally the simplest of all rubrics and the fastest for scoring performance. Checklists are clear for students and instructors, resulting in less arbitrary and more consistent scoring than other rubric types. Checklists can be useful for measuring process-based, in-the-moment performance assessments. They can also be useful for non-performance type of activities or student self-assessments, such as participation in an assignment (e.g., "are all the parts of the assignment present?"). Performance checklists are commonly used for work-based performance evaluations. Employers, supervisors, and internship directors can easily assess student performance in the workplace using simple performance checklists. Table 20 provides an example of a performance checklist that could be used in an animal or veterinary science class or in an animal work-based setting.

A common flaw in rubric development is ambiguity. When developing scoring rubrics, teachers should be certain to provide sufficient descriptive detail to avoid uncertainty in scoring. Students are more likely to meet the requirements of the assignment when there is clarity in the expectations. Additionally, when too many criteria are present in a rubric, it becomes too overwhelming for teachers to assess student performance accurately and effectively. Furthermore, rubrics can be an important guide for students in

Table 20. Example of a performance checklist used in agricultural education

Performance Standard	Satisfactory	Unsatisfactory
Prepares medication for IM administration		
• Selects appropriate medication, syringe, and needle	☐	☐
• Prepares correct dosage	☐	☐
Medication administered correctly		
• Identify correct IM site based on species	☐	☐
• Inserts needle at appropriate angle	☐	☐
• Aspirates	☐	☐
• Removes needle safely	☐	☐
Safety precautions used		
• Utilizes appropriate PPE	☐	☐
• Properly disposes of needle in sharps container	☐	☐
Notes:		

preparation for assessment. Therefore, providing the scoring rubric to students at the beginning of the assignment or project is important.

Constructing Portfolio Assessment

When employing portfolio assessments, rubrics and checklists are the most common measurement tools utilized. General guidelines for developing portfolio assessments are presented in Table 21.

Table 21. General guidelines for developing portfolio assessments

• Ensure that portfolios are perceived by the students as their own selection of work, not just a place to store course assignments. • Student and teacher should work together in decisions of collecting, storing, and evaluating samples. A combination of self-assessment and teacher-led assessment should occur. Student-teacher conferences should occur regularly to assess progress. • Provide clear expectations, steps, and procedures for students in all phases of the portfolio assessment (e.g., selection and storing samples, self-evaluation, etc.). • Provide students multiple opportunities to evaluate their own products. This should be an ongoing process. Each time an evaluation occurs, the evaluations should be dated and attached to the sample.

Developing Assessments in Laboratory Settings

In school-based agricultural education, students are often assessed in classroom, laboratory, and on-the-job settings. Laboratories play a significant role in school-based agricultural education, providing students with opportunities to interact with materials and models and to practice and improve skills. The laboratory offers a unique environment because traditional forms of assessment often do not adequately assess student achievement in such settings. Since laboratory provides a space for authentic and performance-based learning and assessment to occur, the strategies in this chapter for developing and implementing alternative assessments apply directly to laboratory settings.

While laboratory settings provide a rich environment for student learning and development of higher-level thinking and critical skills within the discipline, educators are sometimes reluctant to conduct assessments in the laboratory because of factors related to safety, cost, time, and difficulty in designing authentic assessments. Assessment within the laboratory can be time and resource intensive. It can also be a challenge to develop assessment methods that capture the full range of student learning (e.g., cognitive, psychomotor skills, reflective) that are observable and measurable.

While challenges of accurately assessing students within a laboratory can exist, teachers can employ a variety of laboratory-specific strategies to aid in successful assessment practices within the laboratory. The following is a list of strategies and recommendations for SBAE teachers assessing students within a laboratory:

1. Align learning objectives with laboratory assessment: Teachers must always remember to begin with the learning objectives first, and then design the laboratory assessment (and activities) aligned to those learning objectives. Clarifying the specific learning objectives or outcomes is helpful for the students as well as the teacher.
2. Use appropriate measurements to assess learning: Many times, assessment within a laboratory consists of authentic and/or performance-based assessments. Often, Job Operation Sheets are utilized in laboratories to guide students through learning experiences. These sheets can also be utilized as assessment tools when rubrics, checklists, or traditional assessment items are integrated. Table 22 provides an example of a Job Operation Sheet as an assessment tool for a SMAW welding assignment (Illinois State Board of Education, n.d.). Using reliable and accurate rubrics, checklists, and assessment items are important. See previous sections for more information about how to effectively construct rubrics, checklists, and traditional assessments, which can be applied in laboratory settings as well.
3. Utilize experts to assist in assessment: It can be beneficial at times for teachers to invite experts to assist with assessment, especially when more authentic locations are desired. For example, when assessing industry-specific skills, conducting the assessment at a specific industry location (e.g., meat processing plant, floral shop, local nursery, etc.) or inviting industry professionals to bring supplies and equipment to your own school laboratory can be very beneficial. Finally, when students will be assessed in work-based learning settings, involving the workplace experts in assessment is critical.
4. Traditional assessments: While many authentic and performance-based skills will be assessed in laboratories through alternative assessments, the assessment of some skills and knowledge may be better suited for traditional assessment. For example, when a teacher is interested in assessing students' understanding of safety, or ability to follow laboratory instructions, or the conceptualization of certain scientific principles, a traditional assessment, such a quiz, might be most appropriate.
5. Utilize technology: When appropriate, consider utilizing technology as a way to enhance or replace real laboratory experiences and assessments. Technology can be particularly useful for assessing

students that are absent when specific laboratory activities are conducted (e.g., especially time-intensive preparation) or when pre-laboratory activities or assessments are desired (e.g., flipped instruction). Technology can also be utilized to replace real laboratory experiences using virtual technologies or through simulations. If authentic settings are not practical, simulated environments might be an effective alternative. However, teachers should consider that certain aspects of learning may not be able to be assessed accurately through simulation. For example, assessing agriculture students' ability to identify food samples by conducting a triangle test (i.e., flavor, aroma, texture) would be nearly impossible to do virtually or through simulation. Technology can also support assessment activities by reducing time and workload. For example, electronic laboratory journals or computer-based quizzes can save teachers time in administering, organizing, and scoring student assessments.

6. Use a variety of assessment tools: Since multiple and sometimes complex skills and knowledge are often assessed, utilizing a variety of methods for assessment can be beneficial. Utilizing peer-assessment or self-assessment can also be useful tools for formative assessment within laboratories and authentic workplace settings (see Table 23). Some tools are well-suited and specific to laboratory settings, such as laboratory reports, laboratory notebooks, and laboratory learning logs or journals (UNSW, 2012).
    a. Laboratory reports are another common tool utilized to assess students in laboratories. Laboratory reports can be formatted in a variety of ways and can serve a variety of purposes, but generally they provide opportunities for students to make and record observations, reflect and think critically, and to demonstrate their knowledge or ability related to specific learning objectives. Providing students with templates (e.g., laboratory worksheet) can provide clarity and can be very useful in guiding students through the laboratory activity and assessment (see Table 24).
    b. Laboratory notebooks are utilized by scientists and industry professionals working in laboratory setting, so utilizing laboratory notebooks can be an effective component of authentic assessment. Laboratory notebooks can be physical or electronic in nature and can be an important tool for measuring students' abilities to "do" science like scientists do. Teachers can assess students' ability to properly record their scientific activities, or they can also assess students' knowledge of specific concepts as students respond to questions or prompts within their notebooks (see Table 25).
    c. Learning logs or journals provide students with an opportunity to record personal reflections about their own learning. Like portfolios, learning logs and learning journals can help students assess their own learning progress over time in relation to specific learning objectives. These types of assessments should focus on communication, critical thinking, and other critical skills to be assessed, and not just be a record of routine activities.
    d. Other types of assessment to consider within a laboratory setting might include student presentations, construction of projects or peer-teaching. Regardless of the assessment tool used, teachers should assess the accuracy and consistency of the assessment as it relates to the learning outcomes.

Table 22. Sample job operation sheet as an assessment tool

Shielded Metal Arc Welding, 4x4 beads
Job Operation Sheet

Objective: Students will demonstrate their ability to strike and maintain a SMAW weld for four, four-inch beads.

Laboratory Materials Needed:
Arc Welder with Power Source
Mild Steel Plate 1/8" Thick, Measuring 5 x 7" for Each Student
Masking tape and marker

Materials Needed by Each Student:
Four (4) E6013 Electrodes
PPE:
Welding Helmet- Shade #10
Safety Glasses
Leather Gloves
Other proper protective clothing for SMAW welding
Pliers
Chipping Hammer
Wire Brush

Overview:
After the instructor demonstrates the key steps and procedure for striking and running a four-inch bead, students will gather their required materials as instructed. Working with assigned lab partners, students will practice striking and maintaining an arc to run short beads.

Procedures:
Clean and position the base metal plate flat on the welding table.
Check work connection to table or work piece.
Set the polarity and amperage on the welder appropriately for 1/8" metal using AC current.
Hold the electrode inclined at about a 70-degree angle in the direction of travel.
Strike and establish the arc keeping the electrode about 1/16" to 1/8" distance from the metal plate. Move the electrode across the plate at consistent speeds as demonstrated by the instructor. Run a bead that is four inches in length.
Repeat the process and practice running beads. Students will satisfactorily complete four, four-inch SMAW beads.
Clean each weld by chipping the slag off each bead with a chipping hammer and cleaning with the wire brush. Remember: slag should be chipped away from one's body.
Visually inspect each bead for shape, penetration, and consistency. Identify weaknesses in the beads and reflect on ways in which improvement can be made.
Students will complete a self-assessment and peer-assessment of their own and their lab partner's beads using the SMAW self and peer assessment rubric.
After completing the self- and peer-assessment, students will write their name on the metal by using the masking tape and a marker. The tape should be placed on the opposite side of the beads.
Submit the metal coupons to the instructor as directed to be evaluated.

Evaluation:
Each student will receive a total of three evaluations. One from themselves, one from their lab partner, and one from the instructor. The rubric below will be utilized for all three evaluations

Indicator	Proficient (3)	Approaching (2)	Basic (1)
Width Consistency	Beads are uniform width all along the length of each bead. Each bead has a smooth appearance across the entire length of the bead.	Beads are mostly consistent with one or two uneven areas per bead.	Beads are not consistent. On average, there are three or more areas that are uneven on each bead.
Travel Speed	Beads demonstrate a consistent and accurate speed throughout the entire length of the bead.	Beads have one-two areas (per bead) of inconsistent speed that is either too fast or too slow.	Beads have more than two areas per bead of inconsistent speed that is too fast or too slow.
Arc Length	Spatter is not present or only minimal splatter is present. Less than 10% of the beads show splatter.	Spatter is present, but not excessive on each weld.	Spatter is present on more than half of the beads (each bead).
Penetration	There are well-defined and consistent penetration markings on the underside of the base metal for each bead.	Evidence that the beads penetrate moderately deep. There are penetration markings on the underside of the base metal, but they are inconsistent.	There is no obvious penetration. Beads float on top of the metal. No markings on the underside of the base-metal are present.

Table 22. Continued

Overall Appearance and Safety	Slag is removed and beads are cleaned. No slag is present.	Some slag remains or some beads not fully cleaned.	Beads were not cleaned appropriately. Excessive slag was present.
Safety	All safety criteria and protocols were followed including wearing proper PPE. Followed instructions.	Most safety criteria and protocols were followed including wearing proper PPE or following instructions.	Violated more than one issue with safety protocols not being followed, including not wearing proper PPE or following instructions.
COMMENTS			

Table 23. Laboratory or worksite performance self or peer evaluation

Student/Trainee name ________________			
Evaluator name (if self-evaluation, leave blank): ________________			
On-the-job Skills	High Degree (3)	Medium Degree (2)	Low Degree (1)
Cooperation and Attitude: Observes laboratory or worksite rules, works harmoniously with group members and teacher/ supervisor. Welcomes suggestions for improvement.			
Dependability: Sticks with the job through difficulties, arrives and leaves on time, works hard regardless of whether the supervisor/ teachers is present or not.			
Dress, PPE, and Safety: Wears appropriate clothing and PPE for the laboratory work/ jobs. Demonstrates safe behaviors in the laboratory/ worksite.			
Quality of Work: Shows skill and accuracy with jobs and skills performed. Takes pride in completing jobs with quality.			

Table 24. Laboratory report sample rubric

Level of Performance	Lab Report Rubric Lab Title: ____________________
Advanced (4)	All proficient standards, plus... At least two testable hypotheses have been written and one has been chosen for further investigation. The independent variable, dependent variable, and more than three extraneous variables have been properly identified and all extraneous variables are properly controlled. The conclusion is logically drawn from the data and accurately describes error analysis, applications, and possible further experimentation.
Proficient (3)	The lab report is neatly written, properly titled, and lab headings are identifiable. The question is properly phrased and testable. A testable hypothesis has been chosen for further investigation. The independent variable, dependent variable, and at last two extraneous variables have been properly identified and properly controlled. The procedure is written in a stepwise format and describes how to do a well-controlled experiment. The predicted results are based on logic stemming from the chosen hypothesis and the experiment. The data table(s) is properly formatted and contains accurate, relevant, sufficient data. The computer graph(s) is properly formatted and accurately displays the data. The analysis references the data, identifies patterns and explains the patterns using logic and/or critical thinking. The conclusion is logically drawn from the data and describes two of the following: error analysis, applications, or possible further experimentation.
Basic (2)	The lab report is mostly legible and most lab headings are identifiable. The question is present and testable. At least one hypothesis is present and has been chosen for further investigation. The independent variable, dependent variable, and at least one extraneous variable has mostly been identified and the extraneous variables are somewhat controlled. The procedure mostly describes a controlled experiment. The predicted results are mostly present. The data table(s) is present and contains some relevant data. The graph(s) is present and displays some of the data. The analysis references some of the data and identifies some patterns. The conclusion is mostly logical and drawn from the data.
Below Basic (0-1)	The lab report is not very legible or organized. The question lacks clarity. A hypothesis is not well written or lacks clarity. An attempt to identify the variables has been made. The procedure vaguely describes how to do an experiment. The data table is missing data. The graph displays some of the data. The analysis references the data but needs improvement. The conclusion is present but needs improvement.

Table 25. Laboratory notebook sample holistic grading rubric

Indicator	Mastery Criteria Points Possible (0-4)	Points Earned
Organization and Professional Standards	All required sections are present and are formatted correctly (experiment/ observation title, procedures, and references). All pages are numbered, dated and signed. Notebook is neat and legible. All information recorded in pen not pencil. Proper protocols are followed in regard to errors, empty space, etc.	
Content and Analysis	All methods, observations, and data are recorded completely and neatly presented. Methods section clearly and accurately describes what was done. Results section is written clearly and adequately describes what was observed and what was concluded. Graphs and figures are presented accurately and neatly. Sources of error are discussed.	

Guiding Principles of Assessment

Regardless of the type of assessment, there are some general principles of assessment that should be followed by teachers. Table 26 provides an overview of the guiding principles of effective assessment, regardless of the type of assessment.

Table 26. Guiding principles of assessment

Principle of Assessment	Description
Align Learning Objectives with Assessment	Assessments that are not aligned with the learning objectives defeat the purpose. The point of criterion referenced assessment are to determine how students perform in relation to the learning objective. Student knowledge or skills must be measured and evaluated using the learning objectives as the criteria.
Focus on Validity	Teachers should be certain that the assessment actually measures what it is supposed to measure, and that the interpretation of the results is accurate. This is important in all phases of the assessment process. This chapter provides many examples and explanations related to assessment validity.
Variability	A variety of assessment strategies should be used when appropriate. For example, integrating both performance-based and traditional assessments can be effective. Furthermore, an appropriate balance of formative and summative assessments should be employed.
Minimize Bias and Inconsistency	When scoring student work, teachers should attempt to eliminate unintentional bias. Being overly generous or severe are examples of unintentional (or intentional) bias. While blind scoring is not possible in all situations, eliminating bias whenever possible is a good idea. When possible, teachers should evaluate assessment items without knowing the identity of the student. Evaluating students anonymously, such as blind scoring, can be an effective way to reduce bias. This can be done by using student identification numbers rather than names for example. Additionally, teachers can use rubrics or other scoring guides for objective scoring. Teachers can also evaluate all the students' answers one page at a time/ one question type at a time.
Clarity	Teachers should aim for transparency and clarity in assessments. Students should have a clear understanding of expectations and how they can demonstrate mastery of skills and knowledge. Expectations for student performance should be clear. Assessment items should be clear, concise, and should not cause confusion.
Formative Assessment	Students should receive regular formative feedback before a summative assessment.
Provide Assessment Options	When appropriate, provide students with choices or options for demonstrating mastery. Differentiating assessment by providing options can be an effective way to assess various learners within a class and can increase validity.
Process, Practice, or Apply Before Summative Assessment	Learning activities are different than summative assessment. Learning activities provide students an opportunity to practice or apply learning concepts and skills. Learning activities should serve as practice with formative feedback provided, not as an evaluation of mastery. Teachers should provide students with opportunities to practice, process information, or apply new knowledge and skills before being assessed. This is true with cognitive knowledge as well as psychomotor skills.
Favorable Testing Environment	Distractions can reduce reliability and validity. Testing environments should be quiet and free of distractions. Factors such as room temperature, air circulation, and lighting should be considered and adjusted to meet student needs whenever possible.
Higher-Level Skills and Thinking	If appropriate, teachers should develop assessments that adequately assess higher levels of understanding and complex thinking when possible.
Meaningful Assessments	Teachers should be purposeful in their assessment practices and not just test students just to test students. Furthermore, assessments should be meaningful to students, so they motivate students to perform their best.
Time	Consideration must be taken for the time needed per assessment item type. For example, matching, multiple choice, and completion items: 1 minute per item, drawings, arrangements, identification, and short answer: 2 to 3 minutes per items, essay and performance: 10 or more minutes.

GRADING PRACTICES USED IN SCHOOL-BASED AGRICULTURAL EDUCATION

After assessing students, assigning meaning or a value to that assessment is important. Assigning a grade is one way to assign meaning or value to an assessment or group of assessments. Grades enable students to determine performance, weaknesses, and strengths and enable teachers to evaluate instruction and make instructional decisions. Grades also serve as a motivator for many students to learn. Grades however play a much larger role in today's formal education system. For example, grades are useful for student placement decisions; data for program, school, and district administrators; and communicating student performance in a standardized way to outside audiences. Grades can impact perceptions and decisions of colleges, employers, and other outside audiences towards individual students. Therefore, grading becomes a crucial part of the job of any formal educator. As a result, teachers should consider important factors when establishing their grading practices. The following sections discuss factors to consider when grading.

What Should be Graded?

A common question that teachers are often faced with is what should be graded? While this chapter does not answer that question, it does provide some guiding thoughts to help teachers make that decision for themselves. The first thing to consider is the purpose of assessment and grades. Generally, teachers assess students to document student progress, provide formative feedback to the student, and to inform instructional decisions. Grades should not be used to motivate students to participate in activities, encourage effort, or promote attendance. While these are important, grades should reflect an overall evaluation of student performance, skill, and knowledge in relation to learning objectives. Grading students on attendance or participation in activities generally dilutes the accuracy of evaluations and can negate the overall goal of student evaluation (Vatterott, 2015). For example, what does a grade of an "A" in an agricultural science course actually mean if most of the grade was based solely on the student's willingness to participate in activities? With this in mind, teachers should consider what is essential to be graded in relation to the actual standards the students are being evaluated on.

Within school-based agricultural education, FFA and SAE are important components of a complete program and are often embedded within the state, district, or local learning standards. As a result, the skills specifically stated in the SAE and FFA standards should be treated like other standards students are evaluated on in each course. In other words, grades should generally incorporate learning objectives related to FFA and SAE whenever appropriate.

Other Grading Considerations

While the questions, "What should be graded?" and "Should FFA and SAE be graded?" are important, agriculture teachers should consider other grading situations as well. Table 27 outlines some grading considerations. Many educators hold different opinions regarding the considerations listed below. The list of considerations in Table 27 is not designed to be a rigid guide for the "right way" to go about grading. Rather, this short list is designed to provide recommendations for educators to consider.

Table 27. Other grading considerations for educators to consider

Considerations	Description
Align Assessments to Learning Objectives	Teachers should consider avoiding the grading of things like behavior, attendance, and effort. Teachers should also consider using caution when providing extra credit and bonus points, especially if it is enough to alter the validity of a grade.
Multiple Attempts at Mastery	If assessment is to provide students an opportunity to progress and improve, teachers should consider avoiding the penalization of student for retaking assessments to show mastery. Otherwise, why give formative feedback at all?
Individual Grades	Grades should be based on an individual student's achievement in relation to a learning objective. Teachers should therefore consider avoiding group grades (Kagan, 1996). When cooperative learning occurs, teachers should consider grading students on specific individual tasks related to the learning objective(s) whenever possible.
School-Level Requirements for Grading	For example, a schoolwide grading policy may be set for final exams to serve as x% of final grade in each course. Or no teacher is allowed to give a student a zero grade. It is important for teachers to inquire within their school system to know if any school policies are set on grading.

Common Grading Systems

A grading system is a method of calculating and reporting evaluations of student achievement. The grading system chosen by a teacher should align with what is approved and supported by the local school district. Teachers should understand their school district's policies with grading and reporting grades.

Two common types of grading systems teachers generally employ are weighted grading system and points system. Each system has advantages and disadvantages. In a **weighted grading system**, the teacher calculates each assessment as a percentage of a final grade worth 100%. The maximum number of points of any given assessment can be any value. Each assessment is assigned to a category, which is a percentage (weight) of the final grade. One advantage of using a weighted system is the ability to emphasize certain tasks as more important than others. For example, a final exam where 50 points are awarded can have more weight than a daily quiz of 50 points in a weighted system. Additionally, with the weighted grading system, teachers need not worry about the number of points on each assessment, attempting to get the right number of points to work out. So, a quiz of 30 points and another quiz of 90 points can essentially contribute equally to the final grade in a weighted grading system. This also means teachers need not worry about adjusting points each time a new assignment is added, or one is taken away. Finally, with a weighted grading system, it is easy to change how much weight to assign particular assignments without complex mathematical calculations and conversions.

With a **points system**, a cumulation of points earned is the focus. Final grades are calculated by summating a student's score on all assessments and dividing by the sum of the total possible points from all assessments. One advantage of the points system is how easy it is for students to understand. In contrast to a weighted system, explaining to students and parents how individual assignments contribute to a weighted category of the overall grade can be challenging. Additionally, teachers embracing the philosophy of standards-based grading should consider each academic standard as equal, and consequently there would be no need for weighting any assessment.

SUMMARY

Terms used in educational assessment help to understand the broad scope of assessment in education. Assessment is critical to determine if and at what level students learned the criterion. Effective assessment may involve traditional methods or alternative methods. This chapter discussed techniques for helping teachers effectively assess student learning through various forms of assessment. After assessing students, assigning meaning or a value to that assessment is important. Grading is that value. This chapter discussed common grading practices used in school-based agricultural education and provided thoughts and suggestions for teachers regarding grades and grading practices. In conclusion, assessment is critical, and following the guidelines and recommendations in this chapter can help teachers effectively develop and administer valid assessments, and effectively interpret and report scores and grades.

REVIEW QUESTIONS

1. What is the difference between assessment, measurement, and evaluation?
2. What is validity in assessment and why is it important? What can a teacher do to ensure valid assessments?
3. Explain the key differences between formative and summative assessment. When would a teacher use each of these? Provide specific examples of each type of assessment.
4. Compare and contrast traditional assessments and alternative assessments including the advantages and disadvantages of each. Provide examples within school-based agricultural education when each type of assessment would be most appropriate.
5. Utilizing the guidelines listed in this chapter, construct a traditional assessment utilizing selected-response (alternative-choice, multiple choice, matching) and constructed-response (short answer, essay) type of items.
6. Utilizing the guidelines listed in this chapter, construct an alternative assessment specifying the student task and providing a scoring guide. Construct the scoring guide (e.g., rubric or checklist) to measure the learning outcome specified in the task.
7. List at least three common grading issues to consider as a teacher.
8. Compare the points and weighted grading systems. Which type of grading system do you prefer and why?

REFERENCES

Fellenz, M. R. (2004). Using assessment to support higher level learning: The multiple-choice item development assignment. *Assessment & Evaluation in Higher Education, 29*(6), 703–719. doi:10.1080/0260293042000227245

Illinois State Board of Education. (n.d.). *Example assessment: Shielded metal arc welding*. Illinois State Board of Education Local Assessment Support Project. https://www.isbe.net/Documents/shielded-metal-arc-welding.pdf

International Test Commission. (2006). International guidelines on computer-based and internet-delivered testing. *International Journal of Testing*, *6*(2), 143–171. doi:10.120715327574ijt0602_4

Kagan, S. (1996). Avoiding the Group-Grades Trap. *Learning*, *24*(4), 56–58.

National F. F. A. Organization. (2021). *National FFA career and leadership development events handbook.* Agricultural Sales Career Development Events. https://www.ffa.org/participate/cdes/agricultural-sales/

Popham, W. J. (2020). *Classroom assessment: What teachers need to know* (9th ed.). Pearson.

University of New South Wales. (2012, August 10). *Assessing laboratory learning*. Assessment Toolkit. https://teaching.unsw.edu.au/assessing-laboratory-learning

Vatterott, C. (2015). *Rethinking grading: Meaningful assessment for standards-based learning*. ASCD.

ADDITIONAL READING

Burke, K. (2010). *Balanced assessment: From formative to summative*. Solution Tree Press.

Center on Education and Training for Employment. (2002). *Assess student performance: Skills*. American Association for Vocational Instructional Materials.

Center on Education and Training for Employment. (2003). *Assess student performance: Knowledge*. American Association for Vocational Instructional Materials.

Center on Education and Training for Employment. (2003). *Determine student grades*. American Association for Vocational Instructional Materials.

Fisher, D., & Frey, N. (2014). *Checking for understanding: Formative assessment techniques for your classroom*. ASCD.

Stoughton, A. L., & Myers, B. E. (2008). *Creating and working with rubrics*. EDIS. doi:10.32473/edis-wc069-2008

Tomlinson, C. A., & Moon, T. R. (2013). *Assessment and student success in a differentiated classroom*. ASCD.

Warner, W. J., & Myers, B. E. (2006). *Evaluating learning in laboratory settings*. EDIS.

Wormeli, R. (2018). *Fair isn't always equal: Assessing & grading in the differentiated classroom*. Stenhouse Publishers.

KEY TERMS AND DEFINITIONS

Alternative Assessment: Assessments that are different in form than traditional (e.g., quiz) assessments.

Analytic Rubric: A detailed rubric that provides a description for each level of mastery for each criterion being assessed.

Assessment: The process (the verb) or tool (the noun) used for collecting information to monitor progress and make educational decisions.

Authentic Assessments: Assessments that focus on real-world situations as the context for specific skills and tasks to be performed and assessed.

Checklist: Basic rubrics in which only two performance levels are present (e.g., yes/no; pass/fail) for each criterion.

Criterion-Referenced Testing: Testing to determine a student's level of mastery or performance for a pre-established criterion.

Diagnostic Assessments: A form of pre-assessment that evaluates how much knowledge or skill a student has before a new lesson begins.

Evaluation: The procedure used to judge a student's skills or knowledge in relation to predetermined criteria.

Formative Assessment: Assessment that is conducted during instruction or a learning activity with the goal of improving learning.

Holistic Rubric: General rubrics that consist of a description for only one single level of mastery with a number value corresponding to the level of mastery for each criterion.

Measurement: The systematic assignment of numbers or names to student learning.

Norm-Referenced Testing: Assessments that focus on a student's performance in relation to other students in a group (e.g., the norm).

Performance-Based Assessment: Assessments that enable students to demonstrate their knowledge and skills as they perform a specific task.

Points System Grading: A system of grading in which the points of each assessment are summated to calculate the final grade.

Portfolio Assessment: Assessments that involve a systematic collection of student-developed products in which students highlight individual growth, progress, and accomplishments over time.

Reliability: The consistency of results from assessments.

Summative Assessment: Assessment that occurs at the end of instruction or a learning activity with the goal of making final evaluations on student performance.

Traditional Assessment: Assessments that use traditional methods (e.g., quiz) as a means for evaluating student performance.

Validity: The extent in which the assessment measures what it purports to measure.

Weighted Grading: A system of grading in which the final grades are calculated by summating a student's score on all assessments and dividing by the sum of the total possible points from all assessments.

Chapter 11
Evaluating a School-Based Agricultural Education Program

Jonathan D. Ulmer
https://orcid.org/0000-0002-5078-4822
Kansas State University, USA

Kevin W. Curry Jr.
https://orcid.org/0000-0002-8731-6256
Pennsylvania State University, USA

ABSTRACT

Systems of education intentionally evaluate students and teachers in the teaching and learning process. For many of the same reasons, the case can be made that programs should undergo systematic evaluation to discover where and how growth can occur. The purpose of this chapter is to provide a practical overview of the SBAE program evaluation process. The authors are careful not to be overly prescriptive in the recommendations for how these should be conducted, in part because effective program evaluation is tailored to the local program. Still, the case is made for why program evaluations should be conducted, what the research indicates good program evaluations look like, and how one should go about conducting one at the local level.

INTRODUCTION

In most states, a school-based agricultural education program in good standing must have components of all three circles within the Agricultural Education model (Instruction, FFA, and SAE). While being in good standing is important for funding and participation, there is little accountability on the quality of each circle. A program should be assessed periodically to determine if success is being achieved, changes to be made, and how growth should occur.

The evaluation of an agricultural education program is primarily a tool for the local school district but in some states, evaluations are required. A difficult fact for Agricultural Education is each state has differing expectations. Some states expect specific processes and outcomes and at the opposite end,

DOI: 10.4018/978-1-6684-3420-8.ch011

some states have very little guidance for any Career and Technical Education program. Fortunately, Agricultural Education does have National Quality Program Standards and National Agriculture, Food, and Natural Resources Content Standards. While both sets of standards are important, this chapter will focus on the role of the Program Standards.

The objectives of this chapter are to:

1. Identify the benefits of conducting a SBAE Program Evaluation
2. Discuss the different theories and models on program evaluation in Career and Technical Education as it pertains to secondary SBAE programs
3. Outline the process of conducting a Program Evaluation

BENEFITS OF CONDUCTING A PROGRAM EVALUATION?

It is always difficult for a person to put themselves in a situation where they feel criticized. While a program evaluation is putting oneself out there, it should be seen as a process for improvement and not as criticism. The ultimate goal of program evaluation is to collect information that will assist in the improvement of a program.

Every teacher has multiple roles to fulfill but especially an agriculture teacher. A Google search will lead you to find many lists of teacher roles. One such list of broad roles of teachers from Harden and Crosby (2000) is: Information provider, Resource developer, Planner, Assessor, Facilitator, and Role model.

As a teacher reviews each of these roles, one may wonder how they should know if they are doing their jobs. Are each of these being fulfilled? How could a teacher do each of these better? An evaluation of a program should give insight to determine if roles are being met or if they could be done better. All teachers can see each of the roles of a teacher in their daily instructional life. Interestingly, an agriculture teacher should be able to see each of the roles within the three circles of Agricultural Education.

In school-based agricultural education (SBAE), the development of students occurs in formal instruction, a student organization, and an actual or simulated workplace. Teachers are continually accessing their performance in the classroom through student performance and feedback. Performance in advising and guiding the FFA chapter and Supervised Agricultural Experiences (SAE) is not as easily assessed. To gain insight into the FFA and SAE a teacher must look beyond their students and collect information from multiple sources, many outside of the school itself. The community holds much insight as to the performance of the program and the direction it could go.

Knowing that every SBAE program must reflect the community in which it is held, adds an additional layer of accountability. All three circles of the program should either reflect or impact the community at large. Instruction should be determined by interest and industry, activities of the FFA chapter should have a positive influence on the surrounding people, and SAEs should be conducted in surrounding industries. All of these combined elements demand the need for a program assessment that incorporates all facets of the agricultural education complexity. Therefore, a systematic approach is required to give insight to the program, a *program evaluation*.

Just as an evaluation can provide a program with information for improvement, it can also validate the job a teacher has been doing. A systematic evaluation can reveal that the teacher is leading students to learn and impact the local community. Many evaluations have led to the expansion of teachers' contracts or the addition of teachers. While a program evaluation can show areas of improvement, it can also reveal opportunities for growth and planning for the future.

With the need to assess the SBAE program clearly established, it is important to understand the foundation of a program evaluation.

THEORIES AND MODELS ON PROGRAM EVALUATION

The purpose of program evaluation is for development, to assist in the understanding of the workings of a target program. It is different from research in that it focuses on the workings, structures, and outcomes of a program and not generalizable variables or outcomes that are of common interest across situations. The evaluation of a full program reveals the positives and negatives of the work a program is conducting. It assists people in and out of the program to understand what is being done well and what could be improved. It makes judgments and reveals perceptions while telling the story of the program. All of this should be done in an objective manner that is reliable and minimizes bias.

Countless textbooks have been written about *program evaluation,* each one giving facts, tips, steps, and philosophy. *Program Evaluation* is a discipline that is widely used and sometimes even required in systems such as the federal government. For example, when the United States Department of Agriculture, National Institute of Food and Agriculture awards an external entity, like a university, a grant, the entity must evaluate the developed program and the use of the money. In reality, people evaluate almost everything they do. At lunch, you may evaluate the food you ate and determine if you will order it again. As a teacher, you evaluate every lesson you teach and determine what could be better next time. The evaluation of a program is simply more systematic than your lunch and with larger complexity and potential outcomes.

Program Evaluation is the systematic process of collecting information (data) with the intent of making decisions about a program. Patton (2008) put it in the simple terms of "What? So what? And now what?" (p 5). While textbooks are focused on the *discipline* of program evaluation, this chapter uses their information to help guide the evaluation of an agricultural education program. The information is easily transferable to an SBAE context.

Chen (2015) identified five elements of a program: inputs, transformation, output, environment, and feedback. Inputs are the resources that are put into a program to make change. Transformation is the operation of the program or the work that the program does to turn inputs into outputs. Outputs are the results, the desired changes from the program. The Environment is the elements that influence the program from the outside. Finally, Feedback is the process of getting information from the outputs to make change over the program.

For a practical view, one can visualize an Agricultural Education program across Chen's (2015) five elements (Figure 1). The inputs of a program could be visualized as the faculty, funding, equipment, and other elements needed to offer SBAE. The outputs can be seen as the results of the program, graduates, their knowledge gain, and community impact. In between the inputs and outputs there are transformations. In Agricultural Education one can visualize the transformation occurring through the three-circle model: teaching (classroom and laboratory), work-based learning (SAE), and the student organization (FFA). All of these elements must happen within an environment, potentially defined as the community, school, industry, and other factors that impact from outside the program. Finally, the feedback system in a

Figure 1. A systems view of an agricultural education program (Chen, 2015)

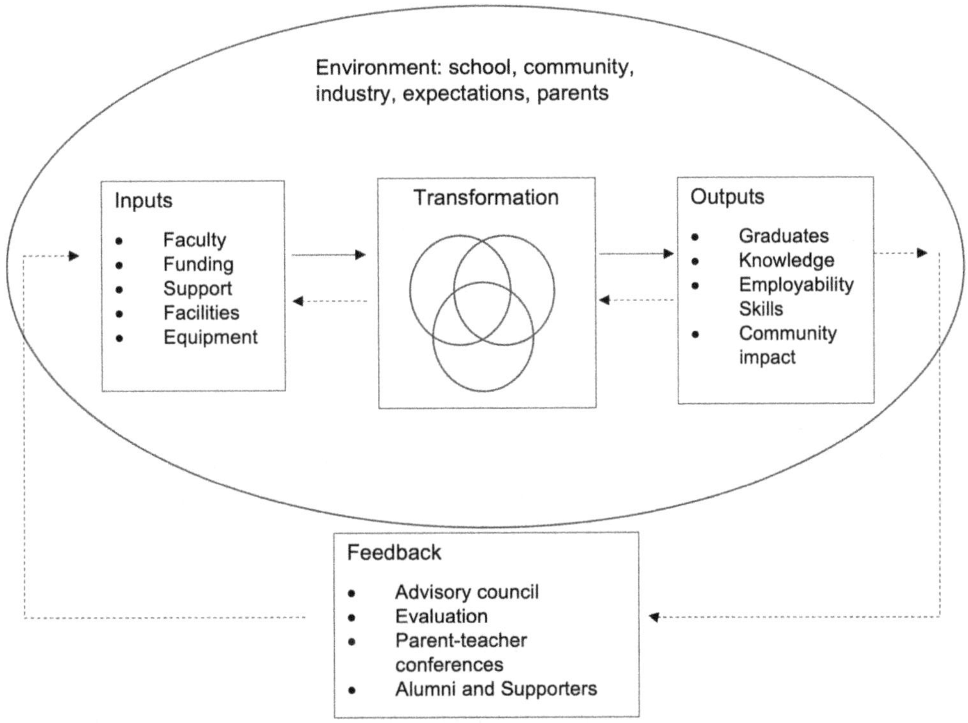

program could be student evaluations, parent-teacher conferences, or the advisory council. To view Chen's model with SBAE it is easy to see how an evaluation can reveal needed information for improvement.

An evaluation of a program generally has two major outcomes; the first is to identify areas of improvement, changes and/or additions, and the second is to identify processes that should be discontinued. Many educators use a stop, start, and continue process with their students to receive input about classes. Evaluating a program is similar; the scope is bigger, and input is from multiple sources.

Regardless of scope and size, thousands of programs have been evaluated over the past several decades and many of the recommendations have gone unused. We can look at social programs that were not effective and when evaluated the recommendation was to end the program and the program lives on. Countless school programs have been evaluated with outcomes that speak to expansion and the programs do not change. The largest challenge facing an evaluation of a program is creating a usable product: an evaluation must have utility.

Utility is one of the four Standards for Evaluation, they also include: Feasibility, Propriety, and Accuracy (Joint Committee on Standards for Educational Evaluation, 1994). An evaluation should be conducted to meet each of the four standards. Accuracy is to be confident that an evaluation collects and conveys information in the truest format. In addition to having confidence in the information, it must be collected in the most ethical way possible and have high regard for those involved. Legality and ethics are held within the standard of Propriety. Finally, any evaluation must be realistic and prudent, in other words, Feasibility. Having a realistic process is also influenced by timing.

While a full program evaluation is often conducted on a rotation of five to seven years, a smaller less extensive evaluation could be conducted on an annual basis. Collecting information from students, administrators, and the advisory committee can help programs identify smaller, simpler changes that can be made on an annual basis. Part of the annual evaluation may be a needs assessment completed by the current or incoming students.

It is important to assess school-based agricultural education programs on a small scale annually or a full evaluation on rotation. Program Evaluation is the discipline of systematically collecting data to make decisions about a program. In Agricultural Education, program evaluation is using the resources to understand if each aspect of the program is meeting the needs of the students and communities while best using inputs and creating strong outputs. Creating a systematic approach is the key to a quality evaluation.

HOW DO YOU CONDUCT A PROGRAM EVALUATION?

While the details and specifics of a program evaluation will vary from program to program, any quality SBAE program evaluation should include the following four phases: Engage, Design, Collect, and Utilize. The rest of this chapter is organized with these phases as a framework.

Phase 1: Engage

The Importance of Engaging Stakeholders

Working with program partners affords many advantages during the program evaluation process. The perspective of those outside of the "program bubble" can identify strengths, weaknesses, and opportunities for growth that can be difficult to see from within. The first phase of engage is rooted in the notion of being open-minded; that information and unique perspective are to be acquired if you are humble enough to ask for it from those who can offer their expertise to the process.

The Ag Teacher

The SBAE teacher plays a unique and pivotal role in a program evaluation. They serve as the communication hub for all partners involved and help move the process along. States vary in how they incentivize and or require program evaluations, if at all, so it is most likely that the responsibility lies with the local agriculture teacher to make this process a reality. The bottom line here is that if external stakeholders are to be folded into the process or program evaluation, it is most likely the agriculture teacher that initiates the relationship.

The Advisory Council

The advisory council is defined locally in diverse ways, but in general, they are composed of industry leaders, parents, community members, administrators, and others who have an interest in the long-term success of a SBAE program. Several states recommend the composition of advisory councils by mandating certain roles in the community are represented. These include, for example, having a local governmental official (i.e., forest ranger), a member of the chamber of commerce, or a member of a nearby post-secondary institution. Whether these advisory council compositions are outlined at the state level or not, having a diverse group of people from a variety of stakeholder perspectives is vital for this group to carry out its purpose. The advisory committee is tasked with providing recommendations pertaining to the goals and objectives of the program. They can be particularly helpful at aligning program goals to local community needs. For example, industry representatives could highlight the local need for biotechnology sector jobs in food manufacturing and the accompanying entry-level lab skills needed to acquire those jobs. This group will play an important role in all three of the upcoming phases, so time should be dedicated to brainstorming potential members of the advisory council and personally asking community members to serve is time well spent.

Greater Community

While the advisory council is rather formal in nature, there are still many individuals in the local community who can provide feedback and assistance to the process. Support from local administration and school personnel, for example, can help bring legitimacy to the advisory council. Other community leaders and industry representatives can review advisory council findings and recommendations to ensure issues like diversity, equity, and inclusion are met.

Phase 2: Design

The core function of the design phase is to craft a plan with intentionality before the evaluation begins. What are the goals/objectives of the evaluation? Where do stakeholders like the advisory council have input? What types of data will be collected? What program standards, if any, will the evaluation be aligned with? These questions need consideration before the collection of data begins and need to be written down in a way that can be followed by all parties involved in the process.

Determine Criteria for evaluation

One of the first tasks of the advisory council is to identify the criteria for evaluation. This group of stakeholders should have a voice in the types of measurements used in the evaluation process. The key point to notice here is that there is outside input in the evaluation process before the evaluation starts. Since program evaluations are only as good as the data collected, it is essential, then that the advisory council aids in informing what types of data are indicators of program success. The criteria this group establishes should involve what they think is most important to program success. Potential criteria include, but are not limited to, program standards, test scores, industry standards, certifications, program completers, etc.

Program vs Academic standards

An important distinction to make in this chapter is the difference between program standards and academic standards (discussed in chapter 10). Although academic standards (i.e., AFNR, Common Core, NGSS) can be referenced and considered in a program evaluation, it is the program standards that are of primary concern. Program standards address the entirety of the complete SBAE program beyond the learning objectives in the classroom. Program standards consider experiential learning (SAE) and leadership opportunities for students (FFA), in addition to alignment to community needs just to name a few. In short, academic standards become one of many things to evaluate when utilizing program standards to conduct a program evaluation.

Current "Program Standards" Models

There are many examples of program standards that exist that can be excellent starting points in the program evaluation process. In Pennsylvania, for example, there is a statewide evaluation process to receive state funding, wherein a program would submit evidence that they are meeting minimum standards for program quality. Several other states have similar "program standards" that are given as recommended structures for program improvement. Authored by the National Council for Agricultural Education in 2009, the National Quality Program Standards (NQPS) is considered the benchmark of SBAE program evaluation standards. The NQPS standard statements are found in Table 1 below. Not surprisingly, the first three NQPS standards are aligned to the three-circle model of agricultural education (Classroom, FFA, and SAE). The other four standards outline ancillary aspects of successful programs including the need for systematic program evaluation as described in this chapter.

These types of program standard frameworks can be excellent starting points to undergird the design and function of program evaluation, and aid in providing external validation as they were developed by state or national efforts.

Phase 3: Collect

Data

Broadly speaking, there are two types of data: quantitative and qualitative. Quantitative data are data that can be easily expressed in numbers and statistics. Examples include information collected on surveys (i.e., mean of the degree to which stakeholders agree the program is meeting community needs) or descriptive accounts of program components (i.e., number of program completers). Qualitative data, on the other hand, are data that describes phenomenon that is not easily measured with numbers. Examples of qualitative data include data retrieved from interviews, focus groups, or document analysis. In the case of interviews or focus groups, there is often a semi structured list of questions to ask the individual (in the case of an interview) or the group of participants (in the case of a focus group). Even though the data that is derived from this process cannot be presented in numbers, the themes that surface can be powerful additions to understanding the nuance of complex issues.

In addition to uncovering basic descriptive data about the program (number of proficiency award winners, hours invested in SAE, percentage of Ag Ed students with active FFA membership, etc.) there are a variety of questions you can ask of stakeholders to gain their perceptions of program impact. Four common types of questions to ask are Likert-type scales, Semantic differential, Rank order, and Open Entry. Likert-type questions include a series of statements that respondents can choose to rate their degree of agreement. Similarly, semantic differential questions ask respondents to rate their position on a dichotomous scale of two adjectives. Rank order questions can be especially useful in determining the preference of respondents among a list of options, as they rank the options first to last. Examples of these types of questions are found in Figure 2. Finally, open entry questions prompt respondents to write their responses in their own words with space for them to write it. An example might be: What specific areas for improvement do you see for the program? With qualitative data collections where stakeholders are asked to write their thoughts, the writings should be reviewed for common ideas or themes. For instance, if 20 of the stakeholders wrote about service-learning, then a theme within your data could be that the program should conduct more service-learning experiences.

Table 1. National quality program standards (The National Council for Agricultural Education, 2017)

Standard 1A: Program Design and Instruction - Curriculum & Program Design
Standard Statement: A standards-based curriculum in agriculture, food, and natural resource education is delivered through programs of study that incorporates classroom and laboratory instruction; experiential, project, and work-based learning through SAE and leadership and personal development through FFA.
Standard 1B: Program Design and Instruction – Instruction
Standard Statement: Programs promote academic achievement and technical skill attainment of all students
Standard 1C: Program Design and Instruction – Facilities & Equipment
Standard Statement: The facilities and equipment support implementation of the program and curriculum by providing all students opportunities for the development and application of knowledge and skills.
Standard 1D: Program Design and Instruction – Assessment
Standard Statement: A system of needs assessment and evaluation provides information necessary for continual program development and improvement.
Standard 2: Experiential, Project, and Work-Based Learning Through SAE
Standard Statement: Student learning (or instruction) is enhanced through continuous experiential, project, and work-based learning through SAE.
Standard 3: Leadership and Personal Development Through FFA
Standard Statement: All students participate in intra-curricular leadership and personal development through FFA.
Standard 4: School and Community Partnerships
Standard Statement: School and community partners are engaged in developing and supporting a quality program.
Standard 5: Marketing
Standard Statement: Key stakeholders are continually asked, involved, recognized and informed about all components of the integrated program.
Standard 6: Certified Agriculture Teachers and Professional Growth
Standard Statement: Competent and technically certified agriculture, food and natural resource teachers provide the core of the program.
Standard 7: Program Planning and Evaluation
Standard Statement: A system of needs assessment and evaluation provides information necessary for continual program development and improvement.

Needs Assessment

A needs assessment is a careful consideration of the opportunities and challenges which exist in the local community. You could make the argument that the needs assessment belongs in phase 2 (design) if you consider the needs assessment as a planning tool. As a planning tool, the needs assessment helps guide the design of collected data. You could alternatively view the needs assessment as a data collection process. It is described here in Phase 3 (collection) because it can serve as a basis to judge whether a program is meeting the needs of the community. The needs assessment is a gauge of the local industries in agriculture and other related postsecondary opportunities as a way to understand demand in order to best design the supply (in this case students of SBAE programs). The type of data collected here as well as the audiences you collect them from, is heavily dependent on the local agriculture industries in the community, but some general questions to consider are included in Table 2.

Figure 2. Survey question type examples

Likert-Type Scales

Classes being taught are reflective of the local industry

Strongly Disagree	Disagree	Undecided	Agree	Strongly Agree
1	2	3	4	5

The agriculture teacher treats all students with respect

Never	Rarely	Seldom	Sometimes	Usually	Always
1	2	3	4	5	6

The annual FFA community clean-up efforts are...

Not Important	Of Little Importance	Somewhat Important	Important	Very Important
1	2	3	4	5

Express the level of satisfaction with the effort of the FFA during the community clean-up.

Highly Dissatisfied	Dissatisfied	Neutral	Satisfied	Highly Satisfied
1	2	3	4	5

Express the level of satisfaction with the effort of the FFA during the community clean-up.

Completely Dissatisfied	Mostly Dissatisfied	Somewhat Dissatisfied	Neither Satisfied or Dissatisfied	Somewhat Satisfied	Mostly Satisfied	Completely Satisfied
1	2	3	4	5	6	7

The instruction of welding techniques is...

Not Effective at All	Slightly Effective	Moderately Effective	Very Effective	Extremely Effective
1	2	3	4	5

The quality of products created by agricultural education students is...

Far Below Average	Somewhat below average	Average	Somewhat above average	Far above average
1	2	3	4	5

How likely are you to recommend taking an agriculture class to your child?

0 1 2 3 4 6 7 8 9 10

Semantic Differential

When I recall my SAE visits, the teacher's supervision was...

Good ___:___:___:___:___:___:___ Bad
Respectful ___:___:___:___:___:___:___ Inconsiderate
Impactful ___:___:___:___:___:___:___ Trivial

Rank Order

Rank the following jobs with the 1 being the highest demand over the next five years and 5 being the lowest in demand.

_____ Laboratory manager
_____ Laboratory technician
_____ Vet technician
_____ Crop scout

Techniques to Improve Responses

Once the types of data to be collected have been established, it is important to determine how the data will be collected. Surveys are commonly collected in three methods: in person, through the internet, or via the mail. The method used is often dictated by the population that is targeted. For instance, if information is needed from students, the survey instrument is often distributed in person. A teacher may

Table 2. Examples of needs assessment data

Questions for Students	Questions for Industry	Data to search for in existing databases
What agricultural content areas are you interested in?	What jobs will be in demand over the next 10 years?	Which agricultural industries have the greatest economic impact locally?
What are your plans after high school?	What skills are needed of new hires?	What are the educational/certification requirements for agricultural jobs in the area?
What job experiences/internships are you interested in	What skills do see lacking in new hires?	Number of jobs open by industry sector

pass out the instrument during a class or FFA chapter meeting, allow the students to complete it, and then collect them at the end of the time. If information is being collected from parents, the school may have email addresses for each family or even parent. An instrument could be converted into an online survey system for distribution. A short list of popular platforms which have free or premium versions to administer surveys in the cloud include Google forms, Microsoft forms, Survey Monkey, and Qualtrics. Finally, businesses in the community may be the target of the data collection. If a particular person or position in the business is the desired participant, a packet including a letter of invitation and description of the process, the instrument, and a self-addressed, stamped envelope for return of the instrument should be sent. When collecting data through the internet and mail, evaluators often struggle to get as many responses back as possible. Research has been conducted to identify techniques that can increase response rate (Dillman, Smyth, & Christian, 2014). Sending a pre-notification that the survey is to follow, two or three reminders, providing incentives, and using different methods such as sending an email after mailing a packet or mailing a packet after an email are all proven ways to increase the number of respondents. All communications should state the importance of participation to the success of the program. No matter what techniques are used, it is strongly recommended to stress the importance and relationship in all communications.

Evaluators may choose to collect data from a sample rather than the whole population. Usually, all students and parents will be asked to participate, but sometimes the community is too large to include everyone. Statistically speaking, there is a number of participants that can represent a population. Several online sample size calculators exist to help evaluators understand the number within their sample they will need to represent their target population. Samples are often used when the population is too big to reach everyone, or the budget is not big enough for the expense of a large population.

Validity and Trustworthiness

It is worth noting here in the collection phase that there should be consideration as to the quality of the data being collected. In quantitative data collection methods, for example, the *validity* of the survey instruments indicates the degree to which the survey measures what it intends to measure and whether the questions asked canvas the complexities of the phenomenon being studied. Likewise in qualitative data collection methods, you would want to ensure *trustworthiness* of your data. Trustworthiness is how confident you are in the truth of the qualitative data after you collect and analyze it. In both cases of qualitative or quantitative data, the goal is to design measures that are of the highest quality so that

Table 3. Examples of accounting items in each circle

Instruction	FFA	SAE
Total number of students	Number of CDE/LDE attended	Number of foundational SAEs
Enrollment in each class	Number of students in each CDE/LDE	Number of Immersion SAEs
Program completers	Hours of service	Number of SAE visits
Student industry certifications earned	Number of events	Number of Proficiency applications
Guest speakers/field trips	Number of meetings	Number and level (local, district...) of Proficiency awards (bronze, silver...)
Professional development hours for teachers	Number of Degrees	Economic Impact of SAEs
Teacher instructional certifications earned	Area/State/National Level leadership positions held	Total hours invested in SAEs

there can be confidence in final conclusions. A good starting point to ensure validity or trustworthiness would be to have several people review the measures in advance of collection.

Program Data

Sometimes in the program evaluation process it is easy to overlook data that is right in front of us. Agricultural educators have access to many forms of data that can give insight to the program. The work within each circle of the agricultural education model has a diverse array of indicators that could be included in the evaluation. Sources of the data could include National Chapter Award Applications, instructional and FFA budgets, student record books, and programs used to track SAEs. Table 3 contains a short list of example indicators.

In addition to the data that can be found in documents, there are many documents within a SBAE program that should be reviewed for information and potential changes. It is advisable to have a committee, maybe the advisory committee, to review documents as external eyes. Those organizational documents could include Constitution and Bylaws, Program of Activities, chapter meeting agendas and minutes, SAE visit reports, and officer and member training manuals.

Report

For the sake of record-keeping for future reference, and to formalize the results of the data collection process for the advisory council, it is highly recommended to produce a written report. This report should be shared not only with the stakeholders engaged in the program evaluation process, but also those in the local school and community. A best practice is to engage the Advisory Council as a partial external point of view. As the Council reviews the data and findings, their perspective may reveal recommendations that are difficult for teachers to visualize. While the report is beneficial for the annual (short term) evaluation, they are particularly important for the full evaluations that occur every five to seven years. The written report can serve as the starting point for future evaluations and helps bring legitimacy to the process. The report serves as the final step in the data collection process as it should reflect the data that were collected.

Phase 4: Utilize

Putting the Evaluation to Work

If you believe the purpose of program evaluation is to fundamentally better the school-based agricultural education program, then phase 5 (utilize) is where the rubber meets the road and makes the careful planning and work described previously worth the effort. In the utilize phase, the process to interpret the findings from the report in practical ways begins. The first step in this phase is to meaningfully interpret the findings from the report. Survey data from the report such as "100 SBAE students", "Course offerings only in Horticultural Science 1 and 2"; coupled with needs assessment data of large numbers of local vocational positions in dairy and veterinary science; and focus group data asserting the desire of local vet clinics to engage more in SAE opportunities, might lead to the conclusion that animal science courses are due to be added to the scope and sequence.

In addition to making some conclusions about data, programs can find it helpful to creatively disseminate the findings of the evaluation with important stakeholders. This might be an executive summary, an infographic, a brochure, or a presentation that highlights the key findings of the report. What this looks like probably depends on the audience, but interested parties like a school board, an FFA Alumni group, or the local farm bureau would all benefit from having an easily digestible form of the data that is collected.

Making Changes to Practice

Once the report has been reviewed and conclusions made, it is advised to create an action plan to ensure there is follow-through. Just as good objectives are SMART in nature (specific, measurable, attainable, realistic, and time-bound), the action plan should be similarly constructed to avoid crafting lofty aspirations with no real commitment to make it happen. It is important, then, to revisit the notion of not being afraid to make change. When the data speaks to the need for change, trust in the process and people you have brought into the evaluation to make the actionable steps necessary to move the program forward. It is also recommended to entertain the notion of making the report and accompanying action plan public. By sharing with your administration and local community, you help ensure accountability for change and help justify the large investment of time from all parties involved.

SUMMARY

In summary, the evaluation of a program is important to assure that the desired outcomes are being met and that the program is being a good steward of the resources. Teachers can identify program elements to change or justify program expansion. A Program Evaluation is a systematic approach to collecting data that can be used to make decisions. An evaluation should consider the inputs, outputs, transformation, environment, and feedback. Each of these elements can lead us to insight to the system. There are four broad phases to a SBAE program evaluation. The first is to *engage* local program partners in the process. Second, is to *design* the process of how the program evaluation will be conducted, including which data will be collected. The third phase is to *collect* data and draft a findings report. Finally, the fourth phase is to *utilize* the findings to enact needed changes via an action plan. It was intentional in this section to not overly prescribe detailed steps on each of the four phases in part because one of the foundational tenants of a successful program evaluation is being locally driven. The actors involved at the local level in pulling this together should have the loudest voices in who is engaged, how the evaluation is designed, the extent to which data is collected, and what to do with the findings. This chapter has provided a framework for how to go about this process of conducting a program evaluation which is centered on isolating recommendations for improvement and based on what the literature tells us good program evaluation looks like.

REVIEW QUESTIONS

1. What are the purposes for evaluating a program?
2. Describe Agricultural Education in the five elements of a program.
3. What are the four standards for Evaluation?
4. What is the role of the advisory council in program evaluation?
5. How should a program define stakeholders when conducting a program evaluation?
6. Describe each of the four phases of a quality SBAE program evaluation.
7. What is the role of the agriculture teacher in the program evaluation?

REFERENCES

Chen, H. T. (2015). *Practical Program Evaluation: Theory-Driven Evaluation and the Integrated Evaluation Perspective* (2nd ed.). Sage Publications, Inc.

Dillman, D. A., Smyth, J. D., & Christian, L. M. (2014). *Internet, Phone, Mail, and Mixed-Mode Surveys: The Tailored Design Method* (4th ed.). John Wiley & Sons, Inc.

Fournier, D. M. (2005). Evaluation. In S. Mathison (Ed.), *Encyclopedia of Evaluation*. Sage Publications, Inc. doi:10.4135/9781412950558.n179

Harden, R. M., & Crosby, J. (2000). AMEE guide no 20: The good teacher is more than a lecturer-the twelve roles of the teacher. *Medical Teacher*, *22*(4), 334–347. doi:10.1080/014215900409429

Joint Committee on Standards for Educational Evaluation. (1994). *The Program Evaluation Standards*. Sage Publications, Inc.

Patton, M. Q. (2008). *Utilization-Focused Evaluation* (4th ed.). Sage Publications, Inc.

Tyler, R. W., Gagné, R. M., & Scriven, M. (1967). *Perspectives of Curriculum Evaluation*. Rand McNally & Company.

Chapter 12
Designing and Implementing Clinical and Field Experiences

Jon W. Ramsey
https://orcid.org/0000-0003-3424-9237
Oklahoma State University, USA

ABSTRACT

The knowledge, skills, and dispositions needed to lead and deliver school-based agricultural education (SBAE) is introduced to teacher candidates through clinical and field experiences. These experiences are designed to introduce teacher candidates to the roles and responsibilities of the agricultural education teacher and FFA advisor. Experiences are intended to expose candidates to the three-component model of agricultural education and introduce them to the agricultural education career. Nationally accredited teacher preparation programs integrate clinical and field experiences into plans of study and degree plans to meet teacher certification and program accreditation requirements. This chapter focuses on the early field experience and student teaching internship components utilized by teacher preparation programs.

INTRODUCTION

This chapter provides insight into the components of the clinical (student teaching) and early field experiences teacher candidates complete when they are introduced to the roles and responsibilities of an SBAE teacher and FFA advisor and as they implement a School-based Agricultural Education Program (see Figure 1). Objectives for this chapter include:

1. Discuss the role of early field experience in teacher preparation.
2. Discuss the meaning and value of the student teaching internship.
3. Describe selection of cooperating teachers.
4. Describe the roles of student teachers, cooperating teachers, and university supervisors related to student teaching experiences.

DOI: 10.4018/978-1-6684-3420-8.ch012

Figure 1. The three-component model of agricultural education

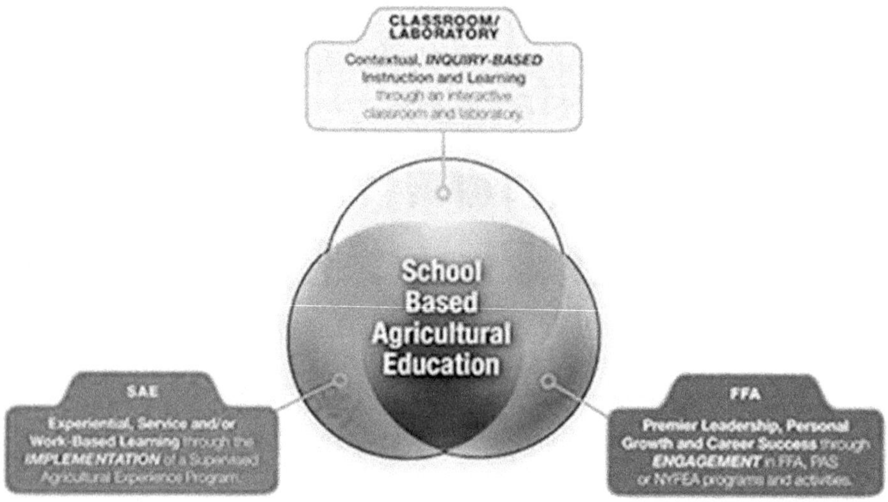

EARLY FIELD EXPERIENCES IN TEACHER PREPARATION

Observation, research, and immersion are common practice for those who aspire to become doctors, lawyers, and teachers. Similarly, early field experiences (EFE) in teacher preparation provide teacher candidates the opportunity to take an active role in developing their knowledge, skills and attitudes regarding teaching and advising prior to the student teaching internship (Retallick, 2005; Retallick & Miller, 2007). Teacher preparation is charged with providing opportunities for students to become involved in their own professional and personal development. Students in teacher preparation programs should be involved as actors, not simply as spectators. Early field experiences provide opportunities for students to become engaged in SBAE programs, work with children, and become involved in the profession (Sagen et al., 2000). These early experiences are guided by national accreditation standards and serve a variety of purposes. Facilitating positive, educative experiences for teacher candidates in the context of teaching and advising students is the aim for teacher preparation programs (Dewey, 1904). To accomplish this experiential learning serves as the basis for EFE in agricultural teacher education (Dewey, 1938). Concrete experiences include activities related to the exploration of the integral nature of the three-component model of agricultural education, applying content knowledge, developing teaching skills, and learning to see themselves as a professional teacher.

To develop their expertise, teacher candidates must observe and practice instructional strategies (know), implement them in a real context (do), gather evidence of learner response to the strategy (use data), reflect upon that evidence (reflect), seek feedback from others like a mentor, coach, peer teacher or observer (get feedback), make adjustments (adjust), and repeat the cycle throughout the three components of agricultural education. To help guide the development and implementation of purposeful experiences, teacher educators can utilize the Comprehensive Model for Secondary Agricultural Education (Baker et al., 2012). The proposed model highlights experiential learning in all components of agricultural education (see Figure 2).

Figure 2. Comprehensive model for secondary agricultural education

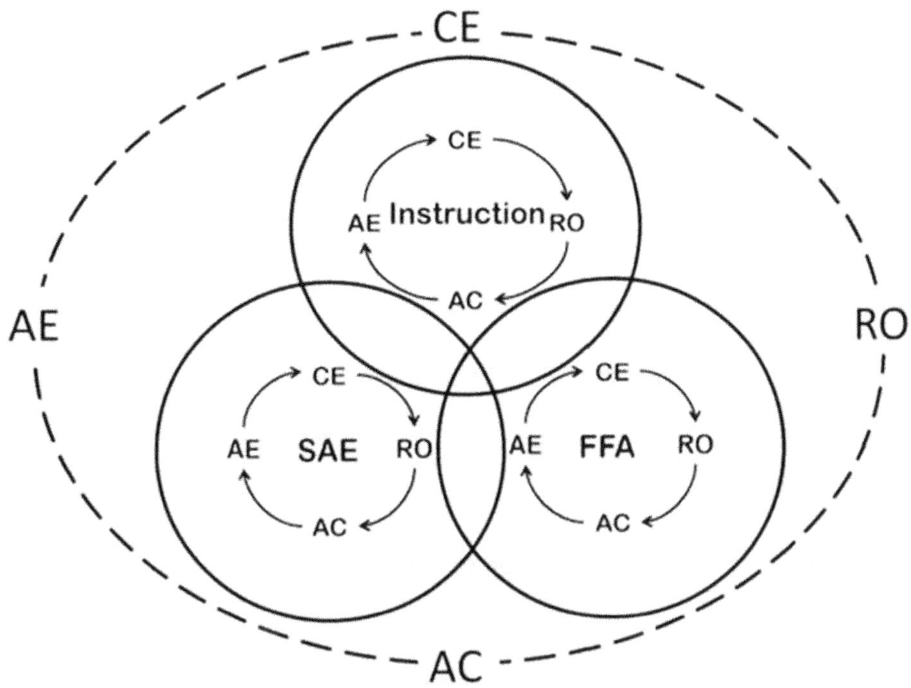

Teacher candidates must be taught how to work through this cycle and should leave their initial preparation with the skill to reflect on their practice alone and with others and evaluate their practices against a framework of developmental growth.

Standards and Accreditation

National Standards

The Council of Chief State School Officers (CCSSO) Interstate Teacher Assessment and Support Consortium (InTASC) developed a model for core teaching standards and learning progression for teachers. Four core areas: the learner and learning, content, instructional practice, and professional responsibility are supported by ten standards which are reflected in accreditation standards and pedagogical exams for licensure i.e., the Praxis Performance Assessment for Teachers (PPAT).

National Accreditation

The Council for Higher Education Accreditation (CHEA, 2013) charged the Council for the Accreditation of Educator Preparation (CAEP, 2021) to create standards which ensure educator preparation providers impart future teachers with the knowledge and skills to support the development of all students. Standard 1 and 2 focus on the learner and learning, content, instructional practice, professional responsibility, and clinical experiences. Teacher educators integrate these standards into EFE and collect data on teacher candidate experiences for accreditation reports.

Teacher preparation programs must look to these standards and accreditation benchmarks when planning, implementing, and facilitating early field experiences (AAAE, 2001). Data is the coin of the realm and teacher educators must consider these standards when crafting program level learning outcomes, data collection, and analysis related to the early field experiences provided for teacher candidates.

Purpose and Activities

Career Exploration

Early field experiences (EFE) help teacher candidates prepare for success as SBAE teachers. Experiences include career awareness, career exploration, and career assessment tied to classroom learning; employability skills training; and work experiences. These experiences have potential to assist candidates with career decisions.

Microteaching Activities

Preparing and delivering prepared lessons in a controlled environment has been an important aspect of teacher preparation. Nearly two-thirds of teacher preparation programs include microteaching activities in their EFE program. Microteaching provides candidates with experience in teaching and skill development while positively affecting their image of themselves as a teacher and their view of the profession. Important elements of the microteaching experience include:

Interaction with peers

Researchers have identified this component of the EFE as "overlooked and undervalued." Teacher candidates have limited experiences, as such, it is difficult for them to comprehend everything expected of them regarding preparing and delivering a lesson. One approach available to teacher educator's is to develop cohorts or groups of students that can discuss and provide support regarding the experience; additional support and encouragement can come from the mentor teacher (Retallick & Miller, 2007).

Observing the mentor teacher

Interactions with the mentor teacher provide candidates a professional contact that can facilitate both formal and informal EFE opportunities. However, it is important for the mentor teacher to recognize the importance of their role and be prepared to have substantial, coaching conversations with teacher candidates.

Observing students

School-based agricultural education programs serve middle school and high school students that are as young as 11 and as old as18 years of age. The physical, emotional, and cognitive development of these students represents substantial differences teachers must consider when preparing lessons and activities. Early field experiences provide opportunities for candidates to observe students in both formal and informal settings which serves as a framework for engagement and motivation strategies as candidates prepare lessons and age-appropriate learning activities.

Microteaching allows teacher candidates to receive immediate feedback on their teaching skills and style (Peters, 1980). It involves teaching a short lesson or selected sections of a prepared lesson plan i.e., the anticipatory set to a group of peers, colleagues, or students. Microteaching sessions vary regarding length of time; however, fifteen minutes is generally allowed for students to demonstrate their teaching skills. These microteaching experiences require candidates to practice their lesson planning skills. An emphasis on student engagement and managing the learning environment are key areas for feedback and reflection for both the mentor and teacher candidate.

Teacher preparation programs introduce the science of teaching and learning to teacher candidates. This introduction often includes educational scholars i.e., Ralph Tyler, John Dewey, Barak Rosenshine, Norma Furst. Tyler's four questions related to curriculum development and Rosenshine and Furst's effective teaching behaviors combined with teaching standards serve as a framework for assessing and evaluating microteaching (see Table 1).

Observation is an excellent way for teacher candidates to become aware of the different components of an SBAE program and the roles and responsibilities of the teacher. An early observation experience for candidates is in essence, a field trip. A structured visit designed to allow teacher candidates to visit and observe the physical surroundings of a program and identify different elements that contribute to the learning environment can be a valuable experience (see figure 3).

Early Field Experience Observation (Example)

Objective: Observe the physical surroundings of the program and determine the climate of the learning environment.

Directions: Carefully analyze the facilities, grounds, meeting rooms, and how the facilities are arranged. Make an unobtrusive observation of the cleanliness, order, graffiti, bulletin boards, magazines, and books in the media center. Try to determine as much as possible about the facilities and the climate without asking questions. Complete the following worksheet as a tool to make an unobtrusive evaluation of the educational climate. Record observations in response to this question, "What was your first impression?"

Candidate observation of classroom and laboratories provides teacher educator's concrete experiences to help guide the reflection process. Guided reflection can help candidates connect principles of student growth and development and the learning environment. Early field experiences are designed to occur early in the teacher preparation program and are cumulative in nature. Certification requirements may require teacher candidates to complete a specific number of hours of EFE to be recommended for licensure. To document the growth and development of candidates, some institutions may require a professional growth portfolio to document artifacts reflecting the early field experience of teacher candidates.

Early Field Experience Portfolio (Example)

Documentation of EFE hours can be done via early field experience portfolios or journals. Teacher candidates are asked to record their thoughts and activities completed as part of their field observation. Separate typed entries (1-2 pages) are required for each visit made to the cooperating school. Each visit must be at least 2 hours long and not more than 8 hours. Each entry must include the following information (see below). Under point number two, the letters are suggested items preservice teachers may observe and prompts for writing.

Table 1. Microteaching Assessment Rubric

	Unsatisfactory	Basic	Proficient	Exemplary
Lesson Planning	Insufficient planning.	Plan meets most elements of the template.	A well-prepared, complete plan is used.	Was of professional quality and could be used by other teachers without explanation.
Introduction /A set	None evident	Attention Getter appeals to limited number of students; few are engaged. Objectives are alluded to.	Attention Getter appeals to some students, most engaged. Objectives are stated/shown.	Effective Attention Getter; All students engaged. Objectives are clearly stated/shown.
Content	Teacher is unsure of content, makes several errors; is "locked" to notes.	Teacher is knowledgeable of content, makes few errors, occasionally uses notes.	Teacher is very knowledgeable of content, no visible errors, infrequent use of notes.	Teacher is very knowledgeable of content, no errors, no use of notes.
Voice	Lacking in energy, variety, and volume	Somewhat expressive; volume is usually sufficient.	Pleasant, enthusiastic, expressive, good volume control.	Voice is very well controlled and used to maintain the class effectively and present material.
Grammar/Use of English	Incorrect grammar used, thoughts disjointed	Some slang, informal English, but clear in communicating thoughts.	Uses standard forms, very clear in communicating thoughts.	Uses standard forms, communicates thoughts clearly, and has broad base of vocabulary.
Preparation	Teacher arrives ill-prepared. Not ready to present at assigned time.	Presents at assigned time, plan and materials are ready.	Presents on time. Plan and materials are well-organized.	Presents on time. Plan and materials are well-organized and polished.
Use of Materials	Teacher/learner materials disorganized, inappropriate.	Teacher/learner materials mostly organized, appropriate.	Teacher/ learner materials well organized, and add to the presentation.	Teacher/learner materials very well organized and promote learning and teaching.
Teacher Mobility	Teacher stays in one spot or in front of room: movement restricted.	Teacher moves around some, but mobility seems to have limited purpose.	Teacher moves around intentionally with purpose.	Teacher moves with purpose, reaches all parts of the classroom.
Pace	Pacing and timing are poor, down time is evident.	Pacing and timing are appropriate, very limited downtime.	Pacing and timing are very good. Lesson moves fluidly and elements are given appropriate priority.	Pacing and timing are excellent. Appropriate priority given to all elements, no down time.
Time	Fails to use time allotted.	Utilizes some time allotted.	Effectively utilizes most time allotted.	Effectively utilizes entire time allotted.
Poise / Presence	Shows anxiety, discomfort, lacks eye contact.	Mostly confident, some eye contact, smooth gestures.	Very confident, good eye contact, gestures enhance lesson.	Extremely confident and appears natural in class, gestures promote learning, but not arrogant.
Teaching Strategies	Strategies inappropriate for lesson and class makeup. Lacking variety and creativity.	Strategies mostly appropriate for situation, some variety and creativity.	Strategies are appropriate for situation, good variety and creativity, addresses some varied learning styles.	Strategies are very appropriate for situation, excellent variety and creativity addresses multiple learning styles.
Technology	Available technology not utilized.	Technology is used but does not enhance instruction and appears forced.	Technology is effectively used by instructor only to enhance instruction.	Uses technology effectively to enhance instruction and involves students in a meaningful way.
Student Engagement / Interaction	Entirely teacher centered, no questioning or interaction.	Some questions and activities that involve students and promote interaction.	Good questioning. Most students actively involved in class. Good interaction between teacher and students.	Excellent questioning. All students actively participating in class. Quality interaction between students and teacher.
Exceptional Learner / Modifications	Modifications are not mentioned or are simply delegated to a para-educator.	Modifications receive limited mention, even if for hypothetical learner.	Modifications are specifically used.	Multiple modifications are specifically used.
Conclusion/ Wrap-up	No closure	Incomplete review of concepts.	Review of concepts covered in lesson.	Review of concepts covered in lesson, includes feedback from all students.

Figure 3. Example Early Field Experience Observation

Observation

Objective: Observe the physical surroundings of the program and determine the climate of the learning environment.

Directions: Carefully analyze the facilities, grounds, meeting rooms, and how the facilities are arranged. Make an unobtrusive observation of the cleanliness, order, graffiti, bulletin boards, magazines, and books in the media center. Try to determine as much as possible about the facilities and the climate without asking questions. Complete the following worksheet as a tool to make your unobtrusive evaluation of the educational climate. Record your observations in response to this question, "What was your first impression?"

Area of Observation	**What did you observe? What was your first impression?**
Example: **The Flag Pole**	• *There was a 25' flagpole in front of the ag building with the US flag displayed. I noticed the flag was faded and torn and there was no lighting on the flag.* • *I sensed that people didn't care about the conditions of the grounds (or at least the flag).*
1. Condition of the Grounds	
2. Organization & Layout of the Facilities	
3. Condition & Cleanliness of the Facilities	
4. Educator Morale	
5. Promotions (what was displayed around the building on bulletin boards or in display cases)	
6. Learners Respect (school=lockers, graffiti)	

1. Date of visit, time of arrival, time of departure
2. Description of the lesson(s)/event(s) observed
   a. What was the teacher doing?
   b. What were the students doing?
   c. Overall objectives
   d. Were objectives met?
   e. Indication of assessment of the objectives?
   f. Engagement
   g. Classroom management
   h. Technology used
   i. How was the lesson brought to a close?
   j. How did the instructor ensure the objectives were met?
3. Description the role of the pre-service teacher i.e., observer

4. Identify the two or three key points learned from the experience
5. A description of how you will use this information as an agriculture teacher
6. Your general thoughts and reaction to the experience

Observation of teacher and student movement patterns is an additional EFE assignment that holds potential for students to observe student engagement, teacher bias and other teaching behaviors demonstrated in the classroom. A simple observation tool can be used to track movement in the classroom.

Observation Tool Teacher/Student Movement Patterns (Example)

Directions: Sketch a layout of the room arrangement using squares to represent the teacher desk, teacher podium, student desks, and learning stations. Use lines with arrows to show which students move about the room during a set amount of time. Do the same to record teacher movements. If you wish to record both during a set time period, use two different colors: one for students, one for the teacher.

When considering early field experience observation, the following areas have potential to provide a comprehensive experience reflective of the comprehensive model of school-based agricultural education.

1. Classroom Observation
2. Laboratory Observation
3. Classroom Microteaching
4. SAE Observation
5. FFA Non-competitive Events
6. FFA Leadership Events
7. FFA Competitive Events

Teacher educators should emphasis the importance of diversity, teaching settings, teaching strategies, SAE types and advising styles in reflection activities.

THE STUDENT TEACHING INTERNSHIP

Student teaching has been described as the capstone experience of the preservice teacher education program and is critical to the process of preparing future teachers. Capstone student teaching experiences require substantial lead teaching responsibilities under the guidance of a cooperating teacher. Through full immersion in a hands-on, real-world experience, teacher candidates begin to master and use professional knowledge such as various aspects of student learning, responsibility, and collaboration (Meyers & Dyer, 2004). The experiences acquired by candidates during student teaching are potentially the most crucial activities involved in the development of prospective agriculture teachers.

Student teaching is a period of guided teaching during which the student, under the direction of a cooperating teacher, takes increasing responsibility for leading the school experiences of a given group of learners over an extended time and engages directly in many of the activities which constitute the wide range of a teacher responsibilities.

The student teacher is a college student who is working under the guidance of a certified teacher in an approved setting. A student teacher, while serving a non-salaried internship under the supervision of a certified teacher, shall be accorded the same protection under the law as that accorded the certified teacher. The student teacher should not be assigned in any situation where he/she is to be a replacement for a qualified teacher.

The cooperating teacher is the regular and certified staff member of a local accredited school to whom a student teacher is assigned.

Value of Student Teaching in Agriculture

Student teaching is perhaps the most dynamic and vital phase of the total curriculum for preparing teachers of Agricultural Education. In no other way can a person quite so effectively develop the competence necessary to assume their role as an agricultural leader and teacher in a local community and the classroom. It is through this pre-employment teaching experience that the prospective teacher is provided the opportunity to bridge the gap between theory and practice under the expert guidance of a cooperating teacher (McIntyre et al., 1996).

During student teaching, through a carefully planned sequence of activities, the student will receive experience and gain competence in:

1. Teaching high school students (classroom and laboratory areas) in a program driven by a strong instructional component.
2. Advising the FFA. Getting the "feel" of a successful well-balanced FFA program.
3. Working with students' SAE Programs: Developing concepts that will enhance a balanced SAE in the local program.
4. Guiding students as they participate in competitive activities. Learning the importance of coaching within a successful Agricultural Education program.
5. Setting an example and working with students in a variety of leadership experiences. Developing a plan for involving "all" students in leadership roles in the local FFA program.
6. Planning activities as students participate in community events and services.
7. Counseling students.
8. Appraising the educational and agricultural needs of high school students.
9. Participating in school and community affairs.
10. Planning and organizing the physical facilities of the Agricultural Education Department.
11. Developing and internalizing the ethical standards and ideals that best characterize the teaching profession.
12. Further clarifying the aims and objectives of public-school programs and the formal and informal patterns of adult education in our society.
13. Developing a better understanding of the important role of related agricultural organizations and businesses to the social and economic welfare of the local community.
14. Serving as a desirable role model for students.
15. Experiencing a broad base of activities in a <u>multi-dimensional</u> program.

Phases of Student Teaching

Student teaching assignments are highly individualistic. Involvement in classroom activities will depend upon the individual's readiness to perform the tasks assigned (Smalley et al., 2015).

The cooperating teacher and university supervisor will assess the student's abilities and determine the student teacher's responsibilities and tasks. **In no case should the student teacher immediately assume total responsibility for all classes.**

Due to the individualistic nature of the assignment, no set timeframe can be attached to the various phases of student teaching. Some student teachers may be involved with observations for several weeks while others will engage in limited teaching by the second week. In general, though, all student teachers will move through four phases during the experience.

1. Orientation and Observation (1 week): The student teacher becomes oriented to the classroom and observes the cooperating teacher. The student should discuss his/her perceptions and observations with the cooperating teacher
2. Progressive Teaching Experience (5-7 weeks): The student teacher assumes responsibility for one or two classes under the close supervision of the cooperating teacher. The student teacher also becomes involved in planning, preparing resources, correcting papers, helping small groups, supervising study periods, tutoring, constructing assessments, preparing instructional materials, and teaching a complete lesson that has been cooperatively developed for other classes. In this phase, the cooperating teacher provides continuity for the classroom and immediate feedback to the student teacher.
3. Extensive Teaching Experience (3-5 weeks): The student teacher assumes responsibility for the full course load, including planning and teaching a unit of instruction, and evaluating the students in the classroom. The cooperating teacher observes all phases of the instructional process and provides feedback regularly.
4. Culminating Experience (1-2 weeks): The student teacher completes the teaching unit and finishes the experience by observing other classes in the building.

Selecting Cooperating Teachers

A successful cooperating teacher creates an enriching experience for both the students in the classroom and the student teacher (Killian & McIntyre, 1983). The cooperating teacher involves the student teacher in such activities as learning students' names, developing seating charts, tutoring, teaching small groups, developing a teaching unit, teaching one class, and eventually teaching multiple classes. The cooperating teacher retains responsibility for the classroom while guiding the student teacher through these activities.

An effective cooperating teacher uses good critiquing skills in guiding the growth and development of the student teacher. In a good student teacher/ cooperating teacher relationship, the student teacher feels free to ask questions and make suggestions while being willing to receive and accept constructive criticism. The two should function as a team working for the development of the pupil. This relationship is developed through actual participation in classroom activities and through conferences involving the cooperating teacher and the student teacher. Conferences are held for planning purposes, discussion of teaching problems, and evaluation of the student teacher's work. While valuable conferences may be held in five or ten minutes, conferences of considerable length will be required periodically; student teachers are required to make themselves available for such as the cooperating teacher deems necessary.

ROLES OF STUDENT TEACHERS, COOPERATING TEACHERS, UNIVERSITY SUPERVISOR AND BUILDING ADMINISTRATOR AS RELATED TO STUDENT TEACHING EXPERIENCES.

Student teaching is an opportunity to learn. Therefore, student teachers should carefully observe not only what is done to manage a program, but also how it is done in the cooperating center. Student teachers should always be discrete about talking to friends and outsiders about what transpires in the cooperating center, especially when discussing personal and confidential information about students, teachers, or parents.

The coming and going of student teachers creates disruptions of the regular school program. At the same time, the student teacher is additional personnel in the school and, as such, should have ideas to present for consideration which can be of value to the educational program. Endeavor to make some permanent, tangible contributions to the department during the student teaching experience. The following is a list of guidelines for student teachers to follow during the student teaching internship.

Student Teacher

1. The student teacher is a guest in the assigned school; therefore, he/she is expected to support school policies and personnel. **All school rules and regulations should be followed by the student teacher.**
2. **Student teaching is a full-time job.** This includes responsibility for the entire class plus any out-of-class responsibilities with student organizations, student SAE supervision, and community activities. Daily attendance and preparation should reflect the seriousness of this responsibility.
3. The student teacher should be part of any regularly scheduled conference with the cooperating teacher and university supervisor.
4. **The student teacher will dress, talk, and act as a professional.**
5. The student teacher is to exemplify professionalism in dealing with confidential information.
6. The student teacher will assume responsibility for the quality of the experience, seek out new involvement possibilities, and ask for new assignments or responsibilities as soon as possible.
7. The student teacher should take part in out-of-class activities such as FFA, faculty and professional meetings, routine teaching and nonteaching tasks.
8. The student teacher is not to be considered as a substitute teacher.
9. The student teacher is to inform the cooperating teacher of any absence prior to missing an assigned experience.
10. The student teacher will communicate with the university supervisor about problems, progress, absences, and visitation times.
11. The student teacher will display self-reliance, desire, enthusiasm, and initiative.
12. The student teacher will develop written lesson plans in advance of the lessons.
13. The student teacher will discuss lesson plans with the cooperating teacher before the lesson begins.

14. The student teacher will observe the instructional strategies of the cooperating teacher and the classroom management techniques employed.
15. The student teacher will learn the names of the students in assigned classes as well as the relevant staff members in the school.
16. The student teacher should test new ideas within the structure the cooperating teacher has established, making them available to the cooperating teacher several days in advance of the lesson.
17. **The student teacher should be involved continually in the self-evaluation process.** This process includes examining the goals and outcomes of each instructional activity, reviewing audio and video tapes of their performance, soliciting feedback with suggestions for improvement, and seeking immediate answers to problems.

Cooperating Teacher

The cooperating teacher plays a vital role in the education of the student teacher and is expected to:

1. Exhibit a whole-hearted willingness to accept student teachers.
2. Ensure adequate continuity, class contact, and supervision of the student teacher program. The student teacher should be part of a teaching team and should not be considered as a replacement for the regular teacher.
3. Define the student teacher's role in terms of his/her duties and responsibilities. To accomplish this, the cooperating teacher should
   a. Schedule a conference with the student teacher early in the semester to clarify the role of the student teacher in the given situation.
   b. Provide many opportunities for the student teacher to observe methods that are appropriate to the teaching profession.
   c. Assist the student teacher in developing a professional attitude in all his/her contacts with school and community.
   d. Assist the student teacher in becoming acquainted with the regulations applicable to pupils and teachers in the school where he/she is working.
   e. Clarify the student teacher's responsibilities with respect to making lesson plans, securing, and organizing appropriate materials, and other necessary activities.
   f. Advise the student teacher well in advance of the time when a lesson or unit is expected to be taught.
4. Attend in-service programs planned for teachers and administrators directly concerned with student teaching.
5. Ensure the student teacher's induction into actual teaching at a rate that seems appropriate for the student teacher. A helpful model is a three-phase process: (1) the cooperating teacher directly models lesson teaching, (2) the student teacher teaches a comparable lesson using the cooperating teacher's lesson plans, and (3) the student teacher plans and teaches a lesson.
6. Make sure the student teacher has planned adequately before being permitted to teach.
7. Keep the student teacher informed of progress, making constructive suggestions and criticisms as necessary.
8. Evaluate the student teacher's performance in writing and provide the student teacher with a copy of suggestions/criticisms.

9. Take the necessary steps to remove the student teacher if work is not satisfactory.
10. Send final evaluation of the student teacher to the Department of Agricultural Education.

University Supervisor

Student teachers should be assigned a university supervisor. This supervisor is an experienced university educator or representative who assumes the responsibilities described below.

Throughout the semester, the supervisor confers with the student teacher and cooperating teacher about the experiences and progress of the student teacher. The supervisor holds regularly scheduled meetings with the student teacher and should provide written documentation about the observations. In these conferences, the student teacher and the supervisor discuss planning activities, learning strategies, classroom management concerns, etc. The student teacher, cooperating teacher, and supervisor will meet individually and together during the semester to discuss the student teacher's progress.

The supervisor will typically visit the student teaching center two or more times during the semester. Generally, one of the visits will be at the beginning of the student teaching experience and one near the end of the experience.

Supervisory Visit by the University Supervisor

Expectations for the teaching interns

1. Have lesson (instructional) plans ready for review upon the university supervisor's arrival.
2. Be prepared to explain the objectives for the classes being taught the day of the visit.
3. Describe how the lesson objectives will be accomplished.
4. At the end of the day (or when appropriate), join the cooperating teacher and the university supervisor in a feedback conference concerning the day's lessons and activities.

Expectations for the cooperating teacher

1. Prior to the arrival of the university supervisor, the cooperating teacher should verify that the intern has a copy of each weekly evaluation.
2. During the first visit by the university supervisor, be present with the university supervisor in the classroom or laboratory to observe the intern teach.
3. Participate in the feedback conference with the teaching intern and the university supervisor at the conclusion of each university supervisor visit.
4. Share any concerns that might strengthen the intern's preparation program.

Expectations for the university supervisor

1. Arrive at the cooperating department in time to participate in a pre–teaching conference and to adequately review the day's teaching plans with the intern prior to start of classes. Sign in at the school's main office.
2. Visit with the cooperating teacher to determine if there are areas in which he/she would like for you to focus your observations.

3. Visit with the intern to determine if any problems exist that you need to address.
4. Review the intern's lesson plan for classes you will observe.
5. Observe teaching with the intent of providing support and encouragement and as reinforcement of successful teaching patterns to the interns. Observe two classes (if applicable). Complete a Formative Assessment of Teaching form (see figure 4) for at least one of the classes. Other observation forms may be used in addition to these forms.
6. Review all evaluations conducted by the cooperating teacher.
7. Conduct post–evaluation conferences with the intern and with their cooperating teacher. Be sure both are clear in your assessment of the intern's progress and recommendations for further growth. Provide intern with objective feedback on the current state of their instructional skills.

Grading and evaluating student teaching interns

The university supervisor and cooperating teacher are expected to provide formal and informal evaluations of the student teacher that will allow for continuous constructive feedback to help the student teacher develop their craft as a teacher, FFA advisor, and SAE supervisor. Teaching should be formally evaluated by the cooperating teacher often, particularly through the early stages of the student teacher's work with a new class. The teaching evaluation should be documented and based on classroom observations and serve as a basis for the student teacher's final grade

Grades for the student teaching internship should indicate the competency of a person to be recommended for initial licensure. Competency should be assessed with reference to the student's peers in the professional setting and as a developing novice teacher. The university supervisor coordinates and is directly involved in student teaching performance assessment. They rely heavily upon the cooperating teacher's input and then forwards a recommendation for the final grade to the university. The teacher preparation program determines the final grade a student receives for student teaching based upon all the data gathered. The terminology used in the grading criteria will assist not only in determining an appropriate grade for a student teacher but also in writing a recommendation or letter of reference.

An individual who earns an A:

- is capable of functioning effectively and independently as an entry level professional.
- has demonstrated outstanding knowledge, skills and attitudes relevant to a wide variety of professional responsibilities and relationships.
- has demonstrated a high level of self-motivation, ability to learn and interest in his/her professional development.

A person who receives an A can be recommended with no reservation.

An individual who earns a B:

- will need occasional assistance to function as an entry-level professional person.
- has demonstrated a good base of knowledge, skills, and attitudes relevant to professional roles responsibilities and relationships.
- has demonstrated an adequate level of self-motivation, ability to learn and interest in his/her professional development.

Figure 4. Formative assessment of teaching

Formative Assessment of Teaching

____________________ Teacher   ____________________ Evaluator

Date: __________ Class: __________ Method: ________________

Instructional Unit: ______________________________
(e.g., biotechnology, leadership, wildlife management)

Lesson: ____________________________________

STRENGTHS OF LESSON

____

SUGGESTIONS FOR IMPROVING THE LESSON

____

TEACHING SKILLS OBSERVED: (check all observed)

Interest development ☐	Professionalism ☐	Appropriate questioning ☐
Enthusiasm ☐	Variation of methods ☐	Application of concepts ☐
Organized/business-like ☐	Creativity ☐	Evaluation of learning ☐
Clarity in communication ☐	Related to student interests/needs ☐	Closure ☐

Curve of Student Class Interest

	Start	1st	2nd	3rd	4th	End
Excellent						
Good						
Average						
Poor						

Motivation

Excellent	
Good	
Average	
Poor	

Involvement

Excellent	
Good	
Average	
Poor	

A person who receives a B can be recommended with confidence for a position after noting strengths and weaknesses.

An individual who earns a C:

- will need considerable assistance to function effectively as an entry-level professional person.
- has demonstrated adequate knowledge, skill, and attitudes in some relevant professional areas, and is making progress in developing a good base level in other relevant areas.
- has demonstrated a need for more initiative and/or guidance in his/her learning and continued professional development.

A person who receives a C can be recommended with reservations.

An individual who earns a D:

- may fall short of overall effectiveness as a professional person, even with extensive assistance.
- has demonstrated limited knowledge, skills, and attitudes relevant to the field and has demonstrated little progress in developing an adequate level of functioning necessary to assuming professional responsibilities.
- has demonstrated characteristics that may preclude success in working effectively with children and others.

A person who receives a D cannot be recommended for licensure and will be counseled out of the program.

An individual who earns an F:

- has inadequate knowledge, skills, and attitudes to function as a professional person.
- has not demonstrated sufficient knowledge, skills, or attitudes for an entry level professional.
- has demonstrated characteristics that could be detrimental to children and/or others.

A person who receives an F cannot be recommended for licensure and will be counseled out of the program.

Building Administrator

The school administrator plays a particularly important role in facilitating and coordinating student teachers. He/she provides a positive environment for student teachers within the building, orients faculty to their role in the student teaching program, orients student teachers concerning philosophy, policies, and regulations of the cooperating schools, assists student teachers in becoming acquainted with the faculty, stays informed about the progress of student teachers, and observes student teachers and provides feedback. The following list outlines the administrator's role regarding the student teacher:

1. To be willing to accept student teachers as part of the total school faculty.
2. To be willing to allow student teachers to work with enough freedom to encourage growth.
3. To be willing to provide facilities that will enhance the training environment.
4. To be willing to give assistance and support to the student teacher in the development of appropriate professional relationships with the faculty, staff, students, and community.
5. To be willing to work with the university supervisor and cooperating teacher in helping to meet the needs of the student teacher.

ADDRESSING PROBLEMS DURING THE STUDENT TEACHING INTERNSHIP

Problem-Solving Process

Most problems that arise between the cooperating teacher and the student teacher are resolved through daily conferences. The following process highlights steps that can be taken by either the cooperating teacher or student teacher to work through problems that may arise during the student teaching internship if the problem cannot be resolved.

1. Either the student teacher or the cooperating teacher should:
    a. Have a conference with the university supervisor to discuss the issues.
    b. Develop a plan to resolve problems with the help of the university supervisor.
    c. Communicate the plan to the principal and to all involved parties.
    d. Work the plan and document progress.
2. If the problem continues, the principal and the university supervisor confer with input from the cooperating teacher and/or the student teacher to remove the student teacher from the assignment.
3. If the student teacher is not satisfied with the decision, he/she may appeal the decision.

SUMMARY

The knowledge, skills, and dispositions needed to lead and deliver School-based Agricultural Education (SBAE) are introduced to teacher candidates through clinical and field experiences. These experiences are designed to introduce teacher candidates to the roles and responsibilities of the agricultural education teacher and FFA advisor. Experiences are intended to expose candidates to the three-component model of agricultural education and introduce them to the agricultural education career. Nationally accredited teacher preparation programs integrate clinical and field experiences into plans of study and develop degree plans to meet teacher certification and program accreditation requirements. Student teaching is the culminating experience for teacher candidates. The immersive nature of student teaching provides opportunities to observe and practice effective teaching with the guidance of a cooperating teacher in a public-school setting.

REVIEW QUESTIONS

1. List three elements of the micro-teaching experience.
2. Identify three specific roles related to the responsibility of the cooperating teacher.
3. Describe the phases of student teaching.
4. Discuss the value of formal feedback provided by the cooperating teacher.
5. Compare the roles of the student teacher, cooperating teacher and university supervisor and identify common themes shared between the triad of individuals involved in the student teaching experience.

REFERENCES

American Association for Agricultural Education (AAAE). (2001). *National Standards for Teacher Education in Agriculture.* Retrieved October 3, 2021, from http://www.aaaeonline.org

Baker, M. A., Robinson, J. S., & Kolb, D. A. (2012). Aligning Kolb's experiential learning theory with a comprehensive agricultural education model. *Journal of Agricultural Education*, *53*(4), 1–16. doi:10.5032/jae.2012.04001

Council for the Accreditation of Educator Preparation (CAEP). (2021). *Initial level standards.* Retrieved November 6, 2021, from http://caepnet.org/standards/2022-itp/introduction

Council of Chief State School Officers. (2013, April). *Interstate Teacher Assessment and Support Consortium InTASC Model Core Teaching Standards and Learning Progressions for Teachers 1.0: A Resource for Ongoing Teacher Development.* Washington, DC: Author.

Dewey, J. (1904). The relation of theory to practice in education. National Society for the Study of Education. Public School Publishing Co.

Dewey, J. (1938). *Experience and education.* Macmillan.

Guyton, E., & Byrd, D. (Eds.). (2000). *Standards for field experience in teacher education.* Association of Teacher Educators.

Killian, J. E., & McIntyre, D. J. (1983). Quality in the early field experiences: A product of grade level and cooperating teachers' training. *Teaching and Teacher Education, 2*(4), 367–376. doi:10.1016/0742-051X(86)90029-6

McIntyre, D. J., Byrd, D. M., & Foxx, S. M. (1996). Field laboratory experiences. In J. P. Sikula, T. J. Buttery, & E. Guyton (Eds.), *Handbook of research on teacher education* (2nd ed., pp. 171–193). Macmillan.

Myers, B. E., & Dyer, J. E. (2004). Agriculture teacher education programs: A synthesis of the literature. *Journal of Agricultural Education, 45*(3), 44–52. doi:10.5032/jae.2004.03044

Peters, J. L. (1980). *The effects of laboratory teaching experience (microteaching and reflective teaching) in an introductory teacher education course on students' views of themselves as teachers and their perceptions of teaching* [Unpublished dissertation]. The Ohio State University, Columbus, OH, United States.

Retallick, M. S. (2005). Early field experiences in agricultural teacher education. *Retrospective Theses and Dissertations.* 17206. https://lib.dr.iastate.edu/rtd/17206

Retallick, M. S., & Miller, G. (2007). Early field experience documents in agricultural education. *Journal of Agricultural Education, 48*(4), 20–31. doi:10.5032/jae.2007.04020

Sagen, H. B., Dallam, J. W., & Laverty, J. R. (2000). Effects of Career Preparation Experiences on the Initial Employment Success of College Graduates. *Research in Higher Education, 41*(6), 753–767. doi:10.1023/A:1007072705601

Smalley, S. W., Retallick, M. S., & Paulsen, T. H. (2015). Relevance of student teaching skills and activities from the perspective of the student teacher. *Journal of Agricultural Education, 56*(1), 73–91. doi:10.5032/jae.2015.01073

KEY TERMS AND DEFINITIONS

Cooperating Teacher: A regular and certified staff member of a local accredited school to whom a student teacher is assigned.

Early Field Experience: Experiences designed to introduce teacher candidates to the roles and responsibilities of the agricultural education teacher and FFA advisor.

Student Teacher: A college student who is working under the guidance of a certified teacher in an approved setting. A student teacher, while serving a non-salaried internship under the supervision of a certified teacher, shall be accorded the same protection under the law as that accorded the certified teacher. The student teacher should not be assigned in any situation where he/she is to be a replacement for a qualified teacher.

Student Teaching: A period of guided teaching during which the student, under the direction of a cooperating teacher, takes increasing responsibility for leading the school experiences of a given group of learners over an extended period of time and engages directly in many of the activities which constitute the wide range of a teacher's responsibilities.

Chapter 13
Assessing Teacher Effectiveness in Agricultural Education

Christopher J. Eck
https://orcid.org/0000-0002-1645-3632
Clemson University, USA

J. Shane Robinson
https://orcid.org/0000-0001-9506-5752
Oklahoma State University, USA

ABSTRACT

Effective teaching is a multidimensional concept that is often considered to be elusive. Those who are deemed effective teachers reflect frequently and often on their practice, continuously develop their human capital through professional development and conversations with others, and regularly solicit feedback and advice on their teaching from experienced and trusted mentors. When considering the development, support, and evaluation of school-based agricultural education (SBAE) teachers, it is essential to consider the sources of feedback along with formative and summative assessments to further develop career-specific human capital to prepare for and sustain a career as an effective SBAE teacher. Therefore, this chapter will identify sources of feedback, describe methods of effective teaching evaluation, and explore opportunities to use evaluations and feedback for continual improvement for pre-service and in-service SBAE teachers.

INTRODUCTION

This chapter investigates what it means to be an effective teacher by exploring the literature related to effective teaching, discussing experiential learning of SBAE teachers such as early field-based experiences and the student teaching internship, and exploring various teaching evaluation tools used by schools. Each of these components leads to the continual development of human capital needed to improve the effectiveness of SBAE teachers. The following learner objectives were established to guide this explo-

DOI: 10.4018/978-1-6684-3420-8.ch013

ration: 1.) describe effective teaching, 2.) describe the various sources of feedback for SBAE teachers, and 3.) explore opportunities for continual improvement.

EFFECTIVE TEACHING

The central question is: What is an effective teacher? All aspiring and current teachers are encouraged to answer this question for themselves. The answer will ultimately become indicative of a teacher's personal teaching philosophy. Yet, for the purpose of this chapter, the attempt is made to define the elusive concept known as effective teaching. Effective teaching is a multifaceted phenomenon that has been described in various ways throughout the literature. It is a term that is used frequently; yet, it is difficult to define. Being *effective* can be look different for every person depending on a teacher's personality, the context by which a teacher instructs, and the community in which a teacher lives and works (especially in relation to stated or assumed expectations), to name a few. Many researchers have conducted complete theses on the topic and devoted their entire careers to defining and describing it. We will attempt to feature a number of those studies in this section as we highlight effective teaching.

Although effective teaching can differ from one context to the next, it appears there are three domains in which teachers need to be competent if they desire to be effective. First, effective teachers must be adept at engaging their students in the content. Effective teachers empower and challenge their students to think critically and solve problems. They motivate students to do more than they believe they are able and scaffold, guide, and support students to increase their expectations of what is possible regarding their learning. In addition, effective teachers prepare students well for the employment sector by equipping them with the necessary skills and experiences needed to obtain and sustain in a career in their chosen field. Effective teachers invest time focusing on ways they can engage their students in each lesson they teach. Such engagement should include a combination of physical, emotional, and cognitive attachment to the material. As an aspiring teacher, how will you seek to engage your students throughout the lessons you teach? The answer to this question is imperative because engagement leads to success (i.e., learning). Therefore, if you want your students to succeed, engage them in the content.

Secondly, effective teachers have a good grasp of pedagogy (i.e., instructional methods and teaching strategies). Simply put, they understand how to teach. Pedagogy is the art and science of teaching. Effective teachers know how to create interesting lessons and present their content clearly. They transition between points easily and make connections throughout and across the lessons they teach. When teaching, they understand when and how to change the pace of a lesson by reading their students' body language. They base their decisions on their understanding of educational learning theory. There are two distinct learning theories that guide peoples' accumulation of knowledge: behaviorism and constructivism. In behaviorism, students learn through reward, reinforcement or repetition of content, and punishment. Oftentimes, this includes students observing a behavior occur and then producing a response to it. Behaviorism assumes that learning occurs by changing a person's behavior and that the instructor is in charge of the lesson. The teacher is perceived as a commander of the lesson and the source of all information in this type of environment. Students are expected to listen to and mimic the behavior of the teacher. This type of instruction is known as teacher-centric pedagogy, relies on mostly lower-order thinking skills (i.e., regurgitation of material), and occurs best in situations where the stakes are high, such as teaching or stressing safety. Imagine teaching students to operate a chop saw, or other dangerous equipment, in the agricultural mechanics laboratory. It might make sense for a teacher to be very prescribed in this situation by laying out a step-by-step procedure for operating such a piece of equipment. Students would be expected to listen carefully to the teacher and follow his or her commands for operating the equipment in a safe and orderly fashion.

In contrast, constructivism assumes that the learner constructs his or her own knowledge by building onto their prior knowledge or experiences. It is an active type of learning where the student is offered choices and is in charge of the lesson. The instructor is perceived as a guide or facilitator in this type of environment. Students are expected to use their prior experiences to solve problems and be active in the learning environment. This type of instruction is known as student-centric pedagogy. It occurs whenever students have familiarity with the subject area and can bring their own experiences to the table and use higher-order thinking skills (i.e., synthesis of material) to solve problems. Imagine teaching a lesson about trouble shooting a faulty small gasoline engine. It might make sense for a teacher to ask open-ended questions to students as a means to set the stage for the lesson, reiterate material learned in the past, and introduce the problem to the students. Then, the teacher becomes a guide and a source of information as the students work (sometimes along with the teacher) to repair the faulty engine.

Although these two examples are on opposite ends of the spectrum of educational learning theories, effective teachers often operate freely along the spectrum to help engage all their students in the learning task at hand. There are times when students need to receive baseline knowledge about a subject delivered through teacher-centric pedagogies. However, there are other times when students need to be challenged using student-centric pedagogies. An effective teacher can read his or her students, make adjustments, and respond to student needs accordingly.

Below, in Table 1, a few examples of teacher-centric and student-centric pedagogies are provided. Teacher-centric pedagogies prescribe to the behaviorism theory. Therefore, when teachers use the methods that fall within this style of teaching, it is assumed they have specific knowledge that needs to be conveyed to students in a particular way. In such cases, there is one correct answer to the problem or way of doing things. The tenets of behaviorism include the teacher being a competent model of delivering the instruction, gaining the attention of the students, and teaching through iteration with a reward structure in place. Examples of this type of teaching include, but are not limited to, the following: Lecture, Direct Instruction, Demonstration, and Simulation.

In contrast, student-centric pedagogies prescribe to the constructivism theory. Therefore, when teachers use the methods that fall within this style of teaching, it is assumed they have problems that need to be solved and are willing to empower students to find answers to these problems. In such cases, multiple correct responses to the problem are possible. The tenets of constructivism include the teacher setting up the problems, asking questions, and gently guiding students to find and be able to defend their answers. Examples of this type of teaching include, but are not limited to, the following: Inquiry-Based Teaching, Case Study, Problem-Based Learning, and Project-Based Learning (see Table 1).

Thirdly, effective teachers possess content expertise. Simply put, they know their content. Effective teachers are those who continuously seek out professional development surrounding the content they teach. They may opt to engage in an advanced degree in a particular subject area or participate in in-service workshops to learn more about their content area. In many ways, they are students themselves. Effective teachers are lifelong learners. They aspire to know the latest, cutting-edge technology and science produced in their field. Once learned, they continuously update their learning materials and content to keep them current so that their students receive the most recent information.

Table 1. Examples of teacher-centric and student-centric pedagogies

Teacher-Centric		Student-Centric
Lecture		Inquiry-Based Teaching
Direct Instruction		Case Study
Demonstration		Problem-Based Learning
Simulation		Project-Based Learning

Effective Teaching Checklists for SBAE Teachers

What does it mean to be an effective SBAE teacher. Teaching SBAE is different than other subject areas. First off, in many states, SBAE is considered an elective. Therefore, not every student in the school must enroll in an agricultural education course at the secondary level. Second, because the SBAE program is intracurricular in nature, it requires a teacher to be effective in areas in addition to the classroom. Specifically, SBAE is based on three areas: classroom and laboratory instruction, FFA, and SAE.

To help SBAE teachers understand what it means to be effective across the three areas, the Effective Teaching Instrument for School-Based Agricultural Education Teachers (ETI-SBAE) was designed (see Table 1). The ETI-SBAE was created to assist aspiring and current teachers consider their effectiveness across all areas of the SBAE program. The complete instrument includes 26-items spanning six components, including intracurricular engagement, personal dispositions, appreciation for diversity and inclusion, pedagogical preparedness, work-life balance, and professionalism. The ETI-SBAE was developed from a nationwide panel of experts, including in-service SBAE teachers, state agricultural education staff, and members from the National FFA board of directors.

Once deisgned, it was then disseminated to SBAE teachers across the country to establish its validity and reliability. As a result, pre-service or inservice SBAE teachers should feel confident in using the instrument (see Table 2) to self-evaluate their preparedness to be an effective SBAE teacher. The ETI-SBAE uses a four-point scale ranging from *very weak* (1) to *very strong* (4).

Aspiring and acting SBAE teachers should consider their effectiveness with each item. As you read through the list, ask yourself: "How effective am I at . . . ?" Based on this self assessment, teachers identify the areas that need to be improved, and then seek out ways to do so. For instance, a self-perception of being weak at advising FFA officers could lead to seeking mentors who have strong FFA officer teams. A teacher educator or state Career and Technical Education (CTE) staff member can help identify specific teachers who develop and use a strong set of criteria for electing their officers, have a formalized officer retreat to develop their annual Program of Activities for the chapter, and use their officers to deliver the assignments for the stations for which they are elected thoughout the year, as examples.

In considering these 26 items, it is imperative that proper reflection occur. At times, teachers can underestimate their effectivenss in a given area. Likewise, it is also possible, and even likely, that teachers often overestimate their effectiveness in a given area. Therefore, objective reflection is critical to becoming a successful teacher.

One potential avenue for advanced SBAE teacher preparation and evaluation can be obtained through video integration. Such technologies can be implemented in SBAE teacher preparation programs, while

Table 2. The effective teaching instrument for school-based agricultural education teachers

Construct	Items
Intracurricular Engagement	Instruct students through FFA
	Advise the FFA officers
	Advise the FFA chapter
	Facilitate record keeping for degrees and awards
	Be passionate about FFA
	Instruct students through SAEs
	Use the complete agricultural education 3-component model as a guide to programmatic decisions
Personal Dispositions	I am trustworthy
	I am responsible
	I am dependable
	I am honest
	I show integrity
	I am a hard worker
Diversity and Inclusion	I value students regardless of economic status
	I value students of all ethnic/racial groups
	I value students regardless of sex
	I care about all students
	I understand there is not an award for all students, but that does not mean they are not valuable
Pedagogical Preparedness	Demonstrate classroom management
	Demonstrate sound educational practices
	Teach every class
Work-Life Balance	I have the ability to say no
	I lead a balanced life
	I am never afraid to ask for help
Professionalism	I have patience
	I show empathy

also being adapted in the growth and evaluation models for in-service teachers. For example, it is entirely possible for a teacher to set up a video camera in the back of the classroom and create a videotape of a lesson taught that can be viewed and critiqued after that lesson ends. When viewing the videotape, use the appropriate items from the ETI-SBAE to conduct a self-assessment of the lesson to evaluate and reflect on the performance based on daily experiences in both formal and informal teaching environments.

The Importance of Teacher Reflection

Regardless of the profession a person chooses to enter, reflection is a key aspect to improving that person's ability to do their job well. It is especially true for teachers, however. Reflection is a key component related to teaching effectiveness and is often included as an integral aspect of pre-service teacher preparation programs. In her book titled, *Mastery Teaching*, Madeline Hunter (1993) stated that effective teaching is "a constant stream of professional decisions made before, during, and after interaction with the student" (p. 3). Therefore, teachers should reflect on their teaching constantly and not just after a teaching experience. Specifically, teachers should reflect *before* students enter the learning environment. This begins with the class syllabus or schedule and the daily lesson plan. Planning for when certain topics will be taught throughout the course of a semester or school year and having rationale for creating a seamless transition between those topics from one week to the next is instrumental for planning a successful class. This includes preparing a variety of assignments and assessments that provide students an opportunity to apply their learning. Once completed, creating a daily lesson plan is critical to the mission of effective teachers. Those who reflect *prior to* students entering the classroom are more prepared to engage their learners at higher levels of cognition than those who do not. Therefore, thinking through the lesson's objectives or instructional goals, how to best develop students' interest in the topic, which teaching methods or instructional strategies should be used to engage the learners, which specific concepts should be taught, and the type of evidence needed to ensure learning occurs are all imperative (pre)flective actions an effective teacher takes prior to engaging with students in the learning environment.

Effective teachers also reflect *during* the lesson. Much can be learned by reading the body language of students. Are students focused on the lesson? Are they attentive and interested in the content being taught? Are they responding to questions? Do they sustain eye contact? Are they on task? If not, the effective teacher will make the decision to pivot the lesson by changing the pace and reengaging students in the learning process. This can be accomplished by asking a question and conducting a think-pair-share activity (i.e., ask a question, have students write out their answer, and then have them pair with a neighbor and share their ideas), using a different learning method or strategy, accessing a different media source or piece of technology, or telling a provocative story or providing an example relating to the topi, to name a few. Regardless of the approach taken, it is imperative that teachers remain attentive to the needs of their students, recognize when they are off task or confused, and respond accordingly. Doing so takes practice and intentionality. It does not come naturally for most teachers. Rather, it is a skill that must be developed over time but can be accomplished when teachers force themselves to reflect during their lessons. Seasoned teachers who excel at reflecting in the moment assume that their students need to work just as hard as they do during the lesson. They ask themselves frequently during a lesson: "What am I doing?" and follow it up with "What are my students doing?" If they determine that they are outworking their students, they will make a conscientious change in their teaching and transition to a student-led activity to empower the students and (re)engage them in the learning process.

Finally, effective teachers reflect *after* they have interacted with students. This usually occurs directly after a class has ended. The greatest question effective teachers ask themselves after a lesson is: "Were the objectives, or learning goals, accomplished?" To answer this question, it is important that the teacher collect evidence each class period that supports whether or not learning occurred. This can be as simple as asking questions at the end of the lesson, administering a quiz, or asking students to perform a *quick write* (i.e., a 60-second exercise where students write out all they learned that class period without the use of their notes), to name a few. It can also include an assessment of authentic products students produced during the class. Examples might include the quality of a weld, the steps taken to ribbon a soil sample, or the effectiveness of delivering a FFA speech. Regardless of the exact measures used, effective teachers understand that reflection is imperative to good teaching and that it must occur prior to, during, and after interacting with students.

Teachers should strive to be reflective practitioners for the following reasons: 1) reflection allows for problem solving and meaning making – when teachers allow themselves to reflect, they are able to formulate creative ideas for improving their teaching, which can relate to better student engagement and learning; 2) reflection allows for teachers to inquire into and understand the needs of their students – this phase of reflection enables teachers to understand the overall feel, vibe, and rhythm of the class and make adjustments where necessary; 3) reflection occurs on a continuum – reflecting on one's teaching can happen in the classroom, during an activity, after school, in the vehicle on the way to check a student's SAE project, after a FFA meeting, or any other time throughout the day or night. It can be an in-depth or quick exercise and can result in wholesale changes, slight modifications, or serve to reaffirm what the teacher is doing currently to help students learn the criterion; 4) reflection occurs best in a social system – reflecting with students, their parents, administrators, or other teachers in the school system is a great exercise that can bring about added clarity to the situation at hand. With agricultural education, it is vital that teachers reflect with students and their parents, especially as it relates to SAE projects when making home, or on-site, visits.

IDENTIFYING SOURCES OF FEEDBACK

Teachers receive feedback from a variety of sources including, school administrators and supervisors (i.e., principals, assistant principals, academic deans), state CTE directors or specialists, university faculty members, local advisory council members, school board members, peers, parents, and students, to name a few. Each of these feedback mechanisms can help inform SBAE teachers of their shortcomings as well as their strengths. One of the biggest challenges of teaching SBAE is knowing which sources are most important to listen to and when. For instance, at times, it is difficult to prioritize the suggestions being given. It becomes imperative that SBAE teachers develop a sense of evaluating their work and having a good understanding of their product – whether it is a lesson they taught, a CDE team they trained, or a SAE visit they made. Knowing one's work and when it is deemed effective does not come easily to most people. In fact, it occurs as a result of years of receiving feedback from people in positions of authority (i.e., faculty in teacher preparation programs, cooperating teachers, peers, and administrators and supervisors) and reflecting on the feedback received.

Preservice Preparation

Coursework Related to Content and Pedagogy

It is imperative that SBAE teachers understand both what to teach (i.e., content) and how to teach it (i.e., pedagogy). Currently, SBAE teachers are expected to be competent in and able to teach across the eight National Agriculture, Food, and Natural Resources (AFNR) Career Pathways. These eight pathways were developed by the National Council for Agricultural Education and include: Agribusiness Systems, Animal Systems, Biotechnology Systems, Environmental Service Systems, Food Products and Processing Systems, Natural Resources Systems, Plant Systems, and Power, Structural and Technical Systems. Therefore, it is vital that SBAE teachers receive instruction in courses that encompass these eight career pathways.

In addition to having a robust knowledge of and exposure to the content, pre-service and inservice SBAE teachers also need to understand pedagogy. Fortunately, teacher preparation courses provide coursework in pedagogy that prepares students to teach. Such coursework includes instructional design, teaching methods, and assessment strategies, to name a few. Often, students have practicums which allow for the application of the content. Microteachings are a simulated teaching experience in which students teach the lesson they designed to their classmates who, in turn, critique them and provide peer feedback. In addition to the presentation of the lesson, microteachings should include a student-made lesson plan outlining the details of the lesson taught and a structured reflection session afterward. Advice, suggestions, and feedback should be shared by the expert instructor of record (i.e., faculty member) on the lesson plan and during the reflective session. Experts should emphasize a coaching model to help the student improve his or her teaching effectiveness.

Early Field-Based Experiences

Another example for acquiring human capital in teacher preparation programs is early field-based experiences (EFEs). EFEs exist as requirements across most teacher preparation programs; however, the expectations ranging across all programs is vastly different. In short, EFEs allow practicing (i.e., pre-service) teachers the opportunity to acquire significant learning experiences by observing inservice teachers in the field. EFEs are usually a collaborative, triadic experience between the student, teacher, and university supervisor. They generally include students collecting data at a specific site. Examples of data collected might include the type of irrigation system installed in the greenhouse, the layout and orientation of the classroom, the number of students enrolled in the courses offered, and an inventory of the hand tools in the agricultural mechanics laboratory. These experiences occur prior to the student teaching internship and are critical for helping students see the bigger picture of the overall program, (re)affirm their intentions to become teachers, identify important learning strategies and outcomes, recognize the connection of and relationship among the total program, and increase their desire to teach.

Student Teaching Internship

In addition to EFEs, preservice teachers also participate in a student teaching internship. Similar to EFEs, the student teaching internship ranges in duration, expectation, and rigor by teacher preparation program. In some states, students may be expected to student teach for an entire academic year. In other states, students may practice their student teaching for one full semester or even part of a semester. Regardless, the student teaching internship is designed to allow the pre-service teacher to practice under the guidance of an expert (i.e., cooperating) teacher. The internship serves as a capstone experience and encompasses the learning acquired throughout the duration of the student teacher's academic career. The teacher preparation program should work to ensure that the student teaching interns learn from the best potential cooperating teachers possible. Such teachers in cooperating centers should be vetted stringently to determine their mentoring ability and perceived fit with the student teaching intern.

Throughout the duration of the student teaching internship, students practice their teaching by participating in all aspects of the SBAE program (i.e., classroom and laboratory instruction, FFA, and SAE). Student teachers invest their time in a variety of ways during their internships. The time invested in activities can vary significantly depending on which semester interns are engaged in student teaching. For instance, interns from one land grant university who student taught in the fall semester spent more time in the classroom (both observing and teaching) than did those who student taught in the spring. In contrast, those who student taught in the spring semester spent more time at FFA events and with SAE projects than did those who student taught in the fall. Interns should be encouraged to participate in a variety of teaching-related activities while student teaching. Although all aspects of student teaching are important to developing an interns' human capital, participating in heavy grading assignments throughout the duration of the student teaching experience have been shown to negatively affect their decision to enter the teaching profession as full-time teachers once the internship concludes.

In-Service Professional Development

Administration and Supervisor Feedback

Feedback from supervisors occurs in a variety of ways. Most school districts across the country implement a formal observation model for providing structured feedback using a defined assessment tool. In most states, once a SBAE teacher is employed in a school district, that school's administrator is required to observe and provide feedback to their teachers once per semester. In addition, in some states, a faculty member and mentor will provide a series of classroom observations to the teacher, especially in cases where a teacher induction program is in effect for early-career teachers. A state CTE director or specialist may also provide a formal observation of the SBAE teacher. These observations should not be looked at as punitive. Instead, the SBAE teacher should welcome these observations as they provide feedback necessary for implementing changes to become, or sustain as, an effective teacher. One of the most frequent forms of feedback for SBAE teachers is that which is provided through a formal observation.

Formal Observations

Teachers, regardless of subject area, should be evaluated using objective data. When teachers are not evaluated objectively, they lose the opportunity to develop and improve their craft as professionals. The main prerogative behind effective teachers is student learning. Although exceptions exist, in the main, if students are learning the content, it can be assumed that the teacher is effective. In these cases, teachers desire student performance data to be included in evaluations because it allows them to be recognized for their efforts.

When such teaching evaluation systems are not perceived as being effective at achieving their goal, however, it is essential that administrators consider new methods and develop a more beneficial evaluation program. Doing so might result in a model that includes additional formal observations from administrators along with follow-up meetings with the teacher to set goals related to their professional growth and improvement. For this model to be effective, there are three keys to success: 1) a mutual bond of trust and respect must exist between teachers and administrators, 2) teachers must make strides to improve their effectiveness and overall performance, and 3) there must be an increase in student performance.

Overall, teachers tend to have a positive perception of evaluation systems when they are implemented effectively. Because SBAE teachers are empowered to teach, advise youth, and supervise students' projects, observations could occur across a variety of the teachers' job duties and expectations. In the end, SBAE teachers should openly invite the opportunity to be observed and receive formal feedback from an administrator. Doing so can improve the relationship between teachers and administrators while allowing teachers to showcase their growth and resilience as an educator and observe their students improve their own knowledge of the agricultural food, fiber, forestry, and natural resources industry.

Formal observations allow teachers to receive important feedback on their effectiveness across the total program (i.e., classroom and instruction, FFA, and SAE). Another form of feedback is that which is conducted through a formal assessment of one's competence.

Assessment

Numerous teaching evaluation systems exist and are used to measure teaching effectiveness on a variety of levels. Unfortunately, there is a lack of consensus related to teaching evaluations and components associated with measurement. Schools often implement the evaluation tool that is mandated by the state or local school district, while others have the opportunity to develop a school/district specific evaluation model. Instruments have been developed to assess the atmosphere of the various learning environments, instructional effectiveness of the teacher, management of students in the program, and student engagement (AIMS). Such instruments attempt to evaluate the complex nature of effective teaching through conducting classroom observations. Other evaluation tools include the teacher performance assessment instrument (TPAI), the teacher assessment and development system-meritorious teacher form (TADS-MTP), and the Florida performance measurement system (FPMS). These measurements have been developed and are viable instruments for teachers at all levels (i.e., pre-service teachers, early career teachers, mid-career teachers, and late career teachers). Although some instruments are designed to fit the needs of inservice teachers, the classroom observation and assessment scale for teaching candidates (COAST) was developed to be an accurate, one-size fits all type instrument. Others are used for teacher certification including the edTPA and PPAT assessments. These assessments vary state to state, discussing certification models with your teacher preparation faculty for your citation is pivotal.

In addition to these instruments, the Marzano model was developed. The Marzano was created to provide "teachers and observers a streamlined, student evidence-based system that ensures standards alignment and helps promote growth in each student and teacher" (Marzano, 2018, p. 2). Therefore, it aims to connect student achievement to instructional styles. Ultimately, the Marzano model was developed to create an evaluation system based on research and classroom observations from over one-half of a decade to meet the challenges of the educational system at the time. This model was developed and implemented on a state-by-state basis beginning in the late 2000's as the federal government provided incentives to those states that adopted teacher evaluations which included measures of student growth.

Soon thereafter, legislation related to the *Every Child Succeeds Act*, that required student performance be included in teacher evaluations, was revoked. Nevertheless, the Marzano model is popular among some school districts.

Teaching evaluations should be used to identify and improve the weaknesses of instructors. Goe et al. (2008, p. 52) suggested six key components to consider when assessing and measuring teacher effectiveness:

- Resist pressures to reduce the definition of teacher effectiveness to a single score obtained with an observation instrument.
- Consider the purpose for the evaluation of teacher effectiveness.
- In considering the validity of various ways of measuring teacher effectiveness, keep in mind that the validity does not lie solely with the quality of the instrument or model but also with how well the instrument measures the construct and how the instrument is used in practice.
- Seek other measures, or create appropriate measures, to capture important information about teachers' contributions that go beyond student achievement.
- Include education stakeholders in decisions about what is important to measure.
- Ensure that data is complete and accurate and that raters are trained and calibrated is essential in order to ensure validity.

It has been determined that teacher performance evaluations are only as effective as the feedback provided to the teachers. Regardless of the metric used to evaluate teachers, the system should relate to accountability and teacher development. Student achievement data itself should only be a component of teaching evaluation, not the sole measurement, as appropriate teaching evaluations can have a positive impact on the development and implementation of new skills. In the end, there are multiple forms of feedback that should inform a teacher of his or her effectiveness.

Student Feedback

Effective teachers also use student feedback to improve their teaching. Specifically, there are two types of feedback mechanisms: 1) Formative assessment and 2) Summative assessment. Formative assessment is feedback which occurs before or during the unit of instruction or activity. It is often used to inform the teacher of the students' current knowledge or understanding of the topic. Formative assessments can be graded, but they usually exist only to serve the instructor with information and a point of reference for where to begin the lesson and the depth and pace at which to teach it. As such, formative assessments are viewed as relatively low stakes tests and are used to inform both students and teachers of the progress, or lack thereof, being made toward the learning goal outcomes.

Oftentimes, instructors can use formative assessment data to improve their teaching and student learning. An example is to ask students to respond to three questions: 1) What am I doing that is helping you learn?; 2) What am I doing that is hindering your learning?; and 3) What do you wished I would do to help you learn better? These three questions have been shown to be effective at helping college teachers reflect and make positive modifications to their teaching. An approach to obtaining more accurate responses from students regarding these three questions is to type them up in a Word document, disseminate them to students, and ask them to anonymously respond and place them in a labeled folder on the teacher's desk or shelf. Once completed, the instructor should review the responses earnestly and seek to make changes where necessary throughout the following weeks. A best practice to show goodwill and sincerity is for the instructor to announce the themes that arose to students after reading the comments and provide ideas for how he or she intends to make changes to better accommodate students' needs and desires.

Summative assessments can range from a unit examination to a semester test. Regardless, they usually include criterion-referenced or standardized test questions in which students choose correct answers from a bank of options or among a set of distractors (i.e., matching and multiple-choice). However, summative assessments can also include authentic evaluations of students' performance, such as projects.

Although a uniform standardized test does not currently exist for students enrolled in agricultural education programs across the country, many states are placing emphasis on industry certifications. These industry certifications vary state by state and by program area, but many school districts have implemented them as a measure of student performance. Serving as a summative assessment of student knowledge related to course content, these evaluations are often linked to teacher performance as well.

Regardless of the type of assessment used, the point with testing students is to determine how much they have learned and their readiness for additional instruction. Students' assessment performances should be a direct reflection of the teaching effectiveness of the instructor. Therefore, teachers should provide their students with the necessary tools for success, such as rubrics, checklists, and exemplar work from previous students. Teachers should then evaluate their students' work carefully and consider areas in which they can enhance their teaching to improve student learning.

Teachers who are serious about improving their effectiveness seek out peers from whom they can be mentored. The school system is full of teachers who can offer insights, suggestions, and mentoring. As such, teachers should feel free to ask for assistance by having a peer observe a class and provide feedback and constructive criticism. It is not necessary that the mentor be another SBAE teacher. In fact, in some cases, having someone outside the discipline who is largely unfamiliar with the content can be better. Creating such a culture for peer mentoring has the potential to improve the entire cadre of teachers in a school system. It also is a way in which teachers can acquire and develop necessary human capital.

OPPORTUNITIES FOR CONTINUAL IMPROVEMENT

Developing Human Capital based on Personal Strengths and Weaknesses

Human capital is an investment a person makes in his or her education, skills, training, and experiences. In turn, various employment sectors seek specific human capital to perform various tasks associated within their chosen industry or profession. This is especially true of education. "Human capital is the largest single investment K–12 districts make" (Myung et al., 2013, p. 3). Determining the specific human capital necessary for teachers who work in school systems is an imperative, yet difficult, task.

Identifying effective teaching qualities of candidates can be difficult for principals who are tasked with making hiring decisions. Some place emphasis on pedagogical knowledge while others require strong evidence of content expertise. However, regardless of certification status, gender, and previous work experiences the candidate had attained in college, the human capital that matters most to hiring officials was found to be their academic rigor, i.e., grade point average. Hiring officials of school systems focus on four aspects of human capital when seeking teachers: 1) acquire the best available teachers possible, 2) develop teaching effectiveness through intense professional development, 3) sustain teachers with a

Figure 1. Teacher human capital framework

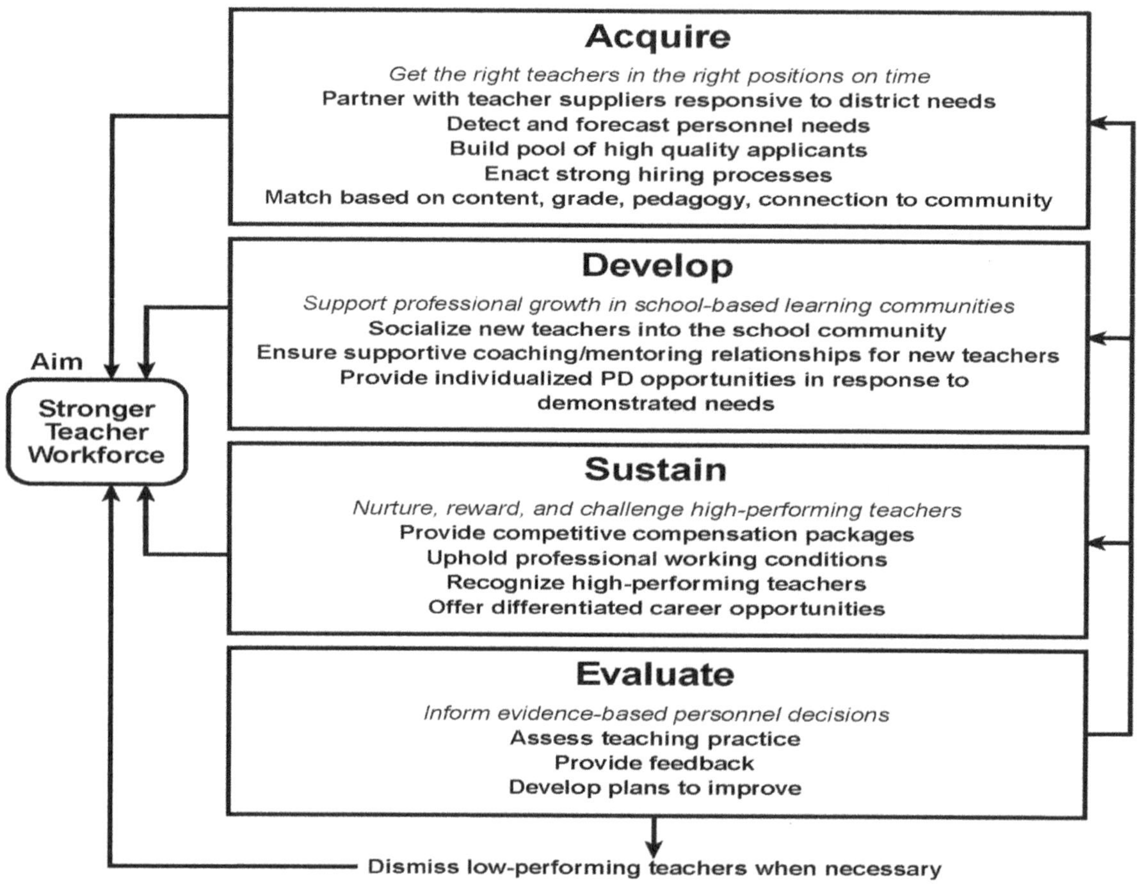

proper range of challenge, and 4) evaluate teachers and help develop professional development plans (see Figure 1).

Specifically, it is imperative that school systems acquire the best teachers at the right time. These individuals should bring unique expertise to the needs of the school district. Their education, experiences, and skill set should be a good fit for the community and match for the school. Because teaching is a lifelong learning endeavor, it is imperative that school systems continue to develop their teachers over time. Support structures should be put in place and enacted. Such structures should encompass coaching and mentoring strategies as well as offering specific professional development to improve the teacher's craft.

Ongoing efforts should exist to sustain teachers once they are employed. Specifically, teaching excellence should be acknowledged and rewarded by the school system. An emphasis should be placed on the school's culture and professional working environment. Finally, after a safe and enjoyable culture has been established, structures should exist to continuously evaluate the teacher with appropriate feedback mechanisms for improvement. The school should work with the teacher to acquire, develop, sustain, and evaluate the necessary human capital over time. To conceptualize this process, the Teacher Human Capital Framework has been developed (see Figure 1).

Figure 2. A conceptual model of effective teaching for SBAE teachers

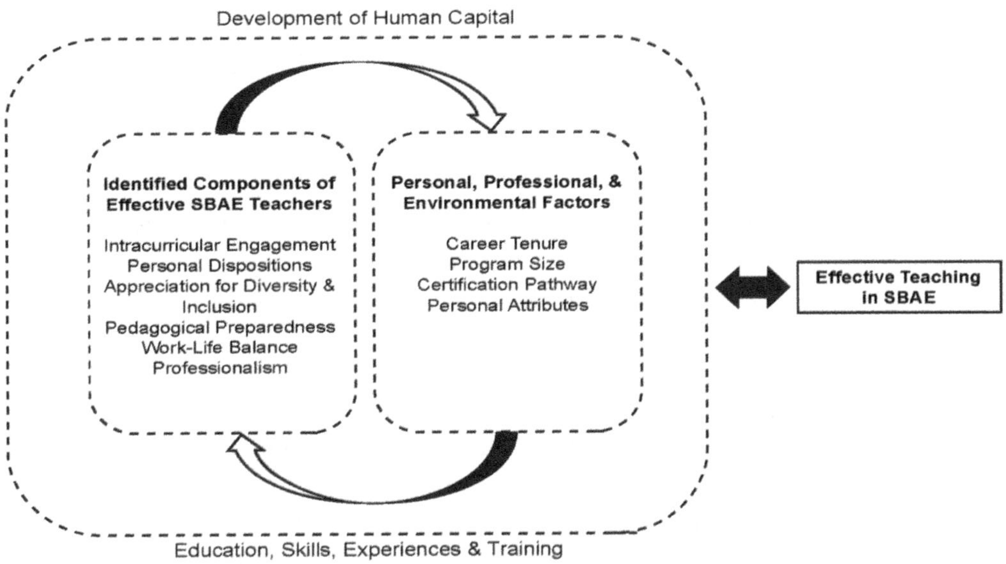

The aim of the figure is to develop a stronger teacher workforce. To do so, school systems should focus on acquiring, developing, and sustaining effective teaching. Creating such a culture is imperative for developing more effective teachers, which directly impact student learning. Evaluating effective teaching does not directly impact the aim. Rather, it exists to inform decision makers of the low-performing teachers that need to be released of their duties. By relieving poor performing teachers of their duties, the aim of creating a stronger teacher workforce is realized. Therefore, the evaluate subsystem has an indirect impact on the overall aim.

Regarding the model, all four subsystems are connected. This implies that none of the four can impact the aim alone. Instead, it requires all four working in tandem to accomplish a stronger teacher workforce (see Figure 1).

The Teacher Human Capital Framework seeks to accomplish a stronger teacher workforce for all teachers regardless of discipline. But how does it apply specifically to SBAE programs, especially when these programs are so diverse and different from other subject areas taught in school systems? SBAE teachers can acquire, develop, and sustain their human capital through their coursework and early field-based experiences received in their teacher preparation program, the student teaching internship, professional development opportunities, and mentoring programs.

The development of human capital related to effective teaching in SBAE should consider the conceptual model of effective teaching for SBAE teachers (see Figure 2). The model encompasses the identified components of the ETI-SBAE along with personal, professional, and environmental factors within the development of human capital (i.e., education, skills, experiences, and training), leading to effective teaching in SBAE.

In addition to the components listed in Figure 2, a criterion for effective teachers is an understanding of when and how to use various teaching methods (i.e., instructional strategies) in the classroom. A focus on such is listed below in section 3.

Effective teachers understand and exemplify the basics of good teaching. They appreciate the importance of building professional relationships and rapport with students. As such, they ask questions before and after class to determine what other interests students have and the activities for which they are involved, attend other school-related events (i.e., football games, school plays, etc.) to show their support to students, and express a genuine and sincere sense of care for their students by showing them proper respect and including them in the learning environment regardless of their differences.

In addition, effective teachers specifically understand when and how to use and navigate among a variety of instructional methods or strategies to engage their students. Simply stated, effective teachers are those who continuously learn. As such, they set goals for their personal and professional improvement and seek out ways to become students themselves.

Goals for Improvement

Both pre-service and inservice SBAE teachers should have a growth mindset. This is accomplished by setting goals for improvement. Such improvement can occur through participation in professional development opportunities, seeking mentorship, and the establishment of an individualized professional development plan. SBAE teachers should work with select local administrators, advisory council, teacher preparation faculty, and CTE staff to create a plan for improvement. This plan should include aspects related to the individual's attempt to be a lifelong learner.

Professional Development

The examples mentioned above include experiences which occur before graduation from a higher education institution. The final example of how SBAE teachers can acquire human capital is through professional development once the teacher has graduated with a degree and accepted a position in a secondary school. Teachers should strive to be lifelong learners, so obtaining professional development is imperative to a teacher's long-term success. Often, individual states will offer induction or onboarding programs and professional development workshops for their SBAE teachers during the summer months. These can be joint activities offered between the state's teacher preparation program's faculty and its CTE staff. In addition, the National FFA Organization offers professional development workshops for teachers each year at the National FFA Convention and through its National FFA Ambassador Program. Effective SBAE teachers should continuously seek ways to improve their craft – both technically, via content knowledge, and pedagogically. One way to continually improve is to seek mentoring from other perceived effective teachers.

Mentorship

Once teachers have been hired, it is imperative that they form a community with other teachers at the school and SBAE teachers across the state. Doing so provides some accountability measures and ensures greater connections, relationships, and overall growth as a person and professional. Ultimately, teachers should determine whether they are a good fit for the position they have assumed. Mentors can help with identifying the fit of prospective teachers. Specifically, teachers should consider the four types of fit: 1) Person-job fit (PJ): this type of fit focuses on how the person matches the demands and requirements of the specific job. In SBAE, it is possible that each local program has certain aspects of its program that it values higher than others. The opposite is true also. There are certainly aspects of the programs that are devalued across a variety of SBAE programs. It is the teacher candidate's job, with the potential help of mentors, to determine the PJ fit in a specific area. 2) Person-organization fit (PO): this type of fit assumes the person understands and aligns with the culture and values of an organization. Regarding SBAE teachers, this might include having a shared philosophy of agricultural education with the administrative leaders of the school. It might also mean that the expectations of the program are compatible between and among the people of interest, such as the teacher, administration, and school board, to name a few. 3) Person-group fit (PG): this type of fit assumes a good working relationship exists between the SBAE teacher and those at the school with whom they will be working most closely. Examples might include an additional SBAE teacher (if it is a dual-teacher program), other CTE faculty, and the school board. 4) Person-community fit (PC): this type of fit is the connection of teachers to their "students' culture, background, and place in the community" (Muyung et al., 2013, p. 17). Assuming these aspects of human capital fit are positive, then the teacher is more likely to remain at the school and be successful.

For SBAE teachers, the perceived fit may include the expectations related to the classroom, FFA, and SAE projects, among others. Teachers may seek to understand the expectations of the administration, students, parents, booster clubs, advisory council, and business owners when determining their fit in the community. For instance, is the community interested in exhibiting livestock at a high level? Is there an expectation, or even demand, of the administration to produce state winning career development teams each year? Do parents expect teachers to provide a robust and high-quality education canvassing all career pathways? Do businesses expect the FFA program to provide frequent community service opportunities? All of these questions and more can help teachers determine their fit and effectiveness in the community long term.

Establishing a Professional Development Plan

Understanding the training needs of teachers is critical, as stakeholders agree that acquiring and sustaining effective teachers is the primary component related to providing quality K-12 education across the country. The Council of Chief State School Officer's (CCSSO) Interstate Teacher Assessment and Support Consortium (InTASC) highlighted the need for teachers to better understand the needs of their students, allowing teachers to customize instruction to meet learners' individual differences, such as students with disabilities or students performing above grade level. Inservice SBAE teachers often need training on 21st century skills, including math, reading, and writing in the context of teaching the agricultural curriculum, as well as student leadership development to foster problem solving skills, and higher-order and critical thinking skills.

To help conceptualize the development process for effective SBAE teachers, the human capital development framework for effective SBAE teachers was developed (see Figure 3). Similar to the Teacher Human Capital Framework, this model includes the need to acquire, develop, sustain, and evaluate SBAE teachers throughout the development of the essential components of effective SBAE teachers. This process is a continual cycle to sustain an effective SBAE teacher workforce. The established framework (see Figure 3) should be used as a guide for the preparation and continual development of SBAE teachers, preparing them to deliver effective instruction throughout a complete SBAE program (i.e., classroom/laboratory instruction, FFA, and SAE).

Figure 3. Human capital development framework for effective SBAE teachers

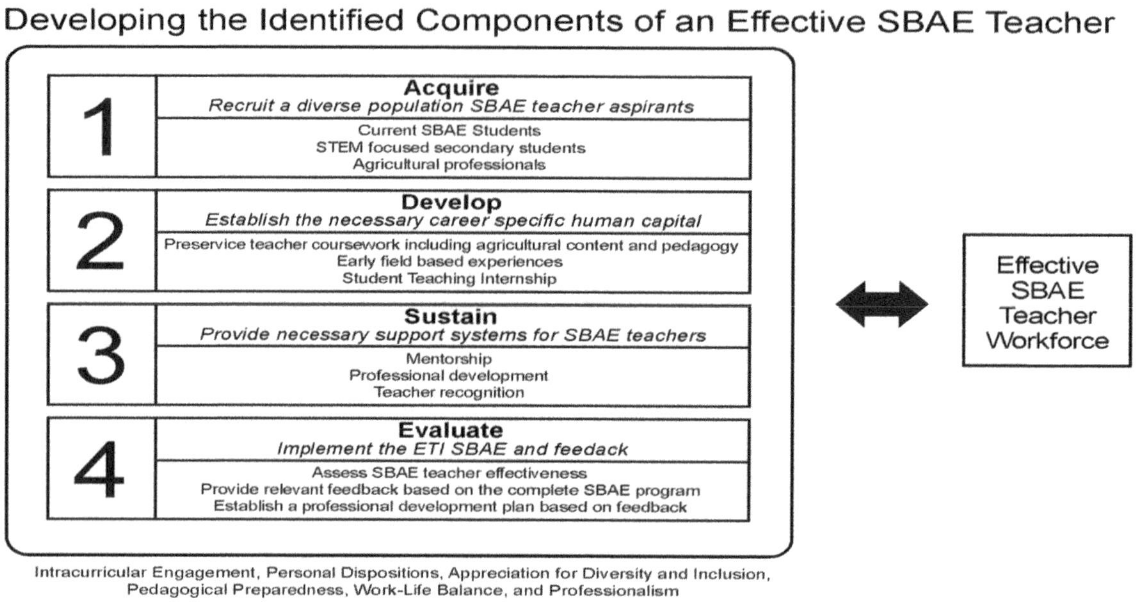

SUMMARY

Being an effective teacher is a lifelong pursuit. Although the needs of SBAE teachers vary nationwide, the identification of these needs early and often can help not only improve the capacity of inservice teachers, but also the preparation of future SBAE teachers. Therefore, it is essential that pre-service and inservice SBAE teachers consider and evaluate their personal and professional needs regularly. This process allows teachers an opportunity to reflect and establish a professional development plan. When establishing an individual professional development plan, SBAE teachers should consider the ETI-SBAE, the teacher human capital framework, and the conceptual model for effective teaching in SBAE as guides to determine personal and professional components in need of refinement or further development. SBAE teachers should work to form connections with other teachers in the school district and SBAE teachers across the state. In addition, they should seek mentors and invite them and their administrators to conduct frequent formal observations of their teaching. Doing so will provide SBAE teachers with data for which they can reflect, make necessary changes, and add to their human capital repertoire all for the goal of engaging their students to succeed at higher levels.

REVIEW QUESTIONS

1. What are the components of an effective SBAE teacher?
2. What are the primary sources of feedback a SBAE teacher should seek to improve their effectiveness?
3. How can pre-service SBAE teachers acquire, develop, sustain, and evaluate their human capital?
4. What pedagogical knowledge and experiences are necessary for effective teaching in SBAE?
5. What content knowledge is important for SBAE teachers to acquire prior to entering a classroom?

6. What steps are necessary to creating a professional development plan for SBAE teachers regarding the improvement of their human capital?

REFERENCES

Goe, L., Bell, C., & Little, O. (2008). *Approaches to evaluating teacher effectiveness: A research synthesis*. National Comprehensive Center for Teacher Quality.

Hunter, M. (1993). *Mastery teaching*. TIP Publications.

Marzano, R. J. (2018). *Marzano focused evaluation models*. https://www.marzanocenter.com

Myung, J., Martinez, K., & Nordstrum, L. (2013). *A human capital framework for a stronger teacher workforce.* White paper. Carnegie Foundation for the Advancement of Teaching.

Chapter 14
Providing Teacher Certification and Professional Development Programs

Rebecca Grace Lawver
Utah State University, USA

ABSTRACT

Agriculture teacher preparation programs, whether traditional or lateral entry, should prepare school-based agriculture teachers with the technical content, pedagogical knowledge, and professional skills to be successful educators in the classroom. This chapter explores the preparation of agriculture and agriscience educators for the classroom through an overview of the types of teacher certification options, describing the components of an agriculture teacher preparation program, including the Standards for School-Based Agricultural Teacher Preparation, exploring opportunities for agriculture teacher professional development, and investigating opportunities to enhance teaching excellence and technical skill enhancement through systematic needs assessment.

INTRODUCTION

Great agriculture teachers are committed to students and their learning, they know the content and subjects they teach and how best to teach it. They can critically reflect on their teaching to improve their practice and are an essential part of the communities in which they serve. To have great agriculture teachers there must be great agriculture teacher preparation programming. Traditional agriculture teacher preparation occurs at post-secondary institutions. On average, 100 institutions across the country prepare 750 school-based agricultural education (SBAE) teachers at post-secondary institutions who are licensed to teach agriculture. While traditionally prepared new hires in SBAE represent the majority of new teachers, alternatively licensed teachers have become necessary to fill teaching shortages across the country. SBAE teachers need to be well educated, acquire a special blend of content and pedagogy, and continue in-service teacher professional growth throughout their careers. Whatever the path, preparing agriculture and agriscience educators for the classroom is a vital component to ensuring that novice teachers are

DOI: 10.4018/978-1-6684-3420-8.ch014

ready to succeed in the classroom and that every student is taught by a great educator (USDE, 2016). The objectives of this chapter are to 1.) differentiate between the various types of teacher certification options, 2.) describe the components of an agriculture teacher preparation program, 3.) identify opportunities for agriculture teacher professional development, and 4.) evaluate teacher technical skills to enhance practice.

DIFFERENTIATING AMONG TYPES OF TEACHER CERTIFICATION PROGRAMS

Teacher certification or licensure is the credential that gives educators the certification to instruct students in a specific content area and at a specific level. All states require teachers to obtain certification to work in public schools. To become certified, you are required to complete a bachelor's degree from an accredited teacher education program, complete a student teaching internship, and pass a background check.

Traditional Teacher Preparation

A traditional agriculture teacher preparation program includes a four-year Bachelor of Arts or Science degree program that incorporates general education courses, technical agriculture content, professional or pedagogical courses, field experiences, student teaching, internship or residency in a school, and state licensure exams. Teacher preparation programs are accredited through either the Council for the Accreditation of Teacher Preparation Programs (CAEP) or other accreditation bodies such as the Association for Advancing Quality in Educator Preparation (AAQEP). Accreditation provides several important benefits to teacher education programs that include outcomes-based evidence that supports learning, provides an accountability system, allows for structured reviews, improved program effectiveness, and third-party validation. Overall, these evidence-based evaluations are grounded in continuous improvement to ensure teacher education programs produce successful teachers.

In agricultural education, traditional teacher preparation programs in the U.S. produce more than 80% of the nation's SBAE teachers (Lawver et al., 2018). For many undergraduate students, the conventional path is to attend a 4-year accredited college or university to pursue a bachelor's degree in agricultural education that leads to teacher licensure. Students may also attend a community or junior college and complete lower division or pre-requite courses before transferring to a baccalaureate institution to complete their degree.

Alternative or Lateral Entry Teacher Preparation

While the traditional route to certification typically requires the completion of a postsecondary degree, many alternative route programs are available for aspiring educators who did not complete a teacher education program. Opportunities exist for individuals who may have a baccalaureate degree, but not necessarily in education, and who have met subject matter competency requirements in agriculture to learn to teach through a district or university alternative certification or internship program (National Council on Teacher Quality, 2021a). These programs place teachers in an agriculture classroom, with supervision by mentors, while they simultaneously complete the requirements for an alternative teacher preparation program. In agricultural education, approximately 20% of all new hires are alternatively licensed.

Many states have one or more alternative routes to certification and to accept students who have a bachelor's degree in another field and deliver the pedagogical education and field experience elements required for certification. While each state requirements for lateral entry or alternative teacher licensure are different, in general, the path may include formal alternative teacher preparation programs, Transition to Teaching, Teach for America, National Board for Professional Teaching Standards certification, Career and Technical Education certification, Teaching residency programs, emergency, and provisional certifications, district-specific training, graduate coursework, and teaching portfolio examinations (National Council on Teacher Quality, 2021b).

Alternative or lateral entry teacher preparation certification programs are an important piece to reducing the agriculture teacher shortage. These programs provide for more teachers and provide flexible certification options to individuals who did not earn a bachelor's degree in agricultural education but who may have experience and content-specific knowledge which brings real-world experience to the classroom.

Graduate Route to Licensure

While the most common path for agriculture teacher preparation is the four-year undergraduate bachelor's degree, some colleges and universities require a two-year graduate program or a bachelor's plus one graduate program for agriculture teacher licensure.

Graduate routes to licensure are programs designed for individuals who have a bachelors' degree in agriculture who desire to become SBAE teachers. The graduate route to licensure may provide students with the coursework necessary to be certified as an agriculture teacher and may include courses in teaching methods, classroom management, curriculum development, etc. These programs may also provide individuals with a master's degree in addition to licensure.

Teacher Certification or Licensure

Once teacher candidates complete the requirements for their undergraduate, graduate, post-baccalaureate or lateral entry program, candidates will be recommended for teacher certification or licensure. Agriculture teacher certification is a process used in all 50 states to ensure that prospective agriculture teachers are adequately prepared to teach at the level and in the subject(s) they wish to teach. While each state has its requirements, the process involves having at least a bachelor's degree or higher, completing a state-approved agriculture teacher preparation program, completing a student teaching experience, passing a knowledge and skills test (i.e., state-approved content assessment, pedagogy assessment, Praxis Core Academic Skills for Educators (CORE), Praxis II Content area exams, etc.) or other professional teacher performance assessment (i.e. state-approved assessments, PPAT® Assessment (ETS, 2021), etc.) and passing a background check (AAQEP, 2022; CAPE, 2022). Many states also require some amount of professional development or continuing education for teachers to maintain or renew their licenses.

Additional Certifications

National Board Certification

National Board Certification is an advanced teaching credential that goes beyond state licensure (NB-PTS, 2021). There are national standards identified for what accomplished teachers should know and be able to. Candidates undergo a rigorous peer-reviewed certification process to demonstrate the skills necessary to be an excellent teacher.

Achieving the National Board for Professional Teaching Standards certification is a voluntary program for teachers who desire advanced certification or where states may provide financial incentives, recognition, educational credits, or leadership opportunities (NBPTS, 2021). In addition, research suggests a positive impact on student learning when students are taught by a board-certified teacher.

COMPONENTS OF AN AGRICULTURE TEACHER PREPARATION PROGRAM

The preparation of individuals who teach middle or high school agricultural education begins in post-secondary agricultural teacher preparation programs. The requirements of agriculture teacher education programs across the country vary greatly depending on college admission, prerequisites, course requirements, experiences, and licensure requirements. The Standards for SBAE Teacher Preparation Programs were developed and adopted by the American Association for Agricultural Education to establish a baseline for agriculture teacher preparation programs across the country and to provide overarching learning goals for graduates of an agricultural education teacher preparation program (AAAE, 2017; Birkenholz & Deeds, 2010). Students who graduate from these programs are expected to earn a license to teach school-based agricultural education. The standards are not considered a curriculum, but as a guide for teacher educators to plan and evaluate their programs.

In addition to the standards for SBAE, teacher education programs should include a broad education for preservice teachers that include additional components comprised of general education and on-and-off campus laboratory and clinical experiences (Cruickshank, 1996).

General Knowledge

All institutions of higher education are required by their accrediting body to offer some form of general education. While students who arrive on campus might want to dig into their major immediately, and sometimes view general education as a checklist of things to do, there are merits to general education courses.

General education courses may consist of foundational concepts in liberal arts and sciences, writing, critical thinking, diversity, and global perspectives (Warner & Koeppel, 2009). They broaden a student's knowledge and understanding of the arts, humanities, and sciences. The general education curriculum focuses on seeing the world through an interdisciplinary lens and is critical for preparing well-educated future teachers with a foundation that supports behaviors that lead to lifelong learning.

While preparing pre-service teachers with specialized knowledge for employability is a fundamental role of agriculture teacher preparation programs, it is the acquisition of general education that ensures that they become well-rounded college graduates (Myer & Dyer, 2004). Further, instilling lifelong learning skills in agricultural education graduates is vital for our rapidly competitive and changing world. General education provides students with essential skills including critical and creative thinking, communication, collaboration, problem-solving, and leadership skills (Rateau, 2015).

Content and Pedagogical Knowledge

Content knowledge is gained by agriculture students through their technical agriculture courses. It refers to the facts, theories, principles, ideas, vocabulary, and skills SBAE teachers must master to be effective. Agriculture teachers should have a deep understanding of the subject they teach and the curriculum they develop.

Yet, agriculture teachers need more than content knowledge; they need a special blend of content and pedagogy (Cruickshank,1996). Essential components in the development of an effective agriculture teacher include knowledge of the content they will teach and knowing how best to teach it (Ball & Knobloch, 2005).

The standards for SBAE Teacher Preparation Programs (AAAE, 2017), identify the knowledge and performance indicators that a 21st-century student completing an agricultural education teacher preparation program and who will be licensed to teach will know and be able to do. The Standards for SBAE Teacher Preparation Programs are not to be considered a curriculum, only as a guide for agriculture teach educators and institutions to utilize as they plan their programs.

Standard 1: Pedagogical Content Knowledge

Pedagogical content knowledge (table 1) is an agriculture teachers' interpretations and decisions about the agriculture subject matter and the context necessary to facilitate student learning. Essentially, it is the intersection of the subject matter, understanding how students learn the subject, and teaching strategies, materials, assessment of learning, and feedback used to teach the subject.

From the beginning, agricultural education has been experiential in nature; providing students with a variety of rich experiences including supervised agricultural experience (SAE) programs, field trips, project-based learning, and service-learning opportunities (Baker et al., 2016). Preservice teachers should be able to demonstrate the capacity to utilize experiential learning opportunities to encourage student development and construct an instructional delivery process that allows all students access to systematic experiential, service, and work-based experience learning. Teachers must connect to students' prior knowledge and personal interest and facilitate experiences that provide students with the opportunity to observe, hypothesize, experiment, and experience the content. Experiential learning is an engaged learning process where students learn by doing and reflecting on their experiences. Whether preservice teachers are teaching the process and dependence of organisms in the environment or teaching about the six nutrients animals need for good health, they must allow students to learn by doing and create opportunities to reflect on that learning. Agricultural education often provides the opportunity for learning to take place outside of the classroom in relevant agricultural contexts. Where the classroom provides more formal instruction, experiential learning opportunities, and out-of-classroom activities provide opportunities for the application of learning.

A positive learning environment (table 2) is created when teachers design a place where students feel safe, engaged, connected, and supported in their classroom. Before students can learn and succeed academically, they must feel safe. Schools use a variety of methods to ensure student safety. However,

Table 1. Experiential learning opportunities (AAAE, 2017)

Knowledge Indicators	Performance Indicators	Additional Indicators
• Highlight STEM concepts in lesson plan construction. • Implement safe and effective laboratory activities. • Facilitate student centered learning using a variety of approaches. • Communicate the reliability of sources to students.	• Plan lessons for instruction from a local and global perspective. • Implement experiential education. • Implement inquiry-based learning. • Utilize higher order questions. • Demonstrate a variety of teaching methods. • Implement research-based pedagogy and best practices for instruction. • Develop and sequence curriculum framework for SBAE program.	• Instruct students on the appropriate use of technology in the agricultural industry. • Appropriately utilize digital literacy and social media to enhance student learning.

teachers must create a welcoming environment, develop appropriate procedures, utilize effective classroom management, and apply appropriate motivational strategies. Students must also be engaged and connected to teachers, staff, and other students in the classroom. Preservice teachers must also be able to appreciate students as individuals and create a learning environment to encourage positive social interactions, self-motivation, and active learning.

Table 2. Positive learning environment (AAAE, 2017)

Knowledge Indicators	Performance Indicators
• Utilize effective classroom management strategies • Utilize effective age-appropriate motivational strategies • Convey excitement about subject matter to students • Demonstrate basic knowledge on meeting the needs of all students	• Demonstrate and display effective classroom management strategies • Engage all students through effective motivational strategies to ensure a positive learning environment • Develop and demonstrate instruction that fulfills the needs of all students • Implement inquiry-based learning.

Standard 2: Technical Content Knowledge (AAAE, 2017)

Technical content knowledge (table 3) refers to the body of knowledge, information that teachers teach, and students are expected to learn in a specific content area. This refers to the facts, concepts, theories, skills, and principles that are taught in the agriscience classroom.

The AFNR Career Cluster Content Standards provide state agricultural education leaders and educators with a high-quality, rigorous set of standards to guide what students should know and be able to do after completing a program of study in each of the AFNR career pathways. Those pathways include agribusiness systems, animal systems, biotechnology systems, environmental service systems, food products, and processing, natural resource systems, plant systems, and power, structural and technical systems.

Preservice teachers should also be able to recognize and appreciate the influence of diverse cultures on global agricultural production practices, possess global competence of issues relating to food and fiber systems, and realize the impact of agriculture/agricultural practices on global populations. Preservice students should have a broad understanding and awareness of curriculum development of relevant agricultural areas (AFNR) as well as be technologically literate.

Table 3. Technical content knowledge (AAAE, 2017)

Knowledge Indicators	Performance Indicators	Technology Indicators
• Broad awareness of AFNR content standards and quality program standards. • Develop curriculum based on community needs. • Know and possess scientific reasoning skills. • Know common core and academic standards role within agricultural education. • Design rigorous and relevant curriculum that integrates core academic standards • Possess knowledge and skills to operate equipment in Agricultural Education facilities	• Exhibit a broad knowledge of the agriculture industry. • Exhibit foundational knowledge of the basic skills and knowledge of each area in AFNR. • Exhibit knowledge of agricultural equipment found in facilities. • Exhibit awareness and utilization of AFNR Quality standards.	• Demonstrate awareness of cutting-edge technology in agriculture and its use. • Teach students how to use technology appropriate to the agricultural industry. • Appropriately utilize digital literacy and social media to enhance student learning.

Standard 3: Program Planning

The local SBAE program is a great asset to a community, developing a program that best meets the needs of the people living in the community is essential. Program planning (table 4) and evaluation is the continuous process where SBAE teachers, with the guidance and leadership of state staff, National FFA, and other stakeholders determine, analyze, and solve local problems through an SBAE program.

Preservice teachers must be able to meet the needs of program stakeholders, plan instruction based on knowledge of the subject matter, students, the community, and curriculum, integrate leadership and SAE and experiential learning development into the SBAE program. Clearly defining the program philosophy and objectives to the school administration, parents, and students is essential to the program planning process.

The leadership or FFA component of an SBAE program provides students with leadership, personal growth, and career success through agricultural education. FFA is the intracurricular student organization for those interested in agriculture and leadership, it is one of the three components of agricultural education.

The experiential learning or supervised agricultural experience (SAE) component of an SBAE program is designed for students to explore practical agriculture activities that are teacher-supervised, individualized, hands-on, and student-developed. They provide students with real-world experience to help them select a career, secure employment, or prepare for further education in agriculture.

To integrate leadership, FFA, Supervised Agricultural Experience, and experiential learning development into the SBAE program preservice teachers must also be able to do the following:

1. Comprehend the value of Career and Technical Student Organizations (FFA) toward the transfer of learned traits and abilities to real-world settings.
2. Possess a depth of knowledge of FFA and its associated programming.
3. Use FFA to compliment SAE and classroom instruction.
4. Comprehend the value of Supervised Agricultural Experiences and experiential learning to apply classroom knowledge in a real-world setting.
5. Possess a depth of knowledge of SAE and its associated programming.
6. Use SAE to compliment FFA and classroom instruction.

Table 4. Program planning (AAAE, 2017)

Stakeholder needs	Instructional planning	
	Knowledge Indicators	*Performance Indicators*
• Engage, manage, and sustain community partnerships within the AFNR program. • Engage business and industry stakeholders to provide currency to college and career ready skills related to AFNR pathways. • Differentiate the purpose and utility of a program advisory committee from other stakeholder volunteer groups. • Develop strategies to effectively publicize and differentiate student learning opportunities and resulting outcomes of AFNR programs to various audiences. • Interpret the contribution of an FFA Alumni Chapter to the instructional program	• Understand the importance of the FFA Alumni Chapter • Understand how to manage a total program in CTE, integrating the SBAE Model. • Identify the role SAE provides to career exploration and in determining career goals • Understand how to balance the classroom with maintaining an active FFA Chapter • Understand how to effectively communicate with students and parents. • Understand how to develop a positive relationship with community and administration	• Evaluate the effectiveness of a SBAE program. • Engage business and industry to guide and provide essential learning for students. • Exhibit an understanding of how to manage SAEs at all levels. • Exhibit FFA knowledge, participation, and leadership development. • Exhibit ability to work with adult volunteers and stakeholders. • Exhibit knowledge of child labor laws and their impact upon SAE. • Maintain inventory necessary for operation of laboratory, greenhouse, mechanics, and land laboratories used for student learning. • Organize, develop, and use a local advisory committee.

Standard 4: Diversity

SBAE teachers are expected to serve as role models for all students. As such agriculture teachers must also embrace and celebrate diversity by creating environments that build positive rapport with students, parents, stakeholders, and community members ensuring fairness and equity (AAAE, 2017).

Standard 5: Professionalism

Teacher professionalism is critical in helping young people achieve success in the classroom. Therefore, SBAE teachers must understand the significance of their own personal and professional growth to meet the needs of their students, school, community, and profession through dedication to all aspects of the discipline, exhibit professional responsibilities in the school environment, and demonstrate personal leadership traits (AAAE, 2017).

To show their disposition for professionalism, teachers of agriculture must demonstrate that they:

1. Engage in professional development opportunities in content, teaching, and experiences in the agriculture industry
2. Reflect on teaching and identify opportunities for personal and professional growth.
3. Utilize feedback to improve teaching and guide professional growth
4. Interact in a collaborative manner and foster positive relationships with stakeholders
5. Maintain a professional image and display professional behavior through ethical conduct
6. Fulfill professional responsibilities including engagement in professional education organizations
7. Value a service-minded approach
8. Model personal leadership traits involving the investment and empowerment of others

Exhibit professional responsibilities in the school environment.

1. Recognize the role of the agriculture program in contributing to the broad mission of the school
2. Exhibit a Collaborative Spirit
    a. Collaborates with other educators
    b. Makes efforts towards curriculum integration
    c. Engages in professional networking
    d. Maintains a positive rapport with school leaders
3. Decision Making

Standard 6: Personal Dispositions

Being a teacher is more than knowing the content or the pedagogy. SBAE teachers must also demonstrate professional attitudes, values, commitments, and ethics that facilitate student learning and convey a positive image of the agricultural education profession (AAAE, 2017).

An SBAE teacher should demonstrate caring, communicative, creative, and contextually aware dispositions. These include the following:

1. Caring Nature-The following qualities, tendencies, and core behaviors are demonstrated by a teacher education candidate:
    a. Compassion
    b. Patience
    c. Caring
    d. Empathy
    e. Unselfish
    f. Willingness to help others
    g. Inclusive spirit
    h. Embrace and celebrate diversity
    i. Passion
        i. For teaching and learning
        ii. For the agriculture content
        iii. For the entire agriculture industry
        iv. As a lifelong learner
2. Creative
    a. Inquisitiveness
    b. Creative and Critical thinker
    c. Problem solver
3. Contextually aware
    a. Organizational skills
    b. Results focused
    c. Manager of change
    d. Adaptable
4. Communicative
    a. Demonstrate professional communication skills (Speaking, Listening & Writing)

Early Field Experiences

The final component in the preservice agriculture teacher preparation program is early field experiences or those on-and off-campus laboratories and clinical experiences (Cruickshank, 1996). Field experiences are essential in the development of preservice teachers and help in decreasing their anxiety in entering the teaching profession, increase their desire to select a career in teaching, develop self-efficacy to teach effectively, understand the nuances of teaching, and expose them to how students learn through observation and practice.

On-Campus Experiences

In preservice agriculture teacher preparation classrooms across the country, students will assume the role of the teacher in a safe space among university teacher educators and peers. This provides an opportunity for preservice teachers to teach a brief lesson, receive feedback, and then reflect on the teaching they have done and what they have learned about the teaching and learning experience.

Clinical and Student Teaching Experience

A critical component of the traditional teacher preparation program is clinical experiences. These include field-based experiential experiences, including student teaching. These provide students with authentic observational experiences and structured teaching practice in schools. The average length of an agriculture or agriscience student teaching internship is 15 weeks, however, they vary across the country from a minimum of 10 weeks to a maximum of 36 weeks (Lawver et al., 2018). Figure 1 below depicts the typical progression through an agriculture teacher preparation program.

Figure 1. Typical progression through agriculture teacher preparation program

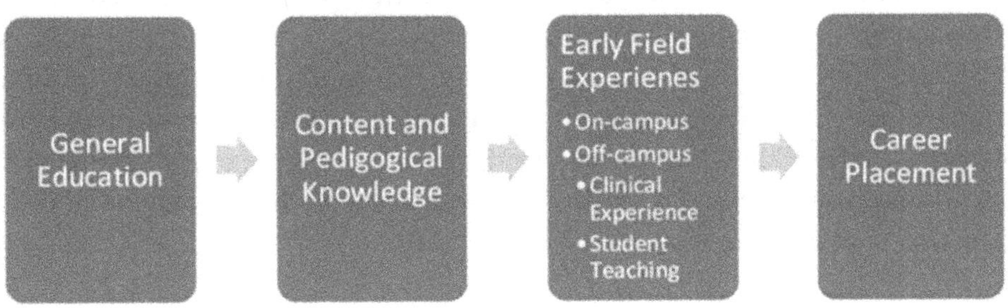

SUCCESSFUL TEACHING ENHANCEMENT

Effective Teacher Professional Development

One important way to address the concerns for improving our schools, teachers, and students is through teacher professional development. Teachers' professional development is a process of enhancing teachers' status through increased awareness and an expanding knowledge base.

While teacher preparation programs and alternative certification programs do as much to prepare agriculture teachers with the most up-to-date skills and knowledge, there is continued learning that must be achieved when agriculture teachers enter the classroom. The extensive range of agriculture content, new instructional methods, developments in technology, changing procedures, and student learning needs present opportunities for continued teacher learning.

One of the most important investments of time and money that local, state, and national leaders make in education is improving teachers' skills and knowledge through professional development opportunities. Professional development should be intentional, ongoing, and systematic (Desimone, 2009). Often, SBAE teachers participate in conferences, seminars, workshops, collaborative learning opportunities, or college classes to continue their learning. Creating unique continuing professional educational experiences that SBAE teachers participate in should build a network of support that is ongoing, embedded in their daily practice of teaching, keeps them up-to-date with changes in the agricultural industry, and provides an avenue for their lifelong learning experiences (McKendree & McKim, 2021).

Professional development programs must also be thoughtful and intentionally planned with purpose, goals, and evaluation measures keeping in mind the multifaceted attributes teachers bring to the table such as whether they were traditionally prepared or if they have alternative certification, what stage they are in their career (e.g., pre-service, early career, mid-career, late-career), their personal attributes (e.g., family support, major life changes or crises, unique aspirations, hobbies, etc.), and their school context (e.g., resources, support, geography, social characteristics, etc.).

Now, widely adopted by school districts, teacher preparation programs, and state departments of education the Danielson framework (1996) suggests effective teaching can be categorized into four domains: planning and preparation, classroom environment, instruction, and professional responsibilities. Domain 4 represents the professional responsibilities and the range of responsibilities outside the classroom that provide opportunities for reflection on teaching, maintaining accurate records, communicating with families, contributing to the school and district, growing, and developing professionally, and showing professionalism. Teachers who demonstrate these competencies are highly valued by their colleagues and administrators, as well as being seen as true professionals.

Foundations for Effective SBAE Teacher Professional Development

The traditional professional development model includes structured professional learning that results in increased teacher knowledge and skills, change in teacher attitudes, beliefs, and instruction, and results in improved student learning outcomes. Agriculture teacher professional development should 1) be focused on discipline-specific teaching strategies, curriculum, or pedagogy, 2) incorporate active learning using authentic and interactive strategies to provide highly contextualized learning, 3) provide opportunities for collaboration and a space for teachers to share ideas and collaborate to create communities of practice, 4) use and model effective teaching practices, 5) provide coaching and support, 6) offer opportunities feedback and reflection, and 7) provide teachers with adequate time to learn, practice, implement and reflect (Darling-Hammond, et al., 2017).

Beginning in the 1980's SBAE research related to teacher needs was common practice (Barrick et al., 1983). These needs assessments began by identifying the importance, knowledge, and ability of teachers to provide the application of specific technical agricultural topics. In the three decades of research conducted on the professional development needs for SBAE teachers, some needs have changed, however many categories have continued to remain the same. Those areas that remain the same include the need for professional development in FFA program management, developing public relations programs, general administrative tasks, SAE development and supervision, and computer technology (Anderson et al., 1992; Barrick et al., 1983: DiBennedetto et al., 2018; Wells & Hainline, 2021). Emerging and new skill attainment includes funding and fundraising, utilizing advisory committees, 21st-century competencies, STEM integration, and working with special populations (Ferand et al. 2020; McKendree & McKim, 2021). It is important to note that professional development needs vary by career stage and by certification type as identified in the research (Roberts et al., 2004; Roberts et al., 2020). Where alternatively certified SBAE teachers may need more assistance in modifying instruction for students with special needs (Roberts et al., 2004 or identifying curricular resources and traditionally certified teachers may need program planning, curriculum development, or writing grant proposals (Roberts et al., 2020). As SBAE teacher professional development needs continue to change, it is important for researchers to continue to identify specific areas for development to provide intentional and purposeful professional development opportunities to support the improvement of pedagogical, technical, or content-specific skills. Further, opportunities for participatory professional development where SBAE teachers serve as the experts and inspire peers through a more collaborative experience (McKendree & McKim, 2021). Outcomes may result in greater perspectives, respect for peers, and a deeper understanding of the topic.

Professional Learning Communities

Professional learning communities (PLCs) are a group of educators that meet regularly, share expertise and work collaboratively to improve the teaching skills and academic performance of students. The purpose of a PLC is to improve the skills and knowledge of educators through collaborative engagement and professional dialogue and to improve the educational achievement of students through stronger leadership and teaching (Darling-Hammond, et al., 2017). Integrating teachers into these communities of practice to meet the educational needs of their students through collaboratively examining their day-to-day practice improves student learning and teaching practice (Vescio, 2008). While the activities of a PLC may differ from school to school the potential to discuss teacher work, student work, student data, and review professional literature to improve teaching and learning is often accomplished collaboratively. PLCs offer opportunities for continuous improvement, innovation, and experimentation with a focus on learning and documented results.

Participation in Professional Organizations

One of the most beneficial and impactful ways to build an agriculture teacher's professional network is to be involved with their professional association. It is impossible to be an agriculture teacher, work in the classroom, and be entirely removed from peers and colleagues. The strength of teachers comes from the connections, networking, mentoring, and relationships with other agricultural education professionals. The most common professional associations for agriculture teachers include the Association for Career and Technical Education and the National Association of Agricultural Educators as well as their representative state-level associations.

The Association for Career and Technical Education (ACTE) mission is to provide educational leadership in developing a competitive workforce. ACTE strives to empower educators to deliver high-quality CTE programs that ensure all students are positioned for career success (ACTE, 2021).

The National Association of Agricultural Educators (NAAE) is a federation of state agricultural educators' associations with more than 9,000 members. Our members are involved in SBAE at any level, from middle school through post-secondary, and state and national agricultural education leadership (NAAE, 2021).

Preservice and inservice agriculture teachers should be informed about the importance of joining their respective professional associations and be familiar with the benefits. Perhaps one of the most important reasons agriculture teachers might join a professional organization is the opportunity to network and learn from other educators. State and national associations provide specific professional development focused on enhancing content and pedagogical skills, improving leadership skills, and providing access to grants, awards, recognition, and teacher leadership opportunities (Desimone, 2009).

ASSESSING TEACHER SKILLS TO ENHANCE PRACTICE

Teacher preparation, whether through traditional or non-traditional routes aims to prepare highly qualified teachers. Yet, qualifications and certifications do not necessarily determine whether a teacher is highly effective or possesses new and modernized skills or competencies to advance student learning. The evaluation of teaching effectiveness and assessment of teaching skills is not only necessary to provide teachers with opportunities to grow, but also critical to improve student success.

Teacher Effectiveness

The simplest definition of teacher effectiveness is the ability of a teacher to improve student learning. Goe, et al. (2008) define teacher effectiveness through the following measures.

1. have high expectations for all students, help students learn, and assess student learning,
2. contribute to positive academic behavior
3. use diverse resources to plan and structure engaging learning opportunities, formatively monitor student progress, adapt instruction as needed, and evaluate learning through multiple sources of evidence,
4. contribute to the development of classrooms and schools that value diversity and civic-mindedness, and
5. collaborate with other teachers, administrators, parents, and education professionals to ensure student success.

Measuring teaching excellence through a variety of evaluations and identifying teacher needs for improvement provides a solid foundation for professional conversations among teachers who seek to enhance their skillset and administrators or teacher education faculty who may be to provide in-service or training opportunities.

Teacher Evaluations

The incorporation of teacher evaluations as part of the preservice and in-service teacher experience provides a way for teachers to document successes and identify opportunities for improvement.

Self-Evaluation

Whether consciously or not, teachers are always aware of their success in the classroom. Periodic self-evaluations provide teachers an opportunity to examine and document their teaching activities to understand and improve their practice. Teachers should evaluate themselves on a range of items including pedagogical content knowledge, technical content knowledge, program planning, diversity, and professionalism.

Student Evaluations

Course evaluations should also be part of identifying areas for growth and success in the classroom. Soliciting end of course student feedback on course content and teaching effectiveness can help refine the content and adjust teaching strategies. Mid-semester student feedback provides the additional benefit of allowing teachers to understand students' concerns during the semester to make appropriate mid-course corrections and changes.

Peer Evaluations

Peer evaluations are an additional component for improved teacher effectiveness. They rely less on the student evaluations and more on the review by a peer in a similar discipline to assess and improve the quality of teaching. Evaluation of a teacher by their peers puts teachers in charge of the quality of their teaching, enables teachers to improve the quality of their teaching, and explores key issues or changes they may want to make. Peer evaluations may be formative and intended to develop or improve teaching, or summative to provide an incremental evaluation.

Teacher Needs Assessments

Whether teachers are traditionally or alternatively certified; in their preservice training or expert teachers, all educators are faced with staying up to date with emerging skills, ideas, or technology. While professional development is an integral component of teacher development, being able to identify what teachers need can be challenging (Desimone, 2009). Agriculture teachers have a strong desire to continue their inservice education, particularly in technical content areas. Identifying appropriate technical content, and meaningful inservice programming can be challenging if left to chance, suggestion, or observation only.

Therefore, conducting a systematic needs assessment can be an important step in identifying any gaps in performance or weaknesses that prevent SBAE teachers from reaching their desired goal of effective teaching. The Borich (1980) needs assessment model offers a systematic process to identify the importance of technical agriculture topics to teachers, their knowledge of specific technical agriculture topics, and their ability to provide the application of the technical content. The results identified through the needs assessments provide opportunities to identify the gaps in importance, knowledge, and application, prioritize them accordingly, and implement evidence-based interventions to target resources and address the needs.

Because agriculture teacher educators and other stakeholders who have a vested interest in providing appropriate and timely professional development, workshops and training should be constructed around results from needs assessments and should focus on specific agriculture teacher needs including but not limited to planning agriculture curricula, alignment to state and national standards, identifying quality curriculum, technical content, student assessment, multicultural and special education, gifted and talented education, STEM concepts, English as a second language, etc.(DiBenedetto et al., 2018; Ferrand et al., 2020; McKendree & McKim, 2021; Rice & Kitchel, 2016; Wells & Hainline, 2021).

With systematic needs assessments and intentional and purposeful professional development, we can assist with the improvement of pedagogical, technical, and content-specific skills of agriculture teachers.

SUMMARY

Agricultural education is an important component of public-school instruction in every state of the United States and five U.S. Territories. There are approximately 1,000,000 agricultural education students in the nation who are taught by nearly 12,000 secondary and two-year postsecondary teachers. It is estimated that the contact hours of in-school instruction in and about agriculture exceed 10 million annually. (National Association of Agricultural Educators [NAAE], 2017, para. 1)

The shortage of qualified SBAE teachers is one of the greatest challenges facing agricultural education. Understanding the various paths toward becoming a certified SBAE teacher may provide essential evidence when recruiting potential educators into the profession. Individuals may pursue traditional four-year baccalaureate degrees, graduate degrees, or alternative/lateral entry into the profession.

When considering traditional routes to licensure through post-secondary education, the requirements of agriculture teacher preparation programs vary across the country (Myers & Dyer, 2004). However, the Standards for SBAE Teacher Preparation Programs establish a reference for institutions and provide overarching learning goals for students who graduate and earn a license to teach SBAE. The standards include:

a) Pedagogical content knowledge
b) Technical content knowledge
c) Program planning
d) Diversity
e) Professionalism
f) Personal dispositions

Professional development is critical for SBAE teachers to learn new knowledge and skills. Given opportunities to improve teaching, leadership, and increase student learning and achievement. Structured professional development that targets the various teacher groups (e.g., career stage, certification level, etc.) that results in changes in teacher practices, improves teachers' professional knowledge, skills, and attitudes, and enhances student learning is essential. Professional development occurs in various settings, including Professional Learning Communities, participation in professional organizations, and structured and ongoing workshops and inservice opportunities.

To effectively identify the professional development needs of SBAE teachers conducting a systematic needs assessment is an important step. This identifies gaps in performance or weaknesses and provides opportunities to deliver prioritized evidence-based interventions to focus on specific teacher needs.

REVIEW QUESTIONS

1. Describe the various types of teacher certification programs?
2. Explain the importance of the general education component of agriculture teacher education preparation programs.
3. In general, what requirements must be met for teacher licensure?
4. Compare and contrast pedagogical content knowledge and technical content knowledge.
5. Explain the importance of program planning in agriscience education.
6. What skills or competencies are needed by preservice teachers to integrate leadership, SAE, and experiential learning in their programs?
7. What role does teacher professional development play on agriculture teacher effectiveness?
8. Explain how often teacher professional development should occur and why.
9. Describe three types of evaluation methods utilized by agriculture teachers.
10. Explain the purpose of a systematic needs assessment to evaluate teacher professional development needs.

REFERENCES

American Association for Agricultural Education. (2017). *Standards for School-Based Agriculture Teacher Preparation.* http://aaaeonline.org/Standards-for-Teacher-Preparation

Anderson, T. J., Barrick, R. K., & Hughes, M. (1992). Responsibilities of teacher education for vocational teacher professional development programs. *Journal of Agricultural Education*, *33*(2), 43–50. doi:10.5032/jae.1992.02043

Association for Advancing Quality in Educator Preparation (AAQEP). (2022). https://aaqep.org/

Association for Career and Technical Education. (2021). https://www.acteonline.org/about/

Baker, M. A., & Robinson, J. S. (2016). The effects of Kolb's experiential learning model on successful intelligence in secondary agriculture students. *Journal of Agricultural Education*, *57*(3), 129–144. doi:10.5032/jae.2016.03129

Baker, M. A., & Robinson, J. S. (2016). Aligning Kolb's experiential learning theory with a comprehensive agricultural education model. *Journal of Agricultural Education, 52*(4), 1–16. doi: .2012.04001 doi:10:5032/jae

Ball, A. L., & Knobloch, N. A. (2005). A document analysis of the pedagogical knowledge espoused in agriculture teaching methods courses. *Journal of Agricultural Education*, *46*(2), 47–57. doi:10.5032/jae.2005.02047

Barrick, K. R., Ladewig, H. W., & Hedges, L. E. (1983). Development of a systematic approach to identifying technical inservice needs of teachers. *Journal of Agricultural Education*, *24*(1), 13–19. doi:10.5032/jaatea.1983.01013

Birkenholz, R. J., & Deeds, J. (2010). National Standards for teacher education in agriculture. In R. M. Torres, T. Kitchel, & A. L. Ball (Eds.), Preparing and advancing teachers in agricultural education (pp. 221-230). The Ohio State University.

Borich, G. (1980). A needs assessment model for conducting follow-up studies. *Journal of Teacher Education*, *31*(3), 39–42. doi:10.1177/002248718003100310

Coleman, B. M., Bunch, J. C., & Thoron, A. C. (2020). Identifying agriscience teachers' instructional practice professional development needs by certification type. *Journal of Agricultural Education*, *61*(3), 86–100. doi:10.5032/jae.2020.03086

Cooper, J. M., & Alvarado, A. (2006). *Preparation, recruitment and retention of teachers*. International Academy of Education.

Council for the Accreditation of Education Preparation (CAEP). (2022). https://caepnet.org

Cowan, J., & Goldhaber, D. (2016). National Board Certification and teacher effectiveness: Evidence from Washington state. *Journal of Research on Educational Effectiveness*, *9*(3), 233–258. doi:10.1080/19345747.2015.1099768

Cruickshank, D. R. (1996). *Preparing America's Teachers*. Phi Delta Kappa Educational Foundation.

Danielson, C. (1996). *Enhancing professional practice: A framework for teaching*. Association for Supervision and Curriculum Development.

Darling-Hammond, L., Hyler, M. E., & Gardner, M. (2017). *Effective Teacher Professional Development.* Learning Policy Institute. https://learningpolicyinstitute.org/sites/default/files/product-files/Effective_Teacher_Professional_Development_REPORT.pdf

Desimone, L. M. (2009). Improving impact studies of teachers' professional development: Toward better conceptualizations and measures. *Educational Researcher*, *38*(3), 181–199. doi:10.3102/0013189X08331140

DiBenedetto, C. A., Willis, V. C., & Barrick, R. K. (2018). Needs assessments for school-based agricultural education teachers: A review of literature. *Journal of Agricultural Education*, *59*(4), 52–71. doi:10.5032/jae.2018.04052

Ferand, N. K., DiBenedetto, C. A., Thoron, A. C., & Myers, B. M. (2020). Agriscience teacher professional development focused on teaching STEM principles in the floriculture curriculum. *Journal of Agricultural Education, 61*(4), 189–202. doi:10.5032/jae.2020.04189

Goe, L., Bell, C., & Little, O. (2008). *Approaches to evaluating teacher effectiveness: A research synthesis.* National Comprehensive Center Quality. Retrieved November 6, 2021, from https://gtlcenter.org/sites/default/files/docs/EvaluatingTeachEffectiveness.pdf

Improving Teacher Preparation: Building on Innovation. (n.d.). Retrieved November 6, 2021, from https://www.ed.gov/teacherprep

Lawver, R. G., Foster, D. D., & Smith, A. R. (2018). *Status of the U.S. Supply and Demand for Teachers of Agricultural Education, 2014 – 2016.* Retrieved November 6, 2021, from http://aaaeonline.org/Teacher-Supply-and-Demand

McKendree, R. B., & McKim, A. J. (2021). Teacher changing the discipline: A case study of participatory professional development. *Journal of Agricultural Education, 62*(3), 72–84. doi:10.5032/jae.2021.03072

Myers, B. E., & Dyer, J. E. (2004). Agriculture teacher education programs: A synthesis of the literature. *Journal of Agricultural Education, 45*(3), 44–52. doi:10.5032/jae.2004.03044

NationalF. F. A. (2021). *About FFA.* Author. Retrieved from https://www.ffa.org/about/

National Association of Agricultural Educators (NAAE). (2017). *Agricultural education in the United States.* Author. Retrieved from https://www.naae.org/

National Association of Agricultural Educators (NAAE). (2021). *Who we are.* Author. Retrieved from https://www.naae.org/whoweare/index.cfm

National Board for Professional Teaching Standard. (2021, November 6). *NBPTS.* https://www.nbpts.org/

National Council on Teacher Quality. (2021a, November 5). *Teacher preparation policy.* https://www.nctq.org/publications/State-of-the-States-2021:-Teacher-Preparation-Policy

National Council on Teacher Quality. (2021b, November 5). *What makes teacher preparation traditional.* https://www.nctq.org/dmsView/NCTQ_-_What_Makes_Teacher_Prep_Traditional_or_Non_Traditional

Rateau, J. R., Kaufman, E. K., & Cletzer, D. A. (2015). Innovative Classroom Strategies that Prepare College Graduates for Workplace Success. *Journal of Agricultural Education, 56*(3), 52–68. doi:10.5032/jae.2015.03052

Retallick, M. S., & Miller, G. (2007). Early field experience documents in agricultural education. *Journal of Agricultural Education, 48*(4), 20–31. doi:10.5032/jae.2007.04020

Rice, A. H., & Kitchel, T. (2016). Influence of knowledge of content and students on beginning agriculture teachers' approaches to teaching content. *Journal of Agricultural Education, 57*(4), 86–100. doi:10.5032/jae.2016.04086

Roberts, R., Wittie, B. M., Stair, K. S., Blackburn, J. J., & Smith, H. E. (2020). The dimensions of professional development needs for secondary agricultural education teachers across career stages: A multiple case study comparison. *Journal of Agricultural Education, 61*(3), 128–143. doi:10.5032/jae.2020.03128

Roberts, T. G., & Dyer, J. E. (2004). Inservice needs of traditionally and alternatively certified agriculture teachers. *Journal of Agricultural Education*, *45*(4), 57–70. doi:10.5032/jae.2004.04057

Shulman, L. S. (1986). Those Who Understand: Knowledge Growth in Teaching. *Educational Researcher*, *15*(2), 4–14. doi:10.3102/0013189X015002004

Teacher Certification Degrees. (2021, August 24). Retrieved November 6, 2021, from https://www.teachercertificationdegrees.com/#guide

The PPAT® assessment from ETS. (n.d.) retrieved November 6, 2021, from https://www.ets.org/ppa/test-takers/teachers/about

Vescio, V., Ross, D., & Adams, A. (2008). A review of research on the impact of professional learning communities on teaching practice and student learning. *Teaching and Teacher Education*, *24*(1), 80–91. doi:10.1016/j.tate.2007.01.004

Warner, D. B., & Koeppel, K. (2009). General education requirements. A comparative analysis. *The Journal of General Education*, *58*(4), 241–258. doi:10.2307/jgeneeduc.58.4.0241

Wells, T., & Hainline, M. (2021). Examining teachers' agricultural mechanic's professional development needs: A national study. *Journal of Agricultural Education*, *62*(2), 217–238. doi:10.5032/jae.2021.02217

KEY TERMS AND DEFINITIONS

Alternative Certification: This is a non-traditional route to teacher licensure.

Content Knowledge: These are the facts, concepts, theories, and principles related to specific academic courses in agriculture. Sometimes referred to as technical knowledge.

Inservice Teacher: Includes all professional teachers who are currently teaching in the classroom.

Needs Assessment: A process of determining gaps in knowledge, practice, or skill.

Pedagogical Knowledge: The specialized knowledge of teachers for creating effective teaching and learning environments for all students.

Preservice Teacher: Includes students in a teacher preparation program who have little to no experience in the classroom.

Professional Development: Continuing education and career training after an individual has entered the workforce.

Standards: The learning goals for what students should know and be able to do after a program or course.

Teacher Certification: The process that which individuals obtain official teaching credentials within a specific field after fulfilling the requirements, coursework, degrees, or other criteria. Often referred to as teacher licensure.

Teacher Preparation: A program designed to prepare students to become certified or licensed teachers.

Chapter 15
Conducting Graduate Education Programs in Agricultural Education

Amanda Bowling
The Ohio State University, USA

Shannon G. Washburn
The Ohio State University, USA

ABSTRACT

Graduate education fulfills a specific need within the professional growth of students but carries different considerations when pursuing a degree or designing a degree program as compared to undergraduate programs. Agricultural education graduate programs possess a nuance in balancing the need to serve a wide audience, use limited departmental resources, and encapsulate the breadth of school-based agricultural education to develop effective future teacher educators and scholars. This chapter serves to provide potential graduate students with an understanding of what to reflect on when pursuing a graduate degree, as well as the various curriculum components and experiences which they might engage in. Additionally, the authors have outlined programmatic decisions for faculty to consider when creating or revising agricultural education graduate programs.

INTRODUCTION

Graduate education is a choice which should not be sought lightly by both the students seeking degrees and the departments which develop and host graduate degree programs. Potential graduate students must understand the rigor and relevance of graduate education and their own personal academic goals as they consider different avenues following their undergraduate or master's degrees. Additionally, as agricultural education academic units design or revise their graduate program they must consider the many philosophies which guide graduate programmatic decisions such as admissions requirements, degree program requirements, curriculum, and professional experiences.

DOI: 10.4018/978-1-6684-3420-8.ch015

Chapter Objectives

1. Describe differences between undergraduate programs and graduate programs in Agricultural Education
2. List criteria to be considered in making graduate admission decisions
3. Explain why specific resources must be considered when offering a graduate program
4. List reasons why graduate programming tends to be more constructivist in design than most undergraduate programs in Agricultural Education
5. Compare Agricultural Education program characteristics for master's degree programs in the following formats: full-time resident programs, teacher certification graduate programs, and online degree programs
6. Describe program considerations for doctoral programs in Agricultural Education for fostering the Scholarship of Teaching
7. Describe program considerations for doctoral programs in Agricultural Education for fostering the Scholarship of Application
8. Describe program considerations for doctoral programs in Agricultural Education for fostering the Scholarship of Discovery

DIFFERENCES BETWEEN UNDERGRADUATE PROGRAMS AND GRADUATE PROGRAMS IN AGRICULTURAL EDUCATION

Transitioning Undergraduates to a Graduate Program

The undergraduate student considering pursuit of a graduate degree has much to consider, such as: academic readiness, application processes, financial preparation, career goals and how advanced education impacts them, degree requirements, institutional fit, etc. University faculty have frequent opportunities to shed light on the world of graduate education with undergraduate students and should engage in these moments with consideration of some of the assumptions or misconceptions the undergraduate may have of the experience and its similarities with and differences from the baccalaureate experience.

Many undergraduates have limited experience with research and often mistakenly assume the research involved in a graduate program is akin to writing a high school or undergraduate "research paper" which is often limited to a review of literature. Care should be taken to help the prospective student recognize the creative and academic expectations of conducting graduate level research since a graduate degree is ultimately a research-oriented degree. One should not view the graduate program is merely a "continuation" of the coursework-centric undergraduate experience. Unlike the behaviorism-oriented approach of many undergraduate programs that center on a tightly prescribed set of curricular expectations, the prospective student should recognize that a graduate program is more constructivist in focus because the student needs to take ownership of the learning experiences and create a tailored program that addresses their individual interests.

As the undergraduate explores their potential interest in graduate education, they would be well-advised to consider that admission to graduate programs is competitive and selective based on the focus of the program and the faculty administering the program. Simply satisfying the printed admission requirements of the graduate program does not ensure admission, as multiple other factors outlined in subsequent sections are involved in admission decisions. Furthermore, admission decisions for graduate programs are frequently made up to a year prior to the program start date, so prospective students should manage their time accordingly. Completing an in-depth self-assessment for graduate school readiness, conducting a thorough investigation of available programs, and preparing well-conceived program applications takes a considerable amount of time and should be approached with both seriousness and realism.

CONSIDERATIONS FOR GRADUATE ADMISSIONS PROCESSES

Academic units contemplating creation of graduate programs in Agricultural Education have many considerations to evaluate including, but not limited to curriculum, program duration, learning outcomes, academic and experiential program focus, faculty preparation and capacity to administer the program, and the institutional processes required for program approval. Several of these considerations will be addressed in sections to follow, but one area that merits careful planning is the program admissions process and expectations of qualification for program entry. Having a clear plan for the processes and policies that govern admission decisions and having philosophical alignment among faculty making admission decisions will result in a more beneficial educational experience for all parties involved. Specific to the expectations sought in the "ideal candidate," faculty should consider the following among the characteristics evaluated in graduate program applicants:

- Students should have a documented record of academic success with a history of meeting or exceeding academic expectations
- Specific academic skill in reading comprehension, vocabulary, and a propensity for critical reflection on academic writing
- In addition to academic aptitude, it is also important that candidates display the academic "maturity" required to embrace learning rigorous content
- Relevant life and work experience to provide meaningful context for new learning
- Students should illustrate a clearly defined research interest with plans for chosen topics of inquiry
- When considering admission of undergraduates continuing into graduate programs at the same institution, students should be evaluated with a focus on how the learning experience will benefit them and a well-defined notion of how the experience will differ from their undergraduate studies

RESOURCE CONSIDERATIONS

While planning for the creation and implementation of an agricultural education graduate program, academic units and their leaders should carefully consider and account for the resources necessary to administer such a program. Certainly, faculty time and expertise to develop and deliver courses at the graduate level will be the most significant resource needed throughout the life of the program, but additional resource demands should not be overlooked.

Beyond graduate course instruction, effective graduate advising demands a significant investment of faculty time. Unlike advising undergraduates who might contact their advisor two or three times per semester, graduate students will need significant and consistent mentoring as they navigate the research process. Regular research consultations, course and career advising sessions, review and feedback of research writing, and relationship building can conservatively require 10 to 12 hours per month for each student. Additionally, each admitted student will need to assemble a graduate committee that meets institutional requirements for faculty members and expertise. Committee service will also require faculty time. Additional faculty engagement in recruiting and meeting with prospective students, reviewing application materials, participating in academic unit discussions about admissions, and providing peer-mentoring support to fellow graduate advisors are time demands that are easy to overlook. These and other expectations of faculty time should be considered when determining the capacity of the program and when making admission decisions.

Administering a graduate program is accompanied by a high volume of correspondence, coordination, documentation, and communication. Staff support to provide essential assistance with these duties is another critical human resource need that should not be overlooked. Staff expertise in coordinating responses to prospective student inquiries, processing multi-component admissions processes, compiling packets for admissions review, assisting admitted students in relocation and settling, helping students and faculty navigate graduate college policies and documentation, and stewarding physical resources are a few of the tasks a capable staff member should be prepared to provide.

Financial resource considerations also require planning – especially when developing a residential graduate experience. Residential students will require a consistent source of funding for graduate assistantship stipends and tuition waivers in compliance with institutional minimums. Several potential sources of support for these finances should be seriously considered including public and private grant programs, college and university graduate fellowships, grants, or allocations, private development efforts to generate endowed support, partnerships with other academic entities and units, and departmental general fund resources. In most cases, academic units must rely on a combination of these and other financial sources to provide residential graduate student support. The program administrator should also consider the time and effort required to develop and steward such resources on an ongoing basis. Financial support for graduate student travel to conduct and present research and participate in extra-curricular professional development activities should also be considered. Some institutions offer competitive mini-grant programs for such opportunities, but seldom will such programs provide adequate funding to address all travel expenses graduate students may encounter.

Online-oriented graduate programs also require dedicated and planned resources. Faculty time for course development unique to online instruction, advising online students, developing student programs of study, and engaging in virtual student committees and capstone experiences all frequently require more time and coordination than similar work in residential programs. Additionally, staff support is perhaps more critical to ensure regular and consistent communication is maintained with current and prospective students – many of whom are also full-time professionals.

Physical resources typically needed for administering a residential graduate program include dedicated student workspace for the completion of assistantship tasks, meeting space for graduate teaching assistants to meet with their undergraduate students, and the space and tools necessary for productive work on research activities. As a social science discipline, the tools necessary for research are often limited to internet access, functional computers with the necessary software to analyze data and occasionally printing services. However, there may be additional resource needs for specialized research projects, so some degree of flexibility for such needs should be anticipated.

THE MASTER'S DEGREE

Driving Philosophies of a Master's Degree

Unlike the baccalaureate experience which tends to align with behaviorist philosophies in both design and delivery, the master's student needs frequent opportunities to learn and practice constructivist thinking. As students transition from the prescribed undergraduate curriculum with its emphasis on developing competency toward the development of scholarship and scholarly thinking more typically associated with graduate programming, their curriculum and learning experiences should reinforce these differences. A hallmark differentiator from the undergraduate experience is the focus on student identification and pursuit of their own academic interests. During this time program administrators should be mindful of the importance that their experience offers support along with the challenges associated with forming one's own knowledge and identifying learning needs. The master's student will likely experience for the first time the expectation to develop a research question based on personal interest and the literature base and to then develop research methodology to address said question. Such experiences are important to the development of scholarly thought and should be reinforced in the way the curriculum is organized to maximize flexibility, in the way courses are structured to blend inquiry with research skill development, and in the advising experiences to which students are exposed. Each of these learning venues structured correctly will enable students to better pursue and unpack their research interests and develop their lines of inquiry.

Fostering master's student experiences for scholarly growth should also be central when considering the philosophical principles underlying creation of student assistantship opportunities. While assistantship funding is most often associated with completion of well-defined tasks in teaching, research, or service project management, the master's student will be better served when these experiences permit and encourage application of active inquiry. The tangible life and workplace application of research skill and scholarly thinking related to teaching, research, and service support roles will better enable master's students to experience application of these skills and thereby be better motivated to pursue them in relation to their own research questions.

Regarding the master's graduate committee, students need access to skilled research mentors knowledgeable in both qualitative and quantitative methodologies who possess a variety of content expertise. Master's students should be encouraged to identify committee members who not only have research interests that align with the intended master's research question, but who will also challenge the student in pursuit of that question. Committee members should be engaged in developing the graduate program of study as well as the graduate research experience to ensure coursework is challenging, aligns with committee member expectations, and best prepares the student for their career and research goals. Care should be taken to satisfy the local graduate college requirements for committee size and member qualifications as well as timelines for reporting the committee make-up to the institution.

Audiences Served by Master's Degree Programs

One of the complexities of designing and delivering a master's degree program in Agricultural Education is the wide range of audiences interested in obtaining such a degree. Exercising caution and being intentional in defining the target audience for the degree program is important in creating and executing the program best tailored to student needs. While many higher education disciplines focus all graduate program efforts on full-time residential students, within Agricultural Education the most likely potential candidates for master's degree programs have a diverse set of nuanced needs. These needs are governed by a myriad of factors including, but not limited to career advancement via continuing education requirements, being place bound in proximity to family agricultural land, financial constraints associated with teacher retirement programs or other investments preventing relocation, affinity for living in smaller communities, desire to continue directly from the baccalaureate program, etc. The following master's program models each have strengths and limitations while also appealing to more specific subsets of prospective students, thereby eliciting need for consideration of learner needs in program design and delivery.

Full-Time Residential Programs

Full-time, residential master's programs tend to be most appealing to academically high achieving agricultural education students continuing their education immediately after completion of the baccalaureate degree; recent undergraduate degree completers seeking to refine career goals or career preparation prior to entering the workforce; experienced teachers with a desire to continue their education - often with a specific career goal in mind; and students interested in pursuing doctoral programs upon completion of the master's degree. While each of these audiences can be well-served by a full-time residential master's program, it can be difficult to adequately serve the needs and expectations of all groups with such disparate experiences and end-goals. Furthermore, admission decisions can be further complicated when comparing candidates with multiple years of professional experience with those wishing to enter the graduate program immediately after baccalaureate graduation.

Program Strengths

The benefits of offering master's programming in a full-time, residential format are numerous for both the student and the academic unit delivering it. For the student, the opportunity to be fully immersed in graduate study with fewer competing professional priorities provides time to engage more deeply in scholarly thought and synthesis of courses and experiences. The proximity to faculty and other full-time students facilitates more frequent opportunities for collaborative reflection and dialog. Since most full-time program participants are supported by graduate assistantship, there is potential for frequent experiential learning related to the program focus and the student's research interest. The likelihood of engaging with graduate student peers or faculty in pre-thesis research is also much greater than with other program formats which can be beneficial for more fully learning the research process. Additionally, it is more common for students to complete the majority of their master's coursework via face-to-face, synchronous delivery, which is often a preferred mode of learning. For the program faculty, greater research productivity and stronger professional relationships are more readily fostered when graduate students are full-time and on-campus.

Program Limitations

Although the benefits of full-time, residential master's programming are abundant, such delivery does come with limitations. Most resident students require assistantship support to offset the cost of education coupled with a lower than typical salary. Such resources are challenging for academic programs to secure and typically must be awarded on a competitive basis, thereby reducing the total number of students who can be served in residential programs. While academic advising can be more rewarding when working with resident students, it can also be much more time consuming than supporting students in other program delivery modes. For students with work experience, the decision to pursue a graduate program full-time likely results in diminished earning potential through the duration of the program even if increased earning potential is a beneficial result of program completion. Such financial limitations can prevent certain socio-economic groups, students with family commitments, and other demographic groups from seriously considering the full-time graduate experience, thereby diminishing the potential for diversity of many forms from being represented in the program.

Teacher Certification Programs

A common question often fielded by Agricultural Education graduate program administrators relates to the potential for earning an initial teaching credential or licensure while also earning a master's degree. In many cases, the full-time, residential master's program described in the previous section is not established for the purpose of delivering requirements for initial teaching credential. Other graduate programs have been tailored at various institutions to serve this specific need and interest. The candidate audiences for such programs include prospective teachers who have completed other degree programs and are seeking teacher licensure; alternatively certified teachers pursuing full licensure; current or potential career changers with industry experience; experienced teachers of other subjects seeking agricultural education credentials; or undergraduate completers of agricultural education degrees in states that require master's level education for full teaching credential.

Program Strengths

The special subset of candidates for master's degree programs accompanied by teacher certification can benefit greatly from such programs. These candidates often find the prospect of credentialing coupled with a master's degree to be more appealing than earning a second baccalaureate degree. As a result, programs delivered following this model hold potential for addressing the research-documented teacher shortage (Foster, Lawver, & Smith, 2020) while providing valuable learning experiences. Students who have completed a baccalaureate degree in another discipline and/or have additional work experiences before choosing to pursue a teaching career bring to their programs a level of maturity and a focus on education that undergraduate pre-serve teachers do not possess.

Program Limitations

Master's degree programs that encompass requirements for teacher licensure require different curriculum and program delivery approaches than other graduate program models. Program administrators and faculty may struggle to balance the constructivist philosophies and expectations for the development of scholarly thinking that are customary for graduate education with the behaviorist philosophies and state curricular requirements more typically found in initial teacher licensure programs. As a result, it can be difficult to strike an appropriate balance between the competing program needs within a curriculum that does not significantly exceed the total number of courses and credits the graduate student applicant may find palatable from time and financial perspectives. Additionally, the notion of delivering a master's teaching certification program concurrently with a more traditional full-time resident master's experience can be challenging and potentially confusing for current and prospective students in each program. Finally, the candidates seeking such programs may not possess the type of prior experiences needed to credibly serve as graduate teaching assistants in undergraduate Agricultural Education programs, or the coursework necessary to serve as graduate research assistants, consequently limiting opportunities for graduate assistantships for students in these programs.

Online Programs

For a variety of reasons referenced earlier in this section, many candidates for master's programming in Agricultural Education may not consider prolonged on-campus learning to be feasible for their personal or professional situations. Such students are likely to seek programs that feature online delivery. Candidates for online master's programs may include working professionals who are place-bound; practicing educators in formal or non-formal teaching occupations who are seeking continuing education; second career teachers or working individuals who aspire to a career in education; and postsecondary educators seeking expertise in teaching. While the curriculum of online programs can closely resemble the full-time residential graduate program, or the combined master's and teaching credential program, special consideration should be given to the unique nuances of online education and the diverse needs of online learners.

Program Strengths

By eliminating barriers to access, online master's programming has the potential to serve a much broader audience and thereby create notable demand for the program. As a result, such programs can be profitable for the institution while meeting a unique student need. Research suggests that enabling educators to secure advanced education while working fulltime has the potential to enhance teacher retention and longevity (De Lay & Washburn, 2013) while also boosting educator enthusiasm for the profession via ongoing learning opportunities.

Program Limitations

Effective delivery of online learning requires a special skill set regardless of the level of instruction. Doing so at the graduate level can be even more challenging. Due to the relational aspects of learning to teach, teacher educators have historically been hesitant to deliver online teacher certification programming at either the baccalaureate or master's level, although some such programs are available at select institutions today. Graduate advising can be challenging in online programs because of the lack of face-to-face interactions and the need to be intentional to schedule and honor virtual advising appointments. The fact that most program participants are also working professionals further complicates opportunities for scheduling advising interactions. The traditional university semester length does not always lend itself well to prolonged engagement in online learning for working professionals – especially those who prefer to enroll in multiple courses each semester. As a result, research suggests many such programs catering to teacher audiences establish shorter course durations with multiple start times throughout the academic year (Roddy, et. al., 2017) to enable completion of multiple courses without concurrent enrollment. While creating arrangements such as these can be effective for online learners, the process of doing so can be cumbersome to navigate within university course approval processes and can also be challenging to schedule for faculty who also deliver face-to-face courses following the regular semester schedule.

Major Components of a Master's Degree Program

The curriculum of a master's degree program should align with the specific needs of the target audience and the program delivery approach outlined in the section above and should provide ample flexibility for the student to construct their own new knowledge in areas relevant to their goals. As a graduate program, however, the curriculum should require development of knowledge and competencies related to research design, methodology, data collection and analysis. These skills should be developed to a minimum level of preparing effective consumers of research and the ability to plan and conduct evaluations of educational programming. Expanding the student's knowledge of theoretical foundations important to the discipline beyond the fundamentals of a baccalaureate program should be an additional goal of the curriculum. As a graduate program in education, most students expect the curriculum to assist them in refining their instructional methods, curriculum development, and learning assessment expertise as well. Beyond these core competencies, the curriculum should allow for student specialization through access to a variety of elective courses at the graduate level. Finally, a culminating synthesis of coursework integrated with research that complies with graduate college requirements should be accounted for in formulating the curriculum.

The *graduate capstone* experience at the master's program level has adopted many forms in agricultural education and these vary from program to program. The most common experience, the *master's thesis*, is often considered the ultimate form of assessing the student's proficiency in knowledge application interwoven with research methodology. Experiences other than thesis research can be valuable learning tools as well and have found a place of service to specific learner career aspirations and learning goals. The *master's project in lieu of thesis* is a prevalent option for many agricultural education graduate programs. The project option often involves action research on curriculum design or implementation, instructional methodology, program evaluation, or another practical application of graduate expertise identified by the student and supported by the graduate committee. A summative master's examination can also be a valuable way for the student to show the graduate committee their expertise of course content and application. When aligned with the goals and intended outcomes of the program and the student pursuing the degree, the master's examination as a capstone experience is a viable option. The *exam option* is often applied in programs leading to initial teacher credential, in online programs, or in situations where the agricultural education faculty workload does not make it feasible for adequate supervision and advising for the thesis or project options.

THE DOCTORAL PROGRAM

Driving Philosophies of a Doctoral Program

Many higher education institutions with undergraduate degree programs in Agricultural Education have access to doctoral program offerings in education in partnership with a school or college of education. Far fewer institutions offer specific doctoral programs in Agricultural Education. The majority of these programs specific to Agricultural Education offer a Doctorate of Philosophy (Ph.D.) as opposed to a

Doctorate of Education (Ed.D.). "Both the Ph.D. and the Ed.D. require rigorous research and scholarship production, but the Ph.D. tends to have a greater focus on these areas" stated DeWitt Scott (2016) of Inside Higher Education. The designation of the program as either Ed.D. or Ph.D. will likely influence the overall philosophy guiding the program.

As a terminal research degree, doctoral programs must be guided by philosophies which engage students in the advancement of knowledge and scholarship. Manathunga and Lant (2006) stated "the fundamental goal of research higher degree programs, particularly doctoral degrees, is to develop independent researchers who are able to adapt to diverse workplaces in academe, industry and the professions" (p. 72). Thus, doctoral programs should create a foundation built on coursework and student outcomes driven by teaching, scholarship, and service experiences. Specifically for agricultural education programs, courses should not only immerse students in the nuances of agricultural education and begin to establish the student's area of expertise but also engage students in the sister disciplines of agricultural communication, leadership, and extension. Consequently, creating future faculty members or industry professionals well versed in both School-Based Agricultural Education and the cross disciplines within agricultural education.

As a discipline, agricultural education is heavily situated in experiential learning and our doctoral programs should be no exception. Professional growth-based experiences should be prevalent in agricultural education doctoral programs to allow students to *learn to do* while also engaging in their coursework. Doctoral student experiences could represent a variety of skills and outcomes including but not limited to research assistantships, teaching assistantships, hosting professional development workshops, presenting at research conferences, or developing extension programming. Assistantships create unique scenarios where departmental needs are met through assigned student workers but may provide students with college teaching and/or faculty mentored research experience. Department administrators and program coordinators should balance assistantship placement decisions between meeting departmental needs and providing doctoral students with a variety of experiences to strengthen their experiential learning outcomes. Faculty advisors and mentors should also be intentional about engaging doctoral students in experiential learning opportunities which align with the student's intended career goals. Potential future faculty members should be provided a balance of teaching, scholarship, and service experiences which also allow the doctoral student to begin to develop and disseminate their line of research. Doctoral students who intend to seek state staff, educational consulting, or other industry-based careers should seek opportunities to engage in the current needs of in-service teachers, extension professionals, and industry educational specialists and trainers.

Developing as a Scholar of Teaching

Most universities require their incoming agricultural education doctoral students to have a minimum of three years of professional, preferably teaching experience, prior to entering their program. Thus, it is likely that teaching is the area of scholarship with which doctoral students have the most previous experience. So, what transitions a person from being a teacher to becoming a scholar of teaching? In the traditional sense, Boyer (1990) suggests this transition occurs through the acquisition of new knowledge and one's ability to think, learn, and communicate. Since Boyer's seminal work the scholarship of teaching has grown in both supporting literature and scope of work (Vithal, 2018). While some argue that teaching lacks peer review, and a publicly disseminated product to be true scholarship, Fincher and Work (2006) suggest that through continued knowledge development and the connection of the other areas of scholarship – application and discovery – one can be a scholar of teaching. Within agricultural education, the scholarship of teaching is the foundation for the classes we teach, the research we conduct, and the service we provide. To fulfill the role of a scholar, doctoral students should seek to continually advance their knowledge through their course work and experiences specifically situated within teaching and learning.

Curriculum Related to Teaching

While agricultural education doctoral students frequently enter their programs with formal training in educational theories, teaching and learning it is crucial that these studies continue to be, "steeped in the knowledge of their field" (Boyer, 1990, p. 23). Early in their program, students should take a course focused on various learning theories to serve both as a refresher from their undergraduate and/or master's degree and to further expand their studies. Next, students should take courses which help to transcend the specific discipline of agricultural education into the broader field of educational psychology. The content of these courses could include but is not limited to adolescent development, working with diverse learners, learning and cognition, emotions, and motivation. Additionally, students should seek sequenced courses or courses with similar content to build their expertise in a specific area of education. This expertise area will serve as the foundation for future faculty members line of inquiry, will have application within agricultural education coursework, and will allow for collaboration opportunities with educational psychology faculty. Lastly, to further expand their teaching and learning expertise, students should consider including courses which center around specific teaching methodologies or learning contexts such as advanced teaching methods, assessment, online teaching, curriculum development or college teaching.

Experiences Related to Teaching

Throughout the doctoral program, students should engage in experiences which either a) aid in their acquisition and application of knowledge and b) connect the scholarship of teaching to the other scholarship disciplines. To further enhance knowledge development and apply newly acquired expertise, students should serve as teaching assistants for faculty led undergraduate courses in agricultural education, general education, and other department courses. As students progress in their programs, faculty should find ways in which doctoral students could be lead instructors of a course so that they are making curriculum, instructional, and assessment decisions. These experiences allow for application of learning theories, educational psychology concepts, teaching methods, and assessment which allows for instant feedback based on student learning. Reflection on and evaluation of teaching are additional experiences future scholars should continually engage in. Supervising student teachers allows doctoral students to evaluate novice educators through their semester long field experience. Whereas, conducting a peer review of another graduate student's teaching allows for a one-time evaluation within the context of a post-secondary classroom. Reflection of one's own teaching can occur through peer reviewed lessons conducted by faculty or other graduate students. Additionally, graduate students should be active participants in faculty discussions around programmatic level curriculum decisions, individual course revisions, and mechanisms for student support. This integration will allow students to see the various factors which influence curriculum, courses, and advising decisions such as accreditation agencies, university and college funding structures, licensure requirements, departmental needs, faculty time, student assessment data, and student needs. To begin to see how teaching, research, and service can intersect students should engage in the Scholarship of Teaching and Learning (SoTL) through which innovative practices are assessed, peer reviewed, and disseminated out. SoTL will be discussed further in the *Developing as a Scholar of Discovery* section of this chapter. Additionally, students should use their depth of knowledge to serve stakeholders and practice intersecting being a scholar of teaching and a scholar of application. Students should provide professional development workshops for preservice and in-service teachers to share the knowledge they have gained through their course work. This would allow the student to engage stakeholders with research supported and innovative educational concepts and teaching strategies.

The Agricultural Education Context of the Scholarship of Teaching

Agricultural education presents a unique context as doctoral students transcend into scholars of teaching through the lens of agricultural education. Being a scholar of teaching often focuses on specific disciplines within educational psychology, teaching, learning, and assessment. However, being a scholar of agricultural education bridges beyond these to also include the nuances of the agricultural education three circle model and Career and Technical Education (CTE). Thus, creating a scholar who can apply educational principles to agriculture classrooms while understanding the role CTE programs play within school systems and the ways in which agricultural education teachers fulfill these roles through various programmatic components. Future scholars will need to incorporate these complex interactions within their teachings to allow preservice teachers to begin to see their future roles as teachers, advisors, and program managers. The application of the scholarship of teaching through the lens of agricultural education into teaching requires a deep understanding of the individual intricacies of and complex connections within the agriculture classroom/laboratory, National FFA Organization, and Supervised Agricultural Experience (SAE) programs.

Developing as a Scholar of Application

Curriculum Related to Application

Courses related to the scholarship of application should be taken which engage students in a variety of content and service areas needed to engage with a variety of stakeholders. To begin, courses could be taken which focus on program planning, implementation, and evaluation. This content would help with creating and implementing single occur or sequenced professional development workshops related to teaching, learning, or disseminating research supported best practices. Additionally, these courses would assist with planning, implementing, and evaluating education or outreach programs as part of grant funded or interdisciplinary projects. Related to teaching, agricultural education doctoral students often have many years of preparation and experience teaching in formal education settings, however taking non-formal teaching courses can help future faculty members as they prepare to teach in a variety of non-formal settings. Lastly, courses which focus on diversity, equity, and inclusion (DEI) should be taken to provide future faculty members with the necessary knowledge and skills to help lead or engage in DEI initiatives with stakeholders.

Experiences Related to Application

Doctoral students should work closely with their faculty advisor and other faculty members as they engage with and serve various stakeholders in formal, non-formal, and informal educational settings. Faculty members frequently find that regular interactions with end-user stakeholders can be mutually beneficial. The stakeholder gains access to current research and discovery while the faculty member benefits from stakeholders identifying research needs as well as sharing examples of relevant experiences that often enrich faculty course instruction.

The Agricultural Education Context of the Scholarship of Application

Teacher educators in Agricultural Education faculty positions have been found to frequently "collaborate with their state department of education and their state agriculture teachers association to provide activities and programs for (in-service) teachers" (Franklin & Molina, 2012, p. 128) Such activities are important contributions to the professional development of teachers and are also important examples of the scholarship of application. Such traditions of collaborative work with statewide "Team Ag Ed" partners also provide impetus and opportunity for doctoral students to engage in the scholarship of application to share both research findings and research-based teaching practice with those practitioners who stand to benefit from such expertise. Scholarship of application opportunities also exist for Agricultural Education doctoral students in serving university faculty peers from other disciplines, extension state specialists, and extension education practitioners with research-based principles of effective teaching. Opportunities for such programming are abundant and can be mutually beneficial to both the developing Agricultural Education scholar and the multiple audiences they may serve. On occasion, these partnerships can also result in assistantship opportunities, open doors for research partnerships, or forge relationships with community partners who can be valuable to subsequent research projects. While it is common for the scholarship of application to be overshadowed by the scholarship of teaching and the scholarship of discovery, it is a crucial part of the Agricultural Education professional culture, and a typical expectation in faculty position descriptions. Consequently, refining the scholarship of application is an important practice for the student to develop in preparation for the professoriate.

Developing as a Scholar of Discovery

As a future scholar, students need to develop knowledge and skills which will help them conduct their own research projects, mentor graduate students through the completion of a research project, thesis, or dissertation, consume and apply empirical findings, and review conference and journal submissions to serve their professional organizations. Doctoral students should take courses which immerse them in research design and multiple methodologies to develop well rounded researchers. Additionally, students should begin to develop their expertise area within a specific line of inquiry which can be pursued through their scholarly career.

Curriculum Related to Discovery

Coursework related to discovery should begin with an introductory research methods and design course to introduce students into the review of literature, conceptualization of a research problem, selection of a theoretical or conceptual framework, and the alignment of the study's purpose to potential methodologies. Following the introductory research course, students should take at least one quantitative and qualitative course, and, if available, a mixed-methods course to provide foundational knowledge to be an informed consumer and reviewer of research regardless of methodology utilized. Doctoral students should seek a quantitative course which focuses on developing and testing data collection instruments and basic data analysis and interpretation. The introductory qualitative course should provide an overview of narrative, ethnographic, case study, grounded theory, and phenomenology methodologies as well as qualitative data collection methods.

Future scholars should then take additional courses which will build their expertise in at least one methodology. Students should reflect on their future line of inquiry or research agenda to align the appropriate methodologies required to address future research questions. Additional quantitative courses could be taken which expand upon specific data analysis procedures including but not limited to multiple linear regression, logistical regression, analysis of variance, factor analysis, and multi-level modeling. Additional qualitative courses could be taken which either explore the use of critical theories and transformative worldviews to address societal issues, dive deeper into one or more of the qualitative methodologies mentioned earlier, or examine different data collection methods such as focus groups, observations, or document analysis. Further, students should consider taking grant writing courses or courses which focus on interdisciplinary research to better situate themselves as a future social scientist.

Experiences Related to Discovery

While courses will engage students in various methodologies and research related concepts, it is important that students seek out experiences to apply this new knowledge. Some of these experiences may come in the form of course assignments but it is how the student approaches these assignments which transforms them from a grade to growth. Students should utilize course assignments to conceptualize and conduct research projects utilizing both the feedback from the course instructor and the guidance of the faculty advisor. Further, doctoral students could serve as teaching assistants for graduate level research methods courses to enhance their understanding of the course concepts. Additionally, students should seek out informal learning opportunities such as seminars or lunch and learns to engage in topics not discussed in their course work such as how to advertise discovery expertise to others, develop research agenda, and conduct interdisciplinary research.

Beyond classroom-based learning, future scholars should engage in a variety of research projects to expand their understanding of and experience within the research process. Early in their doctoral studies, students could serve as a research assistant on faculty developed projects to engage in various research strategies. Additionally, students should attend research and professional conferences to begin to consume innovative research within agricultural education and the broader discipline of education. As their research skills progress, students should be conceptualizing and conducting their own research studies under the mentorship of their faculty advisor. The conceptualization of these studies could lead to students submitting graduate research grant applications. Regardless of whether the research is grant funded or not, upon completion of the project, students should disseminate their findings through research and professional conferences, as well as publishing their work in peer reviewed journals. As the students move towards and through candidacy or comprehensive exams, they should continue to look for opportunities to engage in the research process beyond their dissertation. This could include mentoring undergraduate researchers alongside a faculty mentor or serving as research poster reviewers for their professional organizations and research conferences. Ultimately the culmination of these experiences should boil down to developing a clear line of inquiry or research agenda which can be continued throughout their academic career.

The Agricultural Education Context of the Scholarship of Discovery

With typical heavy teaching loads, it is important to connect teaching and learning with the scholarship of discovery. Often this occurs through the continued pursuit of knowledge through research agendas which align with teaching and learning. However, the connection between classroom teaching and scholarship can also be achieved through the Scholarship of Teaching and Learning (SoTL). Scholars suggest SoTL is the practice of critically assessing student learning to improve teaching practices, to then share out best practices so that others can analyze their own practices (Hutchings et al., 2011). Further, SoTL focuses on "research-led teaching" where educators demonstrate excellence to evaluate the practice of teaching to that of research within higher education institutions (Fanghanel et al., 2016). Shulman (2001) proposes that there are three broad rationales for advancing SoTL: a) professionalism, b) pragmatism, and c) policy. As educators and scholars, we have a professional responsibility to advance education through discovery and application. Regarding the pragmatism of SoTL he further states, "By engaging in purposive reflection, documentation, assessment and analysis of teaching and learning, and doing so in a more public and accessible manner, we not only support the improvement of our own teaching. We raise the likelihood that our work is transparent to our colleagues who design and instruct many of the same students in the same or related programs" (p. 50). Lastly, SoTL can provide measurable evidence of meeting educational goals for accrediting agencies and policy makers. However, SoTL may not be widely accepted as scholarly work or well understood, especially in colleges of agriculture where bench or hard sciences dominate. Thus, it is crucial that agricultural education faculty advocate for and continually engage in rigorous SoTL work. Additionally, as experts situated within the intersection of agricultural concepts, teaching and learning; agricultural education faculty should mentor colleagues in the conceptualization and implementation of innovative teaching activities and assess the effectiveness of these activities through SoTL projects. Researchers indicate SoTL work could also expand into leadership activities outside of traditional institutional leadership roles at influencing departmental or college-wide instructional decisions (Miller-Young et al., 2017).

SUMMARY

The graduate program experience in Agricultural Education plays a key role in the professional growth of future scholars of teaching, application, and discovery in the profession. Because of its research-focused, constructivist-oriented nature, both aspiring students and program faculty should be prepared for a challenging learning environment that encourages individual learning design and active inquiry. Agricultural Education graduate program administrators should anticipate the unique resources required to deliver a quality learning experience and should facilitate a careful focus on the needs of the target audience the program seeks to serve. At the master's level potential exists to serve candidates from a variety of background experiences with a wide spectrum of professional goals guiding their graduate program interest. Faculty involved in advising, mentoring, and planning growth opportunities for doctoral students in

Agricultural Education should be careful to intentionally foster the well-rounded development of future scholars in teaching, application, and discovery to best prepare program graduates for career success.

REVIEW QUESTIONS

1. What considerations should undergraduates take as they consider graduate school?
2. What departmental resources should be considered when implementing a graduate program?
3. Compare and contrast the different master's degree programs.
4. What curricular decisions should be considered within a doctoral program to engage students in the scholarship of teaching, application, and discovery?
5. What professional experiences should agricultural education doctoral students engage in to advance their scholarship of teaching, application, and discovery?

REFERENCES

Boyer, E. L. (1990). *Scholarship reconsidered: Priorities of the professoriate*. Princeton University Press.

De Lay, A. M., & Washburn, S. (2013). The role of collaboration in secondary agriculture teacher career satisfaction and career retention. *Journal of Agricultural Education*, *54*(4), 104–120. doi:10.5032/jae.2013.04104

Fanghanel, J., Prittchard, J., Potter, J., & Wisker, G. (2016). *Defining and supporting the Scholarship of Teaching and Learning (SoTL): A sector-wide study*. High Education Academy.

Fincher, R. M. E., & Work, J. A. (2006). Perspectives on the scholarship of teaching. *Medical Education-Oxford*, *40*(4), 293–295. doi:10.1111/j.1365-2929.2006.02404.x PMID:16573663

Foster, D. D., Lawver, R. G., & Smith, A. R. (2020). *National Agricultural Education Supply and Demand Study, 2019 Executive Summary*. Retrieved from: http://aaaeonline.org/Resources/Documents/NSD2019Summary.pdf

Franklin, E. A., & Molina, Q. F. (2012). Teacher induction programs in Agricultural Education: Description of the role of AAAE Higher Education Teacher Preparation Programs. *Journal of Agricultural Education*, *53*(1), 123–135. doi:10.5032/jae.2012.01123

Hutchings, P., Huber, M. T., & Ciccone, A. (2011). *The scholarship of teaching and learning reconsidered: Institutional integration and impact* (Vol. 21). John Wiley & Sons.

Manathunga, C., & Lant, P. (2006). How do we ensure good PhD student outcomes? *Education for Chemical Engineers*, *1*(1), 72–81. doi:10.1205/ece.05003

Miller-Young, J. E., Anderson, C., Kiceniuk, D., Mooney, J., Riddell, J., Schmidt-Hanbidge, A., Ward, V., Wideman, M. A., & Chick, N. (2017). Leading up in the scholarship of teaching and learning. *The Canadian Journal for the Scholarship of Teaching and Learning*, *8*(2), 1–14. doi:10.5206/cjsotl-rcacea.2017.2.4

Roddy, C., Amiet, D.L., Chung, J., Holt, C., Shaw, L., McKenzie, S., Garivaldis, F., Lodge, J.M., & Mundy, M.E. (2017). Applying Best Practice Online Learning, Teaching, and Support to Intensive Online Environments: An Integrative Review. *Frontiers in Education.* doi:10.3389/feduc.2017.00059

Scott, D. (2016). *Ph.D. vs. Ed.D.: Which terminal degree is right for you?* Retrieved from: https://www.insidehighered.com/blogs/gradhacker/phd-vs-edd

Shulman, L. (2001). From Minsk to Pinsk: Why a scholarship of teaching and learning? *The Journal of Scholarship of Teaching and Learning*, 48–53.

Vithal, R. (2018). Growing a scholarship of teaching and learning institutionally. *Studies in Higher Education*, *43*(3), 468–483. doi:10.1080/03075079.2016.1180350

Chapter 16
Conducting Programmatic Research in Agriculture Teacher Education

Will Doss
Texas A&M University - Commerce, Commerce, USA

John Rayfield
Texas Tech University, Lubbock, USA

ABSTRACT

This chapter focuses on programmatic research in agriculture teacher education programs. Included in this chapter are the definition of research, why agriculture teacher education programs should conduct research, approaches to programmatic research, and practical applications for using research in teacher educations programs. The authors break down various methodologies for conducting programmatic research and offer many practical examples of how these can be easily implemented into any agriculture teacher education program. This chapter walks the reader through developing researchable questions, using the scientific method, choosing appropriate research design and methodology, and simple steps to disseminate and communicate research findings. The chapter also touches on the relationship between research and evaluation and how evaluation is used to assess programmatic effectiveness.

INTRODUCTION

Agriculture teacher education programs generally have three different functions that must be performed: teaching, service, and research. These functions usually benefit program stakeholders such as undergraduate and graduate students, secondary school agricultural education teachers in the field, and other agricultural industry representatives with interests in agricultural education. The teaching function of agriculture teacher education programs is to train new agriculture teachers. This training occurs in undergraduate and graduate courses. The teaching function can include courses in areas such as introduction and history of agricultural education, curriculum design, teaching methods, classroom

DOI: 10.4018/978-1-6684-3420-8.ch016

management, laboratory management, and managing FFA and SAE in agricultural education programs. The service function often comes in the form of providing support for agriculture teachers in the field. In some institutions, particularly land grant universities, extension may be substituted for service. They have similar activities but often involve providing education or support beyond the regular on-campus undergraduate and graduate students. Activities related to service or extension can include coordinating FFA events and competitions, providing professional development in summer workshops for teachers, or simply judging FFA and SAE competitions.

The final function of an agriculture teacher education faculty member and focus for this chapter is research. **Research** is a diligent and systematic investigation used to discover or modify facts, relationships, theories, and applications and to study problems in a disciplined manner, generally using the scientific method. Research is increasingly becoming a requirement for university faculty members in all subjects. For many faculty members, there has been a push for more research productivity than in the past. Agriculture teacher education programs increasingly use research to make informed decisions about their programs. This chapter provides an overview of programmatic research conducted in agriculture teacher education programs. Upon completion of this chapter, the reader should be able to accomplish the following objectives:

1. Determine why programmatic research is conducted and how it informs practice in agricultural teacher education.
2. Formulate plausible programmatic research questions.
3. Compare different programmatic research approaches.
4. Identify potential areas of applied programmatic research in agricultural teacher education.
5. Determine how to communicate research findings with appropriate audiences.
6. Identify programmatic research trends in agriculture teacher education.

Why We Conduct Programmatic Research

Systematic research conducted by faculty and staff members in agriculture teacher education programs with a focus and concentration on an overarching issue, subject, or problem is known as **programmatic research**. Programmatic research can have many different benefits and purposes. Research can take an internal perspective of examining the agriculture teacher program itself; or an external perspective, investigating agricultural education beyond the college or university. Agriculture teacher education faculty and staff members can conduct research on a wide variety of topics going beyond agriculture teacher education. Many investigate problems in areas such as agricultural extension, international agriculture, agricultural mechanics, agricultural leadership, agricultural communication, and higher education. However, this text focuses on research related to training secondary school agricultural education teachers and best practices for managing a secondary school agricultural education program.

Agriculture teacher education programmatic research often provides data related to the effectiveness of teacher preparation programs in the form of a program evaluation. A program evaluation can provide information for future decision making related to program planning or improvement and can serve as a form of faculty and staff member accountability. Programmatic research can also be conducted from a different viewpoint. Rather than critically examining the teacher preparation program itself, research is often conducted related to best practices involved in training teachers and running a secondary agricultural education program.

Information gained from research studies plays a critical role in agriculture teacher education programs. Since agricultural education is often considered an applied science, research on the topic is used to inform practice. For example, a program evaluation of an agriculture teacher education program can identify areas of needed improvement and let faculty members know what areas are being implemented effectively. This information can then be used to make decisions regarding adjustments and future directions of a program to train new teachers more effectively. Studies conducted outside of the college or university program can help faculty members understand what is happening in the field and how to support current practicing agriculture teachers. Information from agricultural education research can also contribute to the broader education field with studies in effective teaching methods, learning, and retention.

Formulating Plausible Research Studies

One of the most challenging aspects for researchers in agriculture teacher education is the task of choosing a topic. A research topic is often chosen based on the researcher's personal interests. Researchers in agriculture teacher education strive to choose a topic that will advance knowledge in agricultural education in general, whether it be on the secondary level or collegiate level. Many faculty members have what is known as a research agenda. A **research agenda** provides a framework for the direction and types of studies that must be completed to address an overarching problem or area of interest. The research agenda can be a roadmap for a career in research, providing a list of studies that need to be completed.

Graduate students are often required to complete a research study and author a thesis or dissertation, depending on the level of study. Their research topics are also chosen on personal interest but could be influenced by their chair or advisor's interests as well. The struggle to identify a topic for research is greater with graduate students because of their novice status in the field. To help graduate students and faculty members in agriculture teacher education identify research topics, the American Association for Agriculture Education (AAAE) has produced and published a national research agenda. This can be referenced to aid in selection of a research topic or direction for a research agenda. Another way to pick a research topic is to review literature to identify areas where information is lacking or to identify a problem that needs solving. On the other hand, some study topics may involve a local issue where there is no intention to apply findings beyond action in a local program.

Once a general topic has been chosen, the beginning of a research project can be thought of as the start of an investigation. For a project to be considered research, a systematic method for conducting the investigation should be employed. This is often referred to as the scientific method. The **scientific method** is a systematic procedure used for the purpose of gaining new knowledge that is recognized by the science community. The scientific method uses the following steps:

1. Identify a problem
2. State the research problem in question form
3. Review information related to the problem
4. Form a hypothesis or prediction to the outcome of a problem to be investigated
5. Design data collection procedures and collect data
6. Analyze data and draw conclusions
7. Communicate results with the scientific community

Table 1. Examples of poorly and well written research questions

Poorly Written (Not Researchable)	Critique	Well Written (Researchable)
Should preservice agricultural education teachers student teach in the fall or spring semester?	Research question is written as an opinion question, can be answered with a yes or no, and is subjective.	Do preservice teachers who student teach in the fall semester have higher levels of self-efficacy in performing classroom, FFA, and SAE duties than preservice teachers who student teach in the spring semester?
Do high school principals like agricultural education programs?	Research question is vague and can be answered with a yes or no.	What is the level of principals' perceived importance of classroom, FFA, and SAE activities in agricultural education programs?
Should secondary agricultural education teachers use experiential learning teaching methods?	Research question fails to identify variables to be studied, can be answered with a yes or no, and is subjective.	Is there a difference in student retention of information when teaching a lesson using an experiential learning technique compared to a passive learning technique?

A general topic or area of interest must first be selected by the researcher so a clear problem or lack of information related to the topic can be identified. The process of identifying a problem can be iterative where the researcher consults literature and refines the problem multiple times. Once the problem has been clearly identified and communicated, the researcher can move forward in the steps of the scientific method.

Determining Which Research Questions are Viable

With the second step of the scientific method, a clear research problem should be written in question form to guide the development of a study. A **research question** identifies the problem that will be investigated, is written clearly enough that the variables under investigation are understood and gives indication as to the research approach that will be used to investigate the problem. A research question should be written in such a way that data can be collected and analyzed to test a hypothesis or form a conclusion. Questions that can be answered with a simple yes or no are not specific enough to conduct a study in most instances and can appear subjective. Problems that are matters of opinion are not researchable. Since agricultural education research often deals with human subjects, the research question posed should also be ethical. Refer to Table 1 for examples of questions addressing research problems.

Reviewing Information Related to the Problem

Once a problem has been identified and a research question has been formed, a review of literature should be conducted. In some cases, the literature may provide a solution to your identified question. In other cases, further research is needed to address the issue. Information previously published in peer-reviewed journal articles, textbooks, white papers, or periodicals may be used to better refine your research question so it can be properly investigated. High quality literature sources should be used whenever possible. Published studies that have been scrutinized and peer reviewed can be thought of as high-quality literature sources and will provide the most accurate information related to a research topic.

Often a theory can be used to explain a phenomenon in agriculture teacher education research. A theory will provide an educated guess for the results of your study and can also be used to justify the design of the study. Theories used in agricultural education research have origins from a wide variety of disciplines including educational psychology, sociology, and marketing to name a few. Since agricultural education is an applied discipline, few theories have been developed specific to the field. Our research generally applies theories from other disciplines allowing for the application and testing of those theories in the context of agricultural education.

Designing a Study

After conducting a review of literature and finalizing your research question, you should have an idea or educated guess for possible answers to the research question. At this point, depending on the nature of the research question, you may also form a hypothesis or hypotheses based on literature reviewed. However, unlike bench sciences, social science studies do not always have a statistical hypothesis to be tested. Either way, a study must be designed so data can be collected and analyzed. The design of a study is directly tied to the research question. How the question is asked and how it identifies variables determines which research approach will be used. The research approach used determines which options are available for methodologies and data collection. Methodologies used and data collected influence conclusions that can be drawn and generalizability to the population being studied. Research approaches and methodologies will be discussed in more detail later in this chapter.

Ethical Considerations

Human subjects are often used to provide data for agriculture teacher education programmatic research. When conducting research with human subjects in social science areas such as education, ethical considerations must be made. Treatment of human subjects can have implications on the study design used and can be a limiting factor. Studies must be designed in such a way that human subjects are protected from physical or psychological harm. In fact, social science research dealing with human subjects is conducted on the principle of beneficence. Ethical studies can ensure beneficence by providing a benefit to those being studied (when possible), by increasing understanding through new knowledge that will benefit the participants, and by promoting advancement and opportunity.

Care must be taken to accurately disclose all risks and benefits associated with a study involving human subjects. In some instances, before research is conducted with participants, written consent may need to be collected from the participant to document being informed of all risks and benefits and agreeing to participate in the study. Extra caution is taken when working with minors under the age of 18, as is often the case in agricultural education research involving secondary school students. When studying minors, parents or guardians grant consent to participate. Since a minor cannot consent, they can agree to participate with what is known as assent. Along with disclosing risks and benefits, the researcher is required to assure participants that data collected will remain confidential and results will be anonymously reported. In some study designs, deception must be used to investigate the variables under consideration. While deception should be avoided, if possible, it may be used with caution if it causes no adverse effects on the participants. Once the study is complete, the participants must be informed of the deception.

Research institutions receiving federal funding are required by law to have an institutional review board (IRB) responsible for the approval of research studies involving human subjects. Researchers must submit to the IRB a study proposal and request for permission to conduct a study. A panel reviews the proposal submitted to ensure no harm will be done to participants and that confidentiality of data will be maintained. All risks and benefits to participants must be disclosed. If the IRB approves the study proposal and finds no harm to participants, the researcher may then begin collecting data.

Honesty in agriculture teacher education research is essential. Care must always be taken to report data exactly as the data were obtained. Data should never be manipulated to show findings that were desired or needed. This can yield research conducted as misleading and meaningless. Temptation to alter data can be experienced when dealing with large sums of grant money or when career advancement is on the line. Researchers must always remain cognizant of the need to have honest results or risk losing reputation and even their career.

The misrepresentation of someone else's work or ideas as your own is known as **plagiarism**. This is another ethical consideration to keep in mind when preparing a research study and writing up results. Full and appropriate reference should be made to the original creator or publisher of ideas and research findings. Plagiarism is another form of academic cheating that can have negative consequences. Self-plagiarism should also be avoided. For example, publishing the results of the same study in multiple journals would be one form of self-plagiarism. In some instances, self-plagiarism is done unintentionally but still results in academic dishonesty.

Research Approaches

Agriculture teacher education programmatic research generally falls under one of three different research approaches: quantitative, qualitative, and mixed methods. Selection of the research approach should be tied to the nature of the question or problem to be investigated. Some questions and problems naturally lend themselves to one particular approach of research based upon the desired data for the study. In order to select a suitable approach to use, a general understanding of what is involved with each approach is needed. Each different research approach has implications on which methodologies can be used, data that can be collected, analysis procedures, and generalizability. Information related to quantitative, qualitative, and mixed methods approaches is summarized in Table 2. A **methodology** is the design and plan for data collection, data analysis, and interpretation used in a research study. The selection of the methodology hinges on the purpose of the study or type of question to be answered. Each methodology provides a particular type of data that can be interpreted in a certain way. It is essential that the methodology selected can provide the correct data needed to answer the research question.

Quantitative

Quantitative research is a research approach that uses numerical data to test theories, study relationships between or among variables, and numerically describe or quantify characteristics. Quantitative research methods are clearly defined before data collection begins with a set plan for data analysis. There are several different methodologies used including experimental, quasi-experimental, survey, correlational, causal-comparative, and single-subject studies. Sample sizes for these types of studies are generally larger in size compared to other research approaches. Larger sample sizes are needed because quantitative methodologies attempt to collect enough data to be representative of a larger population. Data collected are in numeric form and may be obtained from instruments and existing data sets. Analysis of quantitative data uses either descriptive or inferential statistics. Descriptive statistics are simple numeric descriptors such as frequencies and percentages. Inferential statistics are more complex and are used to infer results from a sample to represent the population. Inferential statistics are necessary to conduct hypothesis testing where significance is often reported. The ability to use inferential statistics improves the generalizability of this research approach compared to other approaches, however descriptive statistics should not be generalized beyond what they are describing.

Qualitative

Qualitative research is an approach that explores phenomena in an attempt to understand the complexities and meaning individuals or groups attribute to the phenomena. Qualitative research methods are considered emerging because this type of investigation is exploratory in nature and must remain fluid to gain a deep understanding of a particular phenomenon. The researcher may not be able to fully anticipate which method will be needed in this instance. There are however some recognized methodologies to help shape the qualitative approach including phenomenology, case study, narrative, ethnography, and grounded theory. Each of these methodologies uses a smaller sample size compared to quantitative approaches. Data collected can include textual, auditory, and visual observations. Data are generally analyzed inductively by grouping specific data from individuals into more general themes representing the group. Qualitative studies have low generalizability because of their small sample sizes and inability to statistically infer to a broader population. This research provides a detailed account that may answer "why" questions where quantitative approaches may not be able to collect that information.

Mixed Methods

Mixed methods research is an approach that combines elements from both quantitative and qualitative approaches to gain additional information that could not be provided from one of those approaches alone. Mixed methods studies can be thought of as two studies in one. In that context, they use both pre-determined and emerging methods corresponding with the quantitative and qualitative portions of the study. Methodologies are generally considered either convergent, explanatory sequential, or exploratory sequential. The methodology depends on the timing and purpose of the quantitative and qualitative portions. Sample sizes can be larger for the quantitative portion of a study and smaller for the qualitative portion. Mixed methods studies provide a combination of numeric and qualitative data forms. Data from the two different portions of the study are analyzed separately. Mixed methods studies can have high generalizability for quantitative portions of the study but for the qualitative and combined analysis, generalizability can be low. A key concept to understand with mixed methods studies is that the combined analysis of the quantitative and qualitative portions adds a layer of understanding that could not be found by conducting each study by itself.

Two other research approaches commonly conducted by agriculture teacher educators includes historical and philosophical research. While some disciplines would consider each of these approaches to be qualitative, leaders in agricultural education research have separated them out due to the difference in data collection techniques and final purpose of the studies.

Table 2. A comparison of research method approaches

	Quantitative	Qualitative	Mixed Methods
Method Determination	Pre-determined	Emerging	Both Pre-determined and Emerging
Methodologies	Experimental Quasi-Experimental Survey Correlational Causal-Comparative Single-Subject	Phenomenology Case Study Narrative Ethnography Grounded Theory	Convergent Explanatory Sequential Exploratory Sequential
Sample Size	Larger	Smaller	Multiple
Forms of Data	Numeric	Textual Auditory Visual	Combination of Numeric and Textual/ Audio/Visual
Data Analysis	Descriptive Statistics Inferential Statistics	Induction	Induction Descriptive Statistics Inferential Statistics
Generalization	High	Low	Varies

Historical

Historical research is a systematic attempt to collect and objectively evaluate data related to events, conditions, people, and settings from the past to help explain current events or foretell future events. The location of historical artifacts and documents can be a time-consuming process for this research approach but is necessary to provide the data needed for the study. Once data is found, external and internal criticism of authenticity must be conducted. Analysis of data from historical studies is similar to other qualitative methodologies with generally low generalizability. Generalization can vary though, depending on the breadth of data analyzed related to a particular event, condition, population, or setting. This type of research is important for preserving history in agricultural education and making current and future generations aware of the origins of agricultural education.

Philosophical

Philosophical research is a research approach that attempts to critically examine a philosophy or line of thought in an attempt to gain additional insight that can be applied to establish new concepts, theories, or norms. This type of research is less common in agriculture teacher education programs. However, it is sometimes necessary to encourage others to think differently about different topics within the discipline of agriculture teacher education.

Comparing Programmatic Research Methodologies

Agriculture teacher education programs use a wide variety of research methodologies to solve problems, evaluate situations, and make new discoveries. There are many different methodologies beyond those covered in this chapter. However, we will focus on the more popular and impactful methodologies currently used in agriculture teacher education research: experimental and quasi-experimental studies, associational studies, descriptive studies, longitudinal studies, needs assessments, and program evaluations.

Experimental and Quasi-Experimental Studies

Experimental studies are conducted to demonstrate cause and effect. Experiments require the random assignment of participants to experimental groups and control groups. There must be an independent variable that can be manipulated to cause a change in the dependent variable that can be measured. In educational settings, random assignment of participants to groups may not always be possible for ethical reasons. In this case, a quasi-experimental study may be used. Quasi-experimental studies use the same methods as an experimental study with the exception of random assignment of participants to groups.

Experimental and quasi-experimental studies require statistical hypothesis testing. Care must be taken in selecting the appropriate sample size to give the study enough statistical power to detect differences in dependent variables after independent variables are manipulated. A power analysis can be used to determine the appropriate sample size. Sample size is also largely dependent on the design of the experiment. Experiments are designed based on the number of treatment groups, the presence of a control group, and the stage in which data and measurements are collected. Threats to the validity of experimental studies must be considered. These threats can be both internal and external. Literature related to the study can help the researcher anticipate threats and better design the experiment to control the impact of those threats.

Experimental and quasi-experimental studies in agricultural education have mostly tested the impact of teaching methods on student learning. Many of the studies were conducted with students at the secondary level. There is potential for many future experimental and quasi-experimental studies in agriculture teacher education. Studies related to effective teaching methods at secondary and post-secondary levels will continue to be needed as the profession tests more learning theories in the context of agricultural education. Studies related to long-term retention of knowledge, most effective use of technology, and integrating core subjects into agricultural education all could be explored with experimental and quasi-experimental studies.

Associational Studies

Associational studies are designed to explore associations or relationships among variables. Relationships among variables can be both explanatory and predictive. Common methodologies used to discover relationships include correlational studies and causal-comparative studies. Associational research differs from experimental research because the researcher does not attempt to manipulate any of the variables under investigation. Some classify these types of studies as descriptive, however the types of statistical calculations used for associational research is more advanced.

Correlational research attempts to determine the magnitude and direction of a relationship between two or more variables. Correlation coefficients are used to describe the direction of a relationship as well as the magnitude in explanatory studies. If the purpose of the research is to predict, a regression may be used where an equation is developed to describe the relationship among variables and predict an outcome. Data for correlational studies can come from surveys, instrument data collection, or existing data sets. True correlational research should have hypotheses to be tested, especially when significance is reported. Otherwise reporting correlation coefficients would simply be descriptive research.

Causal-comparative studies, also known as ex post facto studies, are used to strongly suggest causes for differences that already occur in groups. Causal-comparative studies differ from correlational studies because they make a stronger case for causation whereas correlational research does not attempt to do this. Statistics for causal-comparative studies usually involve comparing groups which can be done with analyses such as *t*-tests or analysis of variance. Like experimental research, causal-comparative studies should have hypotheses that are tested because of the interest in statistical significance. Causal-comparative studies do not provide as strong of evidence of causation as experimental and quasi-experimental studies because of the inability to manipulate the independent variable. Causation can be established through repeated experimental and quasi-experimental studies testing the variables.

Associational studies in agriculture teacher education programmatic research are useful for exploring descriptive studies in more detail while laying the groundwork for more rigorous experiments in the future. Previously published studies have examined correlations and compared groups on topics such as test scores, efficacy, and perceptions. Existing data sets have been used in the past for correlational studies. However, surveys and measurement instruments generally provide data for correlational and causal-comparative studies in agricultural education. There is the possibility for many future associational studies that could help describe and explain relationships in agriculture teacher education as additional databases are established. This type of research can be used to highlight the benefits of agricultural education and add information to general education research literature.

Descriptive Studies

Descriptive studies simply attempt to describe current populations, events, conditions, and situations. Research taking place at one point in time is known as cross-sectional research. Methodologies from quantitative, qualitative, and mixed methods approaches can be used for this purpose. However, descriptive quantitative studies are most common. Descriptive studies are limited in their scope and impact. They do not test concepts or theories, but they tend to provide basic information. These studies are often exploratory and provide the groundwork for designing more impactful and complex studies.

Quantitative descriptive research is limited to analysis with descriptive statistics. Descriptive statistics can include numerical indices to describe data such as frequencies, percentages, and measures of central tendency like mean, median, mode, range, and standard deviation. Graphical displays may also be used in descriptive studies. Descriptive research stops short of analyzing relationships among variables. Surveys are often conducted in agriculture teacher education research to collect data for descriptive studies. Public information and existing databases may also be used for this purpose.

Descriptive research has been used extensively in agriculture teacher education research to describe populations such as school-based agricultural education teachers, post-secondary students in colleges of agriculture, secondary agriculture students, and many others who participate in agricultural education. Studies describing FFA, supervised agricultural experiences, classroom instruction, technology usage, and many others have been published. While these studies are simple in nature, they are needed to provide the profession with the most current information on populations, events, conditions, and situations. This will enable researchers to design more complex studies examining relationships among variables and testing theories in the context of agricultural education experimentally.

Longitudinal Studies

Longitudinal studies are generally carried out with the use of a survey. The purpose of a longitudinal study is to document changes that occur over time, particularly with the same group or subjects being repeatedly surveyed. Longitudinal studies can span over the period of a few months to several years, differing from most descriptive studies that are cross-sectional, taking place at one point in time.

Longitudinal studies are not conducted very often in agriculture teacher education research likely due to the extended time commitment needed to carry them out. However, some longitudinal studies examining changes in teacher retention, student teachers, efficacy, and attitude have been published. This type of research is important to accurately reflect changes that occur because it uses the same participants over time or examines a particular situation or phenomenon over time. Longitudinal studies have the ability to examine many of the difficult problems that need solving in agriculture teacher education. With increased commitment to long-term studies, researchers could better describe problems such as chronic teacher shortages, long-term knowledge retention, and effects of agriculture teacher education training on long-term success, to name a few.

Needs Assessments

The purpose of needs assessments is to inform program development or strategic planning. A needs assessment may also be carried out during a program to help decision makers adjust their strategy to meet stakeholder needs more accurately. This form of research helps program personnel prioritize efforts to meet the needs of a population in the most fiscally responsible manner. A needs assessment requires researchers to loosely identify a broad area of needs for a population, develop an inventory or data collection strategy, and evaluate data collected to allocate resources.

In agriculture teacher education research, needs assessments are common. Published studies have determined needs of teachers and preservice teachers related to knowledge and skills and professional development activities. Information gathered from these needs assessments can be used to adjust current training practices employed in agriculture teacher education programs as well as allocate resources for providing professional development for practicing teachers in the field. Service or extension is often a requirement for faculty members in agriculture teacher education programs. Needs assessments can help drive service to meaningfully engage stakeholders.

A needs assessment can also be used to justify funding for grant projects and programs. Grantsmanship falls under the research responsibilities of faculty members. Needs assessments should continue in the future and may become more valuable as requirements for grants increase for faculty members. They are also needed to continually inform the profession of stakeholder needs so useful services may continue to be offered and to allow for program training practices for preservice teachers to be adjusted to have successful and prepared graduates.

Program Evaluation

Program evaluations are generally conducted to provide data used to inform stakeholders in their decision making related to a program or policy. Program evaluations generally answer questions about the effectiveness and efficiency of programs, policies, and projects. Historically, agriculture teacher education programs have used this methodology to determine how well their own programs are performing in the context of effective teacher preparation, how adult education programs and professional development workshops meet the intended outcomes, and how certain secondary education programs perform.

This systematic research methodology can use a variety of approaches such as quantitative, qualitive, and mixed methods. The exact portion of a program to be evaluated will often determine which approach should be used. Areas of programs that can be evaluated include program theory and design, how a program is running and being implemented, measuring program outcomes, evaluating impact of the program, determining program effects and magnitude, and examining the financial efficiency of a program.

Results from a program evaluation are often reflective of the initial program planning process. Evaluations provide information to stakeholders regarding programs intended effects, benefits, and cost efficiency. Data from program evaluations are sometimes tied to funding requirements, especially when the program is funded through government grants. Results from program evaluations can also affect decisions related to the continuation of the program or project.

AREAS OF PROGRAMMATIC RESEARCH

Programmatic research is often carried out in the form of action research. The scope of this type of research can vary depending on college or university size and goals of the institution. Programmatic research related to best practices for training preservice teachers can be used to inform teacher educators on how to tailor their programs to prepare new teachers as well as provide supplemental training for teachers already in the field. In the next several sections of this chapter we will explore different areas of programmatic research in agricultural teacher education including preparedness of new teachers entering the field, retention of agricultural education teachers, best practices for teachers in the field, and effective collegiate teaching practices.

Action Research Related to Agriculture Teacher Education

Agriculture teacher education programs use many forms of action research to evaluate effectiveness of their programs. **Action research** is performed by teachers and other educational professionals to change conditions in a particular situation they may be involved in. The purpose of action research is to solve a problem or obtain information to inform local practice in a school district. An example might be, improving math achievement of students enrolled in agriculture classes or identifying better ways to serve students with special needs through agricultural education courses. Action research utilizes participants or stakeholders more so than subjects in traditional research design. This showcases the "active" nature or involvement in the research process.

The steps and procedures used in action research are similar to formal quantitative or qualitative research. Systematic inquiry occurs with the goal being to solve a local problem. This requires little to no research training to conduct. The intent of this research is usually to identify and correct local problems. The research is carried out by local teachers or educational professionals involved in the local educational organization utilizing instruments developed by teachers or the school district. This is typically less rigorous than formal research utilizing purposive sampling techniques and having some opinion of the researcher considered as data or an outcome. Compared to formal research, the generalizability of action research is very low.

There are many ways to incorporate action research activities into agriculture teacher education programs. Program faculty may choose to evaluate teaching candidates on their preparedness to enter the field. Questions such as how prepared are teaching candidates to perform the roles of classroom teacher, FFA advisor and supervisor of agricultural experiences can be answered through action research. Although formal data could be collected on these specific skills, agriculture education teacher preparation programs can employ action research methodology to quickly assess candidates in these areas.

Programs may elect to study retention rates of their graduates. Studies may address questions such as: What percentage of graduates enter the agricultural education profession? Why are teachers choosing to stay or leave the profession? How can agriculture teacher education programs influence those decisions? It is easy to locate and study teachers who have stayed in the agricultural education field. It is difficult to find teachers once they have left the profession. Employing action research techniques to retention research may be a viable solution to add to the small and dwindling research base in this area of concern.

Another way to utilize action research with agriculture teacher education is to identify best practices for beginning teachers in the field. There are many questions in this area that need answering such as: Which teaching methods are most effective for beginning teachers? How should beginning teachers manage their FFA chapter? What are the best practices for getting all students engaged in supervised agricultural experiences (SAEs)? How should a new teacher approach managing a successful total agricultural education program? What is the teacher education program's role in preparing graduates for their new role? Identification of best practices rarely requires formal inquiry. Action research can easily accomplish this for agriculture teacher education programs.

Often agriculture education teacher preparation programs self-examine their teaching practices. These programs should model the most effective teaching practices as they are producing the teachers who are the next generation of educators. Although these programs are held in high esteem as models of excellence, it is always a good practice for them to evaluate their best practices for teaching, advising, recruiting, and mentoring future teachers. Each of the previously discussed programmatic aspects can be examined or evaluated using action research.

Many of the day-to-day aspects of agriculture teacher education programs can be easily placed on cruise control. If everything is working like it should, then why should we make any changes? Simple and easy steps to program effectiveness can be researched without collecting mounds of quantitative or qualitative data. Action research methods allow practitioners to study and solve these problems at the programmatic level without the constraints of formal research methodology.

COMMUNICATING RESEARCH FINDINGS

The final step in the scientific method is to communicate results with the scientific community. There are several different ways agriculture teacher educators share their research findings. The outlet for this information depends on the purpose of the research, who the research is likely to benefit, and the scope of the research project. Theses and dissertations are traditionally used to communicate research conducted in graduate programs of agriculture teacher education. Due to the lack of visibility of these documents, they often provide the basis for additional publishing such as conference papers and journal articles.

Agriculture teacher educators are often members of professional organizations designed to enhance research collaboration and provide an outlet to share scientific discovery. The American Association for Agricultural Education (AAAE) is the professional organization most closely tied to agriculture teacher education. AAAE has annual regional and national conferences to allow researchers to present findings from studies. Presentation of research studies requires the submission of abstracts or manuscripts that are peer reviewed and either accepted or rejected. Accepted abstracts and manuscripts are then eligible to be presented at conferences.

Poster and paper presentations are the main ways researchers share findings from their studies at conferences. A poster presentation involves the submission of shorter abstracts and the creation of an actual research poster as an aid for short oral presentations. Poster presentations are generally less formal and more appropriate for studies with a narrower scope. Conference paper presentations involve the submission of longer abstracts or manuscripts and the creation of a longer, more formal oral presentation. Paper presentations are more appropriate for research studies with larger scopes. Both paper manuscripts and poster abstracts are published in what is known as conference proceedings for others to reference in the future. Information presented at conferences is then carried back to each attendee's institution and put into practice with training future agricultural education teachers and providing professional development opportunities for practicing teachers.

Another way agriculture teacher educators share their research is with peer reviewed journal articles. Journal articles come from the more rigorous and methodologically sound studies that have the greatest potential impact on agriculture teacher education. While there are several different journals available for the dissemination of research findings in agriculture teacher education, the *Journal of Agricultural Education* is the top tier journal for studies in this area. While journal articles are usually only read by other researchers, it is available for school-based agriculture teachers to read as well.

Since research in agricultural education is often applied in nature, findings should be communicated with stakeholders and practitioners in the field who could benefit from the information. White papers, summaries, and informal reports can be appropriate publications that can be used by practitioners. These reports are generally short and have only the main applicable information needed. Short articles in popular press such as *The Agricultural Education Magazine* can quickly be read by teachers and other stakeholders so they can apply findings from research in their everyday work.

One of the aims of research in agriculture teacher education is to inform practice. Findings from research should help agriculture teacher educators improve their ability to train new teachers. Findings from studies can be put into practice through faculty members' own teaching techniques and procedures. Agriculture teacher educators can also share findings from research studies with preservice teachers so they may use the information in the future. Professional development provided to practicing teachers may also make use of research findings and provide a platform to share information from researchers. Professional development may be conducted at the college or university where teachers can be brought in, or it may be conducted at events like a summer teachers conference.

RESEARCH TRENDS

Periodically, the profession of agriculture teacher education self-examines the status of research in the field. In fact, this can be a form of research in itself. Historically, agriculture teacher education research has been considered an applied science. Research conducted related to the subject usually is done in a pragmatic manner and is concerned with how the results can be quickly applied. By examining trends in agriculture teacher education research, we can determine the direction of research in our field, especially when discussing the types of research conducted, who conducts our research, how our research is funded, and how we identify needed research.

Types of Research Conducted

In a study reviewing research methods used in articles published in the *Journal of Agricultural Education*, it was found that the use of a quantitative approach for studies has declined from nearly 100% in the late 1970s and early 1980s to just under 60% in articles published in 2019 (Doss et al., 2021). At the same time, qualitative approaches have increased in the 2000s and 2010s to nearly 30% of all published studies, while mixed methods approaches have stayed at near 10% during this time period. Of all quantitative methodologies used, survey research has remained steady over the 40-year time period, accounting for nearly 70% of all quantitative studies. Surveys have mainly been used to study school-based agricultural education teachers, secondary agriculture students, college agriculture students, and others involved in agricultural education.

Agriculture teacher education survey research is mostly cross-sectional in nature with few longitudinal studies. Even with qualitative research and other quantitative methods, little research is conducted over long periods of time. Causes for the lack of longitudinal research could include time commitment needed for studies, who is conducting most of the published research, and attrition from study participants. This is an area of research that would be beneficial to agriculture teacher education in the future.

Who Conducts Research

The volume of agriculture teacher education research produced across the country has increased over the last few decades. This can be attributed to increased research requirements for faculty members, especially at land grant institutions and institutions with Carnegie classifications of R1 and R2. Universities with R1 and R2 status are considered doctoral universities with very high and high research activity. The desire to reach this status usually translates into increased research productivity for faculty members. Research productivity can be tied to consideration for tenure and promotions. With the university wide increase needed in research productivity, agriculture teacher education programs must also increase their research output. This leads to faculty members conducting more studies and publishing more findings.

One strategy for increasing research productivity includes recruiting graduate students and graduate research assistants to conduct more research. These students are mostly students studying to earn a master's degree or doctoral degree and are usually novice researchers. A 1982 study reported well over half of all research publications originated from graduate student work; an examination of published journal articles and conference proceedings reveals that this is a practice that has not changed (Miller & Warmbrod, 1982). While conducting research is a crucial skill for graduate students to learn, using them to increase research productivity can have implications on published literature in agriculture teacher education. An implication reported by some is that agriculture teacher education research is often fragmented in nature and lacks depth. As this problem persists, there is abundant opportunity for more focused and impactful research in agriculture teacher education.

In recent years we have seen an increased emphasis on research collaboration across agriculture teacher education programs. This can include collaboration across universities or partnering between universities and smaller colleges. Increased collaboration has also been observed between disciplines. These partnerships are often formed in the interest of receiving grant funding and for the collaboration of resources to tackle more complex research initiatives. The requirements for federal grant funding seem to show no signs of backing down from increasing collaboration. In the future agriculture teacher education research will likely continue to have increased opportunity for partnerships in research.

Funding for Research

Funding research in agriculture teacher education has always been a major barrier. Most research done in this area centers around improving the quality of programs and there are few funding agencies willing to finance this type of research endeavor. Most major federal granting agencies (USDA NIFA, NSF, USAID) do not fund projects related to program improvement. In order to secure funding, programs must strategically target foundations or private donors to fund these types of projects. Funding for this type of project would be considered minimal at most universities, but it can help offset the costs associated with conducting the project or potentially pay a portion of a graduate student's salary.

Needed Research in Agriculture Teacher Education

Identifying what needs to be researched in agriculture teacher education can be a challenge. One of the criticisms of research in this area is that it lacks focus and direction. Beginning in 2006 the American Association for Agricultural Education (AAAE) developed the first national research agenda and periodically revisits the agenda as an organization to provide updates and change the focus (Roberts et al., 2016). The national research agenda suggests areas where work is needed in agricultural education broadly defined. Within the agenda lies overarching problems and questions to be addressed specifically related to agriculture teacher education. This document can serve as a starting point for graduate students to choose a topic and for new faculty members to establish a research agenda of their own. Faculty members who stick to a research agenda over a career can have more focus and direction in the research studies conducted and published.

Researchers in agriculture teacher education should strive to publish studies with conclusions that are data driven and are generalizable to larger populations. Studies often are regionalized in their sampling to one institution, one state, or even a small group of people. If we are to tackle larger problems and have more impactful research, efforts should be made to conduct studies on a national scale.

In order for researchers in agriculture teacher education to produce a product that is useful for the discipline, efforts should be made to identify our stakeholders and determine what they need. Our research should directly provide a benefit to our stakeholders, ensuring relevance of our efforts.

SUMMARY

The future for agriculture teacher education programmatic research is bright and has many different areas needing exploration. The profession of agricultural education is ever changing. If agriculture teacher education programs are to remain relevant and continue to effectively train new teachers, continued research will be needed to inform our practice. This can be accomplished by formulating plausible research studies using steps from the scientific method. Quantitative, qualitative, mixed methods, historical, and philosophical approaches will all be necessary to tackle new areas of study in agriculture teacher education. With careful selection of methodologies used in studies, we have the opportunity to conduct impact research in collaboration with other disciplines in the future. There will continue to be a need for program reviews and action research topics will continually arise. We must continue to strive to communicate our research findings effectively and efficiently with stakeholders in the field while training the next generation of researchers at the same time.

REVIEW QUESTIONS

1. Why is programmatic research conducted in agriculture teacher education and how does the research inform practice?
2. Explain how the scientific method is used in designing research studies related to agricultural education.
3. What are the characteristics of a well-written research question? Provide an example research question written in your own words.
4. How is a research approach selected when designing a study?
5. What are some ethical considerations that must made when designing a study?
6. Differentiate between experimental, associational, and descriptive studies.
7. Describe how research may be published in agriculture teacher education.
8. What trends have been observed in agriculture teacher education research?
9. How can agriculture teacher education research projects be funded?

REFERENCES

Doss, W., Rayfield, J., Burris, S., & Lawver, D. (2021). A quantitative content analysis of survey research methods over a 40-year time period in the *Journal of Agricultural Education. Journal of Agricultural Education*, *62*(3), 310–328. doi:10.5032/jae.2021.03310

Miller, L. E., & Warmbrod, J. R. (1982). Research in teacher education in agriculture. In A. L. Berkey (Ed.), *Teacher education in agriculture* (pp. 247–262). The Interstate Printers & Publishers.

Roberts, T. G., Harder, A., & Brashears, M. T. (Eds.). (2016). *American Association for Agricultural Education national research agenda: 2016-2020*. University of Florida Department of Agricultural Education and Communication.

ADDITIONAL READING

Ary, D., Jacobs, L. C., Sorensen Irvine, C. K., & Walker, D. A. (2019). *Introduction to research in education* (10th ed.). Cengage.

Creswell, J. W., & Creswell, J. D. (2018). *Research design: Qualitative, quantitative, and mixed methods approaches* (5th ed.). SAGE publications.

Dillman, D. A., Smyth, J. D., & Christian, L. M. (2014). *Internet, phone, mail, and mixed-mode surveys: The tailored design method* (4th ed.). Wiley.

Fraenkel, J. R., Wallen, N. E., & Hyun, H. H. (2019). *How to design and evaluate research in education* (10th ed.). McGraw-Hill.

Mertler, C. A., & Charles, C. M. (2011). *Introduction to educational research* (7th ed.). Pearson.

Mills, G. E., & Gay, L. R. (2019). *Educational research: Competencies for analysis and applications* (12th ed.). Pearson.

KEY TERMS AND DEFINITIONS

Action Research: Research performed by teachers and other educational professionals to change conditions in a particular situation they may be involved in.

Historical Research: A systematic attempt to collect and objectively evaluate data related to events, conditions, people, and settings from the past to help explain current events or foretell future events.

Methodology: The design and plan for data collection, data analysis, and interpretation used in a research study.

Mixed Methods Research: A research approach that combines elements from both quantitative and qualitative approaches to gain additional information that could not be provided from one of those approaches alone.

Philosophical Research: A research approach that attempts to critically examine a philosophy or line of thought in an attempt to gain additional insight that can be applied to establish new concepts, theories, or norms.

Plagiarism: The misrepresentation of someone else's work or ideas as your own.

Program Evaluation: A research approach conducted to provide data used to inform stakeholders in their decision making related to a program or policy.

Programmatic Research: Systematic research conducted by faculty and staff members in agriculture teacher education programs with a focus and concentration on an overarching issue, subject, or problem.

Qualitative Research: A research approach that explores phenomena in an attempt to understand the complexities and meaning individuals or groups attribute to the phenomena.

Quantitative Research: A research approach that uses numerical data to test theories, study relationships between or among variables, and numerically describe or quantify characteristics.

Research: A diligent and systematic investigation used to discover or modify facts, relationships, theories, and applications and to study problems in a disciplined manner, generally using the scientific method.

Research Agenda: A framework for the direction and types of studies that must be completed to address an overarching problem or area of interest.

Research Question: Specifies the problem that will be investigated, is written clearly enough that the variables under investigation are understood and gives indication as to the research approach that will be used to investigate the problem.

Scientific Method: A systematic procedure used for the purpose of gaining new knowledge that is recognized by the science community.

Chapter 17
Comparing Agricultural Education Programs Around the World

James J. Connors
University of Idaho, USA

ABSTRACT

Agricultural education programs are offered in many countries around the world. Students are taught hands-on vocational agricultural skills that allow them to enter the agricultural workforce or advance to higher education. Interpersonal communication and leadership skills are offered in some programs but are not a common component of secondary agricultural education curriculum in other countries. Agricultural teachers come to the profession after working in agricultural industry, completing a college degree in technical agriculture or general teacher education, or by completing a formal teacher education program in agriculture. A wide variety of different youth organizations is offered in other countries for students studying agricultural education.

INTRODUCTION

Agricultural education programs have been included in schools around the world for centuries. These programs have ranged from individual lessons related to agriculture in primary grades to full units of courses for secondary school students. In many parts of the world the only agricultural education that future farmers receive is through their secondary school program. Wherever agricultural education courses are taught there is a need for trained teachers knowledgeable in technical agricultural science, pedagogy, and youth development.

The inclusion of school-based agricultural education instruction around the world varies greatly from country-to-country. Some countries have very formal secondary agricultural education programs while other countries offer limited instruction in either the elementary or secondary levels. Teacher education programs in institutions of higher education also vary greatly. Some universities offer full baccalaureate degrees complete with formal methods courses and field-based practice teaching. In other countries

DOI: 10.4018/978-1-6684-3420-8.ch017

any higher education graduate with a degree in an agricultural-related field can obtain a job teaching agriculture in an elementary or secondary school.

In addition to the classroom and laboratory instruction, many countries have intra- or extra-curricular youth organizations that supplement the students' instruction in agriculture. These organizations may be based on the non-formal 4-H youth development model, the formal future farmers program attached directly to their classroom instruction, or an industry-based young farmer organization model prevalent throughout the world. These agricultural education clubs and organizations often offer skills competitions, leadership development activities, professional development and networking opportunities, and social events for their members.

This chapter will provide examples of agricultural education programs, teacher education programs, teacher associations, and youth organizations in:

- Australia
- Brazil
- Egypt
- India
- Ireland
- Japan

These countries were selected because they represent quite different cultures in five regions of the world: South Pacific, South America, Africa, Asia, and Europe. They also represent major agricultural producing regions of the world.

Agricultural industries around the world have become very inter-connected through agricultural production, marketing, and trade. This also makes agricultural education programs, teacher education programs, and agricultural youth development particularly important in these regions of the world.

Future agricultural teachers should be aware of the types of agricultural education programs that exist in other countries. Agricultural education programs are not unique to the United States. There are primary, secondary, and post-secondary agricultural education programs in numerous countries in every corner of the globe. Every school-based agricultural education program in the U.S. should incorporate instruction in global agriculture, food security, and agricultural leadership from other countries. Agricultural teachers could also connect with their counterparts in other countries to share information, resources, or teaching ideas. Youth in agricultural education programs could connect with agricultural students in other countries on social media to share their agricultural knowledge, experiences, or leadership development activities.

The importance of a global awareness of agricultural education will continue to grow as we progress into the 21st century. The ease and availability of worldwide travel will allow future agricultural professionals the chance to interact, work, and even live in other parts of the world. A knowledge of agricultural education programs around the world is essential to all future agricultural teachers across the U.S.

The learning objectives for this chapter are: 1) Describe formal school-based agricultural education systems utilized around the world, 2) Compare and contrast agricultural teacher education programs in various parts of the world, and 3) Explain the role youth and adult agricultural education organizations play in worldwide agricultural education.

AUSTRALIA

Agricultural Education Programs

Secondary agricultural education in Australia dates back to the early 1900s. The first half of the 20th century witnessed numerous agricultural high schools established across New South Wales, Western Australia, and South Australia (Falvey, et al., 2017). Some of these agricultural high schools across Australia are included in Table 1.

Table 1. Agricultural high schools in Australia

State	Agricultural High School	Year Established
New South Wales	Hurlstone	1907
	Yanco	1922
	Farrer	1939
	James Ruse	1959
South Australia	Urrbrae	1913
Victoria	Ballarat Agricultural College (now Ballarat High School)	1907
	Colac Agricultural High School (now Colac Secondary College)	1911
	Leongatha Agricultural High School (now Leongatha Secondary College	1912
	Mansfield Agricultural High School (now Mansfield Secondary College)	Unknown
	Mildura Agricultural High School (now Midura Senior College)	1912
	Sale Agricultural High School (now Sale College)	1907
	Shepparton Agricultural High School (now Shepparton High School)	1909
	Wangaratta Agricultural High School (now Wangaratta High School)	1909
	Warragul (now Warragul Regional College)	1911
	Warrnambool Agricultural High School (now Warrnambool College)	1907
Western Australia	Denmark	1942
	Harvey	1953
	Cunerdin	1959

In addition to the four agricultural high schools in New South Wales, there were over 300 secondary high schools teaching agriculture in 1997-2013 in the state. These schools include both government, independent, and Catholic schools. Some trends that have been observed in agricultural education across New South Wales have been a reduction in the number of schools teaching horticulture and an increase in the schools offering instruction in aquaculture (Pratley, 2013; Spencer, 2000).

Hurlstone Agricultural High School

The first formal school for agricultural education in New South Wales was Hurlstone Agricultural High School which opened in 1907. When the school opened on April 11, 1907, it included just five teachers, a principal, and seven students. However, enrollment grew quickly and by the end of the year there were 30 students studying agriculture at Hurlstone. Figure 1 is the Hurlstone Agricultural High School crest.

Figure 1. Hurlstone Agricultural High School crest

As with other agricultural high schools in Australia, Hurlstone was a residential school that provided a hostel for housing students who came from a wide geographical area to study agriculture. A school farm was operated by the students to provide food for the hostel. Students, under the supervision of faculty, constructed poultry pens, a dairy complex, hay shed, tool shed, and seed store. With the purchase of two pedigreed Ayrshire dairy cows, the school established the Hurlstone Ayrshire Stud.

Hurlstone continues to serve as a magnet high school for urban, rural, regional, remote, and isolated students in the New South Wales region. It offers a challenging STEAM curriculum which includes focus areas of Science, Technology, Engineering, Arts, and Math. The school currently enrolls over 900 students. It is associated with the Hawkesbury Agricultural College in Richmond NSW. Hawkesbury Agricultural College eventually merged with another university to become the new University of Western Sydney in 1989.

Yanco Agricultural High School

Located about 400 miles west of Sydney, Yanco Agricultural High School is in a very remote part of south-central New South Wales (NSW). It was established in 1922 adjacent to an Agricultural Experiment Station operated by the NSW Department of Agriculture. Instruction at the school was originally provided by the agricultural specialists at the Experiment farm. The curriculum offered at Yanco included coursework in agriculture, botany, entomology, farm mechanics, woodwork, and practical agriculture. Figure 2 is the Yanco Agricultural High School crest.

Figure 2. Yanco Agricultural High School Crest

Yanco Agricultural High School remains a highly successful agricultural boarding school for students in rural New South Wales. Agriculture is a required subject for students in years 7 to 10. The school offers instruction in cattle, sheep, swine, poultry, and equine science. Students interested in equine management can board a horse at the school and participate in the school's equestrian team. The school also maintains White Suffolk sheep stud, and Shorthorn and Limousin cattle studs.

Farrer Memorial Agricultural High School

As enrollments in Hurstone and Yanco agricultural schools grew, there was a need for a third agricultural high school in the northern section of New South Wales. Farrer Memorial Agricultural High School opened in 1939 with six teachers and 72 students. Subjects taught included Agriculture I and II, Agricultural Botany, Manual Training, and Practical or Field Agriculture. The school farm grew wheat, oats, barley, hale and silage. In its first year Farrer organized both a Field Day and a Speech Day for the local community. It was written that, "the boys receive ample work on the farm and they enjoy it. They

Figure 3. Farrer Memorial Agricultural High School Crest

gain a sound education, besides the ordinary high school curriculum, learning manual training which is invaluable on a farm, and the rudiments of agriculture." Figure 3 is the crest for Farrer Memorial Agricultural High School.

The school also operated a dairy herd. An original herd of 19 Australian Illawarra Shorthorn cattle was established to provide milk for the school and local community. The school also organized a Berkshire swine herd and a poultry flock consisting of White Leghorns, Rhode Island Reds and Autralorps. After the success of the school's Field Day, an Agricultural Science Society was formed to promote and provide educational tours for students to visit agricultural operations in the community.

Farrer administrators also organized an advisory council called the Livestock and Agriculture Committee to provide strategic planning for the agricultural education program and farm at the school. As the school grew over the decades it became evident that an increasing number of boys entering the school had no interest in pursuing agriculture as an occupation. Regardless, the school continued to provide a high-quality hands-on education for all students even if they did not intend to continue in the agricultural industry after graduation.

Farrar continues to offer in depth instruction in a variety of agricultural subject areas. Students supplement their classroom instruction with work in the different production agriculture facilities at the school. The school maintains a 70-cow dairy cattle herd, a 400 head sheep flock including 50 head of Merinos, a poultry flock, and a 240 head Angus cattle herd.

James Ruse Agricultural High School

The newest agricultural high school in New South Wales is James Ruse which was established in 1959. The school is located in a suburban area on the outskirts of Sydney. It is the only agricultural high school that is for day students only and does not operate a hostel at the school. Figure 4 is the James Ruse Agricultural High School crest.

Figure 4. James Ruse Agricultural High School Crest

Being located in a suburban area, the school grounds are only about 10 hectares in size. However, the school farm does include livestock, sheep, poultry, an orchard, and vegetable garden areas. The school maintains a small herd of Angus beef cattle. Leadership activities available to students at James Ruse include a debating and public speaking club, large animal handlers club, and poultry squad.

Teacher Education

From the very beginning of secondary agricultural education in Australia in the early 1900s, hiring qualified, trained teachers was a challenge. The first agricultural high school, Hurlstone often relied on graduates of the Hawkesbury Agricultural College to fill agriculture teaching positions. However, these graduates, while possibly having an agricultural background or coursework, had no formal teacher training and were offered the teaching position based on their diploma alone. By 1921, the Sydney Chamber of Commerce urged the Department of Agriculture to provide trained agriculture teachers for the country high schools in New South Wales.

In 1936, the first Conference of the Association of Teachers of Agricultural Sciences was held. The Armidale Teachers' College was offering a two-year training course for teachers of agriculture. One of the recommendations from the conference was that "The Agricultural High Schools and Hawkesbury College to be the source of agriculture teachers for county schools."

The problem of preparing agriculture teachers continued for decades. By the 1970s, less than 5% of all the agriculture teachers in New South Wales were university graduates. Most of the university graduates were employed at the three agricultural high schools in the state. The Armidale College of Advanced Education (formerly Armidale Teachers' College) was offering a 3-year program in agricultural education that resulted in a Diploma in Teaching. In 1975 Hawkesbury and Wagga Agricultural Universities merged into a College of Advanced Education (CAE) and began awarding their own graduate teaching certificate in agricultural education. As a result, the teacher education program at Armidale was closed.

The Hawkesbury CAE became part of the University of Western Sydney in 1989. Today, the University of Western Sydney offers a degree in Sustainable Agriculture and Food Security. Students in that major can also enroll in the Pathway to Teaching or the Masters of Teaching to become a secondary agriculture teacher.

In 1994, the Armidale CAE became part of the University of New England. The University of New England (UNE) offers several degrees in agriculture including agricultural business, animal science, and rural science. Students interested in agriculture can select agricultural education as one of their teaching areas. Students must complete coursework in agricultural science, agricultural technology, horticulture, plant or animal sciences, agricultural enterprises, or chemical safety.

In the state of Queensland, the University of Queensland offers a Specialist Teaching Area in Agriculture degree. Students enroll in either the Bachelors of Education (Secondary) or the Graduate Diploma of Education (Secondary) program and complete the two-semester program in agriculture to become a secondary agriculture teacher.

Teachers Associations

Agricultural teachers across Australia actively participate in professional teachers' associations in agriculture. These associations include:

- Agricultural Educators Association of Western Australia (AEAWS)
- New South Wales Association of Agriculture Teachers
- Agricultural Teachers Association of South Australia

These associations offer agriculture teachers professional development activities, annual conferences, and networking opportunities for agricultural teachers in their respective states. The state teachers' associations are affiliated with the National Association of Agricultural Educators (NAAE). The members of NAAE are primarily agriculture and/or vocational education teaches working in state and non-government high schools and colleges across Australia. This is a separate and distinct organization not affiliated with the National Association of Agricultural Educators (NAAE) in the United States.

Youth Organizations

There is no single agricultural youth organization for students studying agriculture in Australia. Some individual schools created their own student organization to offer experiential education opportunities outside of the classroom. In 1944, students at Farrer Memorial Agriculture High School formed the Agricultural Science Society to promote educational field trips in the community.

The largest provider of youth activities in Australia is the Royal Agricultural Society of New South Wales (RAS). To promote leadership development through livestock judging activities, the society offers the Young Judges program. The RAS Young Judges program offers competitions for students 15 to 25 years old in judging alpacas, goats, horses, pigs, poultry, and horticulture displays. Each program is subtitled Horse Young Judges, Pig Young Judges, and Horticulture Young Judges. The RAS also offers school competitions for agricultural programs in New South Wales. Students from agricultural schools exhibit poultry, swine, goat, sheep, cattle showmanship, honey, and scarecrows during the annual Sydney Royal Easter Show.

The Royal Agricultural Society (RAS) also organizes a Youth Group. The Youth Group is comprised of young agricultural professionals who aim to educate, entertain, and engage youth through RAS programs, events, and competitions. These activities include Agrichats, the Young Farmer Challenge, and All Roads to the Royal at the Sydney Easter Show. For young adults, the RAS offers the Rural Achiever Award. The Rural Achiever Award is a state-wide leadership program for young agricultural leaders aged 20-29 who wish to contribute to their local communities and rural Australia. Networking activities for Rural Achiever members and agricultural professionals are offered during the Sydney Royal Easter Show each spring.

BRAZIL

Agricultural Education Programs

As the largest country in South America, Brazil has a truly diverse population, geography, agricultural and educational system. Being a largely rural country, Brazil has had rural vocational agriculture training for over a century. Agricultural education started in Brazil in the late 1800s. It has been supported by federal legislation.

The Organic Law for Agricultural Education was adopted by the Brazilian Ministry of Education in 1945. The goal of this law was to reorganize rural and agricultural education that would focus on students learning by doing. The law established five levels of agricultural schools for educating students (Mendonca, 2006). The five new types of schools included:

- Agricultural Initiation Schools – introduced elementary students to agricultural education subjects.
- Agricultural Schools – Continuation of agricultural education in elementary schools with more advanced agricultural topics.
- Agrotechnical Schools – Offered technical and pedagogical courses in agricultural topics to students in secondary education (grades 9-11).
- Technological Schools – Advanced technical education and extension courses in agricultural and veterinary areas.
- Agricultural Training Centers (CTs) – Agricultural training centers to teach rural works in advanced methods of efficient performance in agricultural activities.

Education in Brazil is divided into two tracks, either General Education or Professional Education. The General Education track prepares students to continue to higher education in colleges and universities. The Professional Education track prepares students in technical subjects like agriculture (Leite, et al., 2004). Students who complete the basic level of education (grades 1-8) in agricultural initiation schools can advance on to the technical level in grades 9-11 in agrotechnical schools.

There are approximately 250 agricultural schools at the technical level across Brazil. These mainly rural schools, enroll over 50,000 students (Leite & Radhakrishna, 2004). The schools are supported and administered by the federal, state, and local governments or private organizations. Federally supported schools make up about 50% of the agricultural schools in Brazil. These technical schools offer instruction in a variety of agricultural subjects including agriculture, horticulture, zootechnics, livestock raising, veterinary practice, industrial agriculture, agricultural mechanics, and dairying (Faust, 1959).

Brazil, a predominantly Catholic country, has numerous Catholic schools that teach agriculture. These are called First Family Farming schools (Escolas Famílias Agrícolas – EFA's) and Rural Family Houses (Casa Familiares Rurais – CFR's) (Goris et al., 2021). The aim of these schools is to prepare students for improved rural livelihoods in order to stem the exodus of youth from the rural areas to the urban centers of Brazil. Students in the First Family Farming Schools spend one hour each morning and afternoon completing farming tasks including caring for the crops, livestock, and in some schools, beehives. The remainder of the day is spent in the classroom being taught the theory and practice of agricultural subjects (Osava, 2009).

The federal law that governs education in Brazil states that curriculum and teaching methods in secondary schools should be of interest to rural youth, and aligned with the agricultural calendar, seasonal conditions, and work schedule in the Brazilian countryside. Brazilian education officials have suggested that agricultural extension, secondary agricultural education, and higher education in agriculture should give priority to six major areas which include: community development, economic plurality, leadership and life skills development, social welfare, environment and sustainability, and family agriculture.

Agricultural education in Brazil continues to evolve with the introduction of distance education methods. In 2004, the National Confederation of Agricultural Works (Confederação Nacional dos Trabalhadores na Agriculturua – CONTAG), the largest labor union of agricultural workers in Brazil, established the Jovem saber (youth knowledge) program. The goal of this agricultural education program is to enhance the agricultural skills of young farmers between the ages of 16 and 32. Local study groups of 5-10 people, of which 30% must be female, meet to complete the free on-line training courses. Eight training modules covering different agricultural topics including: family farming, rural health, labor laws for agricultural workers, agricultural credit, agricultural associations, and cooperatives. The study groups review each module on the internet and must complete assignments before moving on to the next module. Additional face-to-face training and educational support activities are also offered by the CONTAG organization (FAO, 2014).

Teacher Education

There are many higher education institutions across Brazil that offer degrees in various agricultural subjects. In the 1970s the Federal University of Rio Grande do Sul in Brazil operated a Vocational Teaching Training Center to prepare teachers for the different agricultural schools in the country. However, that center no longer exists. There is no comprehensive system of teacher education in place in Brazil. This is especially true for vocational educators including agriculture teachers. Teacher training programs are sometimes located in the secondary education level and sometimes in the higher education level. Mandatory university level training for future teachers is a recent requirement in Brazilian education system. Approximately 50% of Brazilian teachers have only a secondary level training. The National Education Guidelines and Framework Law of 1996 requires teaches in all educational levels to have a higher education degree (de Sa Barretto, 2015).

Secondary technical schools do offer vocational programs for students interested in a career as a teacher. The students in grades 9-11 in the technical schools take teacher training courses which include a supervised student-teacher internship consisting of 300 hours of practice teaching in a local school. Students certified to teach through these programs however can only teach grades 1-8 in the basic level (elementary) of Brazilian schools. Upon completion of the secondary teacher training program, they would receive a vocational teaching certificate.

Students who wish to teach at the secondary level (grades 9-11) are required to complete a degree from college or university. Universities do offer a degree in Licenciatura for students wishing to become secondary school teachers. Students interested in teaching agricultural subjects in agricultural high schools will often complete an agricultural degree along with their licenciatura program at a Brazilian agricultural university. There has been a concerted effort recently by the Brazilian government to implement a new education plan to improve the teacher education program for all teachers, general education as well as professional-vocational education, in all levels of education in Brazil.

Youth Organizations

There are limited youth organizations for students interested in Agriculture in Brazil. In 1940, students studying agriculture in Catholic schools developed the Agrarian Catholic Youth (JAC) organization. This was mainly a political organization focused on improving the lives of rural people across Brazil. Also in the 1940s, the Ministry of Agriculture promoted the formation of agricultural clubs for rural youth to promote the training of rural workers after World War II. In 1945, the Ministry of Agriculture reported that there were 886 agricultural clubs in Brazil. Many of these agricultural clubs were based on the American model of 4-H clubs. Members of these agricultural clubs were called Little Ruralists. The goals of these Brazilian agriculture clubs included:

- Instill in students the spirit of love of the land and the gifts it produces.
- Introduce students to productive agricultural activities.
- Orient students to a life of agricultural work as productive citizens in rural Brazil.
- To dignify manual labor and ennoble the farming profession.
- Develop the spirit of cooperation in school, family, and collectivity.
- Encourage polyculture.
- Teach economic habits.
- Organize a cooperative for sales of agricultural products from crops.
- Create partnerships to improve agricultural production.

After the establishment of the Peace Corps in the United States, the Brazil 4-H Peace Corps program began in 1961. Peace Corps volunteers in Brazil, El Salvador, Uruguay, and Venezuela worked to implement new 4-H clubs and improve existing agricultural/4-H clubs in rural communities in Brazil.

In 1983, the Pastoral da Juventude Rural (PJR) was organized. Local affiliates of the PJR formed rural areas to improve the lives of rural peasants involved in agriculture. The PJR works with the Sindicato dos Trabalhadores na Agricultura Familiar (SINTRAF) peasant union to conduct educational workshops and events that introduce students to the topic of agroecology for rural Brazil (Goris et al., 2021).

The PJR conducts both local events and a national meeting. Their annual themes have included the formation of cultural awareness to strengthen the rural peasant identity and help members to be proud of their rural agrarian roots. They have also introduced activities that expose Brazilian youth to agricultural activities that involve the head, heart, and hands that link them to their rural livelihoods.

EGYPT

Agricultural Education Programs

Agriculture has been the foundation of Egyptian society for thousands of years since the time of the Pharaohs. In the 21st century, Egypt is still a very agrarian society with over half the population living in rural areas. These rural communities rely on agricultural education to prepare youth for work in the agricultural industry in Egypt.

After students complete their basic education, they complete a national exam that will determine which track they will follow in secondary school level. Secondary education in Egypt includes two tracks: General education or technical education. The technical schools emphasize industrial, agricultural, or commercial themes for students interested in these vocational areas.

The technical track can last either three years or five years depending on the curriculum selected. The three-year program includes both general education subjects, vocational subjects, and electives. After completing the program and passing a final examination, students are awarded the Technical Secondary Education Certificate. The five-year program provides a more in-depth instruction in a vocational specialization leading to the Advanced Technical Diploma. Students in either the general education or technical education tracks can matriculate to tertiary education in institutions of higher education.

Egypt operates 117 secondary-level agricultural technical schools (ATS)(Barrick et al., 2011a). These are usually larger schools that have over 2,500 students and 150 teachers. The technical agricultural subjects taught in the ATSs include field-crop production, livestock production, animal health, horticulture, agricultural economics, and agricultural mechanics.

In 2008, a USAID program titled *Reforming the Agricultural Technical School (AST) System in Egypt* introduced key innovations and reform activities into the agricultural schools (Barrick et al., 2011b). The activities included:

- Improving the school learning environment and engaging students in the learning process.
- Developing students' technical skills through supervised internship programs (SIP) in partnership with the private sector.
- Enhancing students' career and personal skills through new career development events.
- Connecting schools with their communities by forming external advisory committees involving local agribusiness leaders.

Teacher Education

The preparation of quality teachers for Egyptian schools has long been a problem. This is due to the low salary for teachers and the profession being seen as a "fall back" position for those students who performed poorly in technical fields. Over the last several decades, educational professionals and governmental officials have worked to improve the teacher preparation program in Egypt.

Teacher education programs are now offered at 15 colleges of education in universities in Egypt. A bachelor's degree is now required to teach at all levels of education in Egyptian schools. Teacher education programs are divided into either integrated preparation or continuing preparation programs.

The integrated teacher preparation program includes two years of instruction in the principles of education and psychology, principles of teaching, social and historical foundations of education. At the end of their second year of study, students must pass an examination to advance to years three and four. Coursework in years three and four include methodology, educational psychology and technology, educational philosophy, and curriculum development. In addition to their educational courses, students also enroll in general studies in the area of curricular specialization in which they wish to teach, such as agriculture. Students also complete a student-teaching field experience in a school system in Egypt.

Students who complete an undergraduate degree from a non-education faculty (college) and wish to become a teacher can select the continuing preparation program. After completing their degree in an agricultural subject area, students would then enroll in education courses full-time for one year. At the end of their continuing preparation program, they are awarded a General Diploma in Education (Education Encyclopedia – StateUniversity.com, 2022).

One challenge facing agriculture teachers in Egypt has been the type of instruction they provide. In the past, few teachers were trained in pedagogy and different methods of instruction. The teachers would teach the way they were taught in school, primarily through lecture. They were not prepared to utilize the problem-solving approach to instruction, or incorporate hands-on, critical thinking, or decision- making approaches in their classrooms.

Projects to improve the methods of instruction in agricultural technical schools have been conducted by the Midwest Universities Consortium for International Activities (MUCIA). The project has trained Egyptian agriculture teachers to utilize lesson planning, teaching aides, active learning strategies, and learning by doing activities on school farms located at the agricultural technical schools. The MUCIA project used a six-step process to improve the teaching methods and introduce active learning strategies in Egypt's Agricultural Technical Schools (Maguire, 2021). The steps included:

1. Training the Agricultural Technical School teachers in active teaching-learning methods.
2. Develop instructional materials for ATS teachers.
3. Develop lesson plans for each ATS course.
4. Conduct a study tour for ATS Headmasters to visit vocational schools in the Netherlands.
5. Refocusing ATS school farms for practical skill training.
6. Training ATS teachers in the use of lesson plans and instructional materials.

Youth Organizations

One aspect of agricultural education in Egypt that is severely lacking is youth organizations. During the 1940s, an effort was implemented to improve education through charitable and service societies. Many of these benevolent associations also offered instruction in vocational subjects. These organizations conducted community development activities including efforts to improve the social behavior, morals, and free time of participants.

The Egyptian Ministry of Social Affairs (MOSA) opened rural centers and established clubs for youth and adults across the country. Rural poor could obtain assistance from the MOSA or from the private charitable organizations operating throughout Egypt. The MOSA rural centers also offered education, training, and cultural activities with the goal of reforming rural life in Egypt (Pollard, 2014).

Often the MOSA centers were referred to as Rural Welfare Centers or Rural Social Centers. They offered educational programs throughout rural Egypt to improve skills for rural youth in agriculture, handicraft areas, carpentry, and weaving. Eventually these Rural Welfare Centers transitioned into Com-

Figure 5. Future Farmers of Egypt Emblem

munity Development Associations (CDAs). These CDAs continued the practice of offering non-formal education, skill development and social activities. Unfortunately, there were no agricultural oriented organizations available to rural youth attending the Agricultural Technical Schools in Egypt (LaTowsky, 1997).

A key innovation of the USAID project in 2008 was the establishment of the Future Farmers of Egypt (FFE) youth organization. The goal of the organization was to teach Egyptian agriculture students agricultural marketing, problem-solving skills, and leadership skills including interpersonal communications methods. The emblem of the Future Farmers of Egypt is shown in Figure 5.

INDIA

Agricultural Education Programs

India is one of the most populated countries in the world with almost 1.4 billion people. The educational system in India serves approximately 250 million students in both government and private schools. India is also a very rural country that relies heavily on agriculture to support millions of people, rural communities, and the country's economy.

Because of the importance of agriculture to India, agricultural education is also an important component of the Indian educational system. After India obtained its independence from Great Britain in 1947, the new government placed a strong emphasis on establishing agricultural schools in rural areas of the country.

In 1947, the government in Bombay established the first Manjri School, or agricultural school, in Manjri, Poona District, Bombay State. The goal of the Manjri agricultural school was to educate the sons of Indian farmers in scientific agriculture. The curriculum in the agricultural school included:

- General agriculture
- Animal husbandry
- Dairying
- Veterinary science
- Horticulture
- Marketing
- Agrarian legislation
- Village industries
- Methods of extension
- Forestry

The students at the school had to perform all operations on the school farm including working at the dairy and poultry operations. They also had to maintain accurate records of the agricultural enterprises and visit local villages to demonstrate improved agricultural methods to local farmers. Enrollment in the school was targeted to farmer's sons between the ages of 16 and 30 years old.

It was the goal of the Indian government to establish one agricultural school in each district in the state of Mumbai (Bombay). By the early 1960s, 28 agricultural schools were in operation. The agricultural schools often included farmland for producing crops, a garden, dairy, science laboratory, and school library. To encourage hand-on learning, each student was given an individual plot of land to cultivate and raise crops or produce.

The Indian government also created a very specialized agricultural school for blind students. Because a majority of blind people resided in rural areas, they often came from an agricultural background. The Royal Commonwealth Society for the Blind in England worked with the National Association for the Blind in India to establish the Tata Agricultural and Rural Training Centre for the Blind. Admission was open to blind students who possessed land and/or livestock or had a possibility of employment in the agricultural sector.

When the government of India organized the Basic Education scheme, agriculture was included as part of the curriculum. Mahatma Gandhi envisioned a school system built on the philosophy of developing the capacities of the child through a hands-on craft. Because most of the children came from rural communities, the curriculum in the rural schools was based around agricultural and the rural environment. The aims of the Senior Basic Schools established in 1954 were to:

- Develop a well-integrated personality through agriculture/craft.
- Make education productive.
- Portray the school to be the centre of community life and spearhead progress.
- Provide training in citizenship for a democratic society.
- Provide leadership training and self-government through youth organizations.

As the 1950s progressed, a new type of school was developed by the Indian government. Multipurpose or Diversified Secondary Schools were opened to provide secondary school agricultural, commercial, and other practical courses to help students prepare for vocational pursuits after graduation. The India Secondary Education Commission stated that "Considering that over 75 percent of the population is employed in agriculture…we recommend that all States should provide much greater opportunities for agricultural education in rural schools so that more students may take it and adopt it as a vocation."

These new higher secondary schools included seven different curriculum "streams" including agriculture, commerce, fine arts, home science, humanities, science, and technology. The agricultural stream offered courses for students in their 9th, 10th, and 11th year of education. The objectives for the agricultural stream included:

- To provide the boy a terminal education and with the occupational competency that will fit him to return to the village farming operations.
- To provide him with a practical and diverse training which will serve as a foundation to advanced training in an agricultural college.
- To give him training which will broaden his knowledge and appreciation of community life so that he might assume the responsibilities of village citizenship.

The philosophy of the secondary agricultural schools was to provide education at the "doing level" as it was commonly called in India at the time. Thus, the instruction must be vocational in nature to develop the occupational competencies of the students. Agricultural educators believed that students must learn the theory and "learn to do by doing" at the same time. The curriculum for the three-year training period included (Naik, 1961):

- Agricultural Biology and Chemistry
    - Morphology of vegetative plant organs
    - Plantation campaigns
    - Crop pests and diseases
    - Locust and rat campaigns
    - Fertilizers – nitrogenous and others
    - Soil salts
- General Agriculture including Soil Management and Crop Culture
    - Soils
    - Rainfall and climatology
    - Soil management
    - Irrigation
    - Field and garden crops – culture and classification
    - Fruit crops
    - Preparation of fruits and vegetables
- Farm Management including Animal Husbandry and Dairying
    - General animal science (parts of all animals including poultry)
    - Feeding
    - Breeds
    - Diseases
    - Dairy milk
    - Management records

Agricultural education is still taught at numerous schools throughout India. The following are some examples of vocational agricultural education in India schools.

Rajghat Besant School

The Rajghat Besant School is located on the banks of the Ganges River in Varanasi, India. Part of the school is the Rajghat Rural Centre which consists of two rural primary schools, a vocational training center for women, and a 400-acre school farm. Students at the school work on the farm and raise wheat, corn (maize), pulses, potatoes, tomatoes, vegetable crops and fruit trees including mango. The school has phased out the use of chemicals and adopted sustainable organic farming methods. Along with the crop production, the school also includes an 80 head dairy operation. Milk from the dairy herd is used in the school cafeteria to feed all students. Students also participate in regular field trips to local farmers to better understand modern agricultural practices (Rajghat Besant School, 2017).

Vivekanand Uchchatar Uttar Buniyadi Vidyalaya High School

This rural residential high school enrolls approximately 400 students including some from area foster homes in the district. Most students come from farming families so agricultural education makes up an important part of the school curriculum. Agricultural courses include scientific methods of farming, animal husbandry, dairy farming, computer science, and agro-processing. Students from the high school often advance to additional agricultural training offered at the Muni Seva Ashram Vocational Training Center located nearby in Gujarat, India (Vivekandand Uchchatar Uttar Buniyadi Vidyalaya, 2022).

Muni Seva Ashram Vocational Training Centre

Tribal children who pass high school can enroll in four-to-six-month training courses offered at the Muni Seva Ashram Vocational Training Centre. The center takes pride in the fact that 75% of the instruction is practical training based on the educational philosophy of learning by doing. Agricultural courses offered at the centre include:

- Tractor Operator
- Organic Grower
- Micro-Irrigation Technician
- Paddy Farmer
- Floriculturalist – Open Cultivation
- Quality Seed Grower
- Seed Processing Worker
- Dairy Farmer/Entrepreneur
- Vermicompost Producer

Vigyan Ashram School and Agricultural Training Centre

The Vigyan Ashram School offers the Introduction to Basic Technology (IBT) program which is a multi-skill foundation course (MSFC). It is also considered a Work Centric Educational Program in Secondary Schools in India. This means that the curriculum ascribes to a hands-on, learning while doing, project -based learning, philosophy of education in the school. Students are involved in socially useful productive work (SUPW) and conduct community service projects in the local community as part of their education (Learning While Doing, n.d.).

The agricultural curriculum at Vigyan Ashram focuses on Agriculture and Animal Husbandry. The objectives of the program include:

- To train the students the philosophy of "Learning While Doing."
- To train students for income-generation through self-employment.
- To train students in multi-skills.
- To train students in different technologies and transfer these technologies to the society through them.
- To involve students in various rural development activities as a project work, thus integrating rural development and education.
- To make available various services to the community at a modest cost and giving real-life training to students.

The specific topics covered in the Agriculture and Animal Husbandry course are quite diverse. The topics include:

- Modern agriculture techniques
- Dairy Goat production
- Poultry farming
- Irrigation
- Vermi-Composting
- Nursery production
- Seed production
- Rice Cultivation
- Feeds and Silage Production
- Soils
- Pest Control

Lend-A-Hand-India

The Vigyan Ashram school works closely with an organization called Lend-A-Hand-India (LAHI). LAHI is a non-profit organization created in 2003 by young professionals in New York City, USA with the goal of improving vocational education for youth across India. Both the Vigyan Ashram school and LAHI aspire to the educational philosophy of learning while doing. The Lend-A-Hand-India logo is shown in Figure 6 (Lend A Hand India, n.d.).

Figure 6. Lend A Hand India Logo

LAHI focuses on youth development through various programs that provide vocational training, career development, employment, and entrepreneurship opportunities to rural and urban youth across India. The aim of LAHI is to integrate vocational education with mainstream education at secondary and higher secondary education levels (Grades 9-12, Age 14-18). The following are some of the programs offered by Lend-A-Hand India:

Project Swadheen, which means "self-motivation," provides multi-skill vocational education and career guidance to young boys and girls in secondary and higher secondary schools in urban and rural communities. Students are provided with foundation skills in a wide variety of areas including gardening, landscaping, food processing, welding, and other vocational areas.

Karigar – School of Applied Learning – is a "skills hub" school often described as the "Disney Land of Skills." The objective of Karigar is to make working with your hands popular among Indian students. The Karigar school facilities are also used for teacher training programs for vocational teachers in India. The school uses a "train the trainer" model to provide hands-on training in vocational skills for teachers and other professionals who then share their knowledge and skills with students in rural schools in India.

Project Saad – Beyond Skill Education – provides education in soft skills for students in vocational education programs. Teachers are trained how to incorporate these leadership or soft skills into their vocational instruction. Components of the soft skill instruction includes self-awareness, self-acceptance, confidence, emotional intelligence, interpersonal relations, resilience, increased wellbeing, and happiness.

Skills on Wheels (SOW) is a fully equipped vocational training laboratory on wheels that travels to schools to promote hands-on learning of vocational skills. The vocational trainers travel to small rural schools that have limited laboratory facilities. The trainers deliver vocational skill training for both teachers and students during their visits to the rural schools. Training is provided in welding, plumbing, electrical wiring, food processing, basic healthcare, and mechanical repairs.

Lend-A-Hand-India Junior Fellowship (LAHI-JRF) is a program that offers six-month fellowships and three-month internships to students for on-the-job training. Students can complete the fellowship and then follow it up with an additional internship. After a student completes both programs, they can continue with a one-year long mentorship program with a professional in the industry. The program also assists students with advancing to higher education and additional training in communications, personality development, self-study, and self-reflection.

While vocational education is a part of numerous schools in India, there has been a recent call to introduce agricultural education in all schools in India. An editorial in the Hindu Business Line, Mokkapati (2018) stated,

Generally, students select a career aligned with what they studied. In the case of agriculture, the subject is not formally learned at school. Unless one grows up on or near a farm, one does not gain exposure to farming. Thus, it is important for kids to have the chance to get their hands muddy in primary schools.

The author concluded that, "Bringing agriculture to mainstream education has the potential to feed the world." Indian Prime Minister Narendra Modi has stated that to impart knowledge about agriculture and farming practices among young minds, agriculture should be included in the curriculum at the middle-school level. Prime Minister Modi stated that, "This will develop agriculture-related understanding among students and enable them to relay the information about agriculture, modern farming techniques and marketing to their family members" (PM Narendra Modi pitches for taking farm education to middle schools level, 2020, August 29)

As in other countries, the target audience for secondary agricultural education in India was sons of farmers. The theory was that agricultural education could educate these young men on improved agricultural practices and encourage them to remain on the farm. The idea of offering formal education to young girls, especially in agriculture, was never considered. However, the future of agricultural education in India may be changing.

A new program introduced in Indian schools encourages all students to participate in planting school gardens adjacent to their local schools. These School Nutrition Gardens are being promoted by the Department of Agriculture and Horticulture, Food and Nutrition Boards, agricultural universities, and Krishi Vigyan Kendras, an organization that promotes improved agricultural practices and technology in India. The goal for the program is to teach all students in India the importance of growing their own nutritious food. The food grown in the school gardens is incorporated into school meals for the students. The aims of the program are to:

- Address malnutrition and micronutrient deficiencies by improving the consumption of freshly grown vegetables.
- Give children first-hand experience with nature and gardening.
- Increase knowledge among students about the nutritional aspects of vegetables and the harmful effects of junk food.

Government guidelines suggest that every class of students should spend one to two hours per week working in the school garden. The program is not just targeted to rural schools. Schools in urban areas are encouraged to grow fruits and vegetables in pots on terraces and rooftops or grow plants that will climb walls and need a minimal amount of space.

The importance of educating all students about agriculture, not just males, is also being promoted. The Good Harvest School, in the Indian state of Uttar Pradesh, was established in 2016. It is the first agricultural-based primary school for girls. Many families stress education for boys so that they can leave the rural agricultural areas and find gainful employment in urban areas. Unfortunately, the education of girls has not been promoted because girls tend to stay at home, help their parents, and work in agriculture.

The school uses 50,000 sq. ft. of land to teach the girls about good farming practices through hands-on experience with planting crops, maintaining a school nursery, seed bank, and raising cattle. The school also organizes social gatherings between the students and local farmers and hosts regular farmer's meetings in the surrounding community. The Good Harvest School is one example of the expanding emphasis on educating all students, boys and girls, rural and urban, about the importance of agriculture to the country of India.

Teacher Education

The issue of finding qualified agriculture teachers is a problem in India as it is in other countries around the world. After India gained its independence over 3,000 Senior Basic Schools were established in rural areas across India. Thousands of teachers were recruited to teach agriculture in the Basic Schools. The teachers were provided with a course on Practical Agriculture and Elementary Pedagogics to prepare them to teach agriculture in the Senior Basic Schools. As is the case in other countries individuals re-cruited to teach agriculture had the following backgrounds:

- Individuals with technical agricultural experiences but no formal teacher preparation.
- Individuals with an undergraduate degree but no practical agricultural experience.
- Individuals with a bachelor's degree in agriculture or related field but no practical agriculture experience.
- Individuals who hold a Bachelor's of Teaching (B.T.) or a Licentiate of Teaching (L.T.) degree

Senior Basic Schools and Multipurpose Secondary Schools had difficulty in finding teachers who possessed technical agriculture skills or a degree in agriculture and had completed a teacher preparation program. Many universities across India do offer agricultural degrees and teacher preparation programs, but there is no teacher education program specifically for individuals who wish to teach agricultural science in primary or secondary schools.

Youth Organizations

Because of the size and diversity of education in India there are no organizations for youth studying agriculture. Two relatively new organizations do offer programs for rural youth in India.

Sarvodaya Youth Organization (SYO)

The Sarvodaya Youth Organization (SYO) was established in 1993 to promote and enhance sustainable rural, tribal, and urban development and to enhance the socio-economic status of vulnerable and disadvantaged youth in India. SYO's vision is to work for sustainable human and natural resource development and management. The mission of the organization is to alleviate poverty and protect the environment through promoting participatory principles. Specific objectives of the SYO include:

- To alleviate poverty and protect the environment.
- To improve and bring changes in the community, social economic and cultural aspect.
- To help and enhance the rural poor men and women and differently abled persons in their development process.
- To reduce/stop the various incurable diseases and social misbehaviors
- To promote livelihood and food security.
- To act as a resource and knowledge base centre.

The emblem for the Sarvodaya Youth Organisation is shown in Figure 7.

Figure 7. Sarvodaya Youth Organisation Emblem

The focus areas of the SYO include:

- Agriculture
    - Promotion of organic farming
    - Non-pesticide management
    - Sustainable agriculture
    - System of Rice Intensification (SRI) paddy culture
    - Formation of sustainable community institutions
    - Community-based tank management
- Health
    - Integrated nutrition and health programs

    - Reproductive child health
    - Urban health center
    - Prevention of HIV/AIDS
    - Health care for truck drivers
- Promotion of Solar Energy
    - Distribution of solar lanterns
    - Installation of solar streetlights and solar LED lights
- Women Empowerment
    - Reduce violence against women
    - Economic empowerment of women
    - Vocational training for women
    - Skill development training for women
    - Sustainable livelihood for widows of farmers
- Promotion of Livelihood
    - Natural resource management
    - Plantation for sustainable livelihood
    - Skill development trainings for adolescent youth
    - Income generation programs for youth and landless labours
- Education
    - Promotion of girl child education
    - Literacy centers for women
    - Rehabilitation of child labour

Some of the achievements of the Sarvodaya Youth Organization included converting 10,000 conventional farmers into organic farmers, the formation of 70 farmers clubs and 20 youth clubs, and offering interest free loans to women farmers to support sheep, goats, and poultry operations.

Global Youth Organization

The mission of the Global Youth Organization is to empower individuals, families, and social communities for a better life with healthy relationships and to serve less privileged fellow beings with love and compassion through implementation of holistic education and social empowerment with sustainable programs. The Global Youth Organization works in the following areas:

- Health Hub
- Women Empowerment
- Civil and Political Action Committee
- United Nations Sustainable Development Goals
- Human Rights Education
- Education Hub

The Food and Agricultural Organization within the United Nations reports that the Global Youth Organization in India also works in the areas of agricultural and rural legislation, education, and food and nutrition.

IRELAND

Agricultural Education Programs

Formal agricultural education began in Ireland in the early decades of the 19th century. The Bannow Agricultural School was opened in County Wexford in 1821 and the Templemoyle Agricultural School was formed in 1826 in County Londonderry, Northern Ireland. In 1846, during the height of the Irish potato famine, there were calls for establishment of a system of agricultural schools in each parish in Ireland. After the famine subsided, agricultural instruction was included in the workhouses that were established as part of the Poor Law Act. Agricultural programs in the workhouses grew from seven in 1850 to a high of 80 in 1856. There was a total of 173 agricultural education programs across Ireland by 1856. In 1866, the Limerick Model Agricultural School included a school farm of 12 acres of pasture for the 15 dairy cows, and 33 acres for crop production including wheat, barley, flax, potatoes, carrots, turnips, rape, hay, and vetches. By 1870, agricultural lessons were added as academic subjects in all national schools across Ireland, not just in agricultural schools (Mangione, 2017).

The early decades of the 20th century brought a dispute about the control of agricultural education in Irish schools. Local vocational education committees (VECs) controlled agricultural education, or rural science classes, in vocational schools. However, the Department of Education did not list agricultural education as a function of their department. At the same time, the Ministers and Secretaries Act of 1928 stated that "education in agriculture and in subjects related thereto" as a function of the Department of Agriculture. Decades later, it was finally agreed that while the Department of Agriculture has a special position and responsibility in regard to agricultural education, the Department of Education and the vocational education committees (VECs) have a responsibility to ensure that instruction bearing on agriculture is provided in secondary and vocational schools in Ireland (Daly, 2002).

By the end of the 20th century agricultural education was being offered as an academic subject at the senior level of secondary schools across the country. In 2017, about 7,500 students enrolled in agricultural science courses in Ireland. These students completed the agricultural science examination for

Table 2. Technical agricultural colleges associated with Teagasc.

College	Location	Affiliation	Focus
Ballyhaise Agricultural College	Ballyhaise Co. Cavan	Teagasc	Forestry Plant Science
College of Amenity Horticulture	National Botanic Garden - Dublin	Teagasc	Horticulture
Clonakilty Agriculture College	Clonakilty Co. Cork	Teagasc	Dairy
Kildalton Agricultural and Horticultural College	Piltown Co. Kilkenny	Teagasc	General Ag. Horticulture
Gurteen College	Ballingary Co. Offaly	Private - Church of Ireland	Agronomy/Crops Equine Science
Mountbellow Agriculture College	Mount Bellew Co. Galway	Private - Franciscan Brothers - Catholic	Dry Stock (Beef, Sheep, Swine)
Pallaskenry Agricultural College	Pallaskenry Co. Limerick	Private - Salesian Fathers of Don Bosco - Catholic	Agricultural Mechanics

their Leaving Certificate. This was approximately 14% of the total student population completing their Leaving Certificate in 2017 (Irish agricultural science teachers' association, n.d.).

Agricultural education is now offered in both national and Catholic schools across Ireland and is mainly taught as a science topic and not a hands-on vocational subject. Students who complete the Leaving Certificate at the end of their secondary education have numerous options to continue their post-secondary education in agriculture. Teagasc, the Irish Agricultural and Food Development Authority (pronounced: cha-gus) conducts agricultural research, provides advisory (extension) services, and operates technical agricultural colleges across Ireland. Teagasc operates four agricultural colleges and is affiliated with an additional three privately operated colleges. These technical agricultural colleges provide classroom instruction and hands-on experiential learning activities for students interested in a career in agriculture. Table 2 lists the seven technical agricultural colleges associated with Teagasc.

Teacher Education

Higher education in agriculture is available at several institutions in Ireland. Various agricultural majors are offered at both University College Dublin (UCD) and University College Cork (UCC). However, neither institution offers a specific degree in teacher education in agriculture.

University College Dublin (UCD) does offer graduate degrees in education including Master of Arts in Education, Master of Education, Professional Master of Education, and a Master of Education, Children, and Youth. A student who is interested in teaching agriculture can complete an undergraduate degree in agriculture from the UCD School of Agriculture and Food Science then advance to complete a graduate degree in education.

A similar pathway to teaching agriculture is available at University College Cork (UCC). Students can complete an undergraduate degree in an agricultural subject and a graduate degree in education. UCC offers a Professional Master of Education (Secondary School/Post-Primary Teacher Training). However, agricultural education is not considered a separate subject and is therefore included under biological sciences.

University of Limerick (UL) does offer a specific degree related to agricultural education. UL offers a Bachelor of Science with concurrent Teacher Education (Biology, Chemistry, Physics, or Agricultural Sciences). The University of Limerick shares that "The aim of the programme is to educate young teachers and help them develop the skills and aptitudes to confidently face the challenges of science teaching." Students who complete the program are qualified to teach the following subjects:

- General Science to Junior Certificate Level (Junior High School level)
- Biology, Chemistry, Physics, or Agricultural Science to Honours Leaving Certificate level

The curriculum for the program combines courses in biology and agricultural science so students are prepared to teach both subjects in secondary schools in Ireland. While the University of Limerick does not include a college of agriculture, students interested in teaching agricultural sciences do complete a wide variety of coursework in agricultural subjects including:

- Plant Physiology
- Animal Feeding & Digestion
- Crop and Grassland Science

- Ecology
- Animal Production Systems
- Horticulture
- Agricultural Biotechnology

Students also complete pedagogical courses in educational technology for teaching and learning and two courses in the methods of teaching science. Students in the teacher education program complete two teaching practice placements (student-teaching) in a secondary school in Ireland. The first placement takes place during Year 2 and lasts for six weeks. The second placement occurs in Year 4 (final year) and lasts for 12 weeks. The student-teachers prepare lessons and teach junior and senior level science topics while being supervised by their cooperating teacher and a University of Limerick faculty member.

Many secondary agriculture teachers in Ireland have technical, hands-on, personal or professional experiences in agriculture. However, many entered the teaching profession as a general science or biology teacher but gravitated to an agricultural science teaching position.

Youth Organizations

Currently, there are no organizations specifically for youth studying agricultural education in secondary schools in Ireland. Because agriculture is taught as a science subject in Irish schools, students often conduct agricultural science research projects on different agricultural topics taught in the classroom.

Young Scientists and Technology

Each year secondary students across Ireland are invited to participate in the BT Young Scientists & Technology Exhibit competition. The event is the equivalent to a science fair open to all secondary students in Ireland. Approximately 2,000 projects are submitted by students. About 500 projects are then selected to participate in the face-to-face presentation held at the Royal Dublin Society (RDS) in Dublin (BT Young Scientist & Technology, n.d.).

Science projects completed by the students can be entered in four different categories:

- Technology
- Social and Behavioral Sciences
- Biological and Ecological Sciences
- Chemical, Physical, and Mathematical Sciences

Agricultural subjects fall within the Biological and Ecological Sciences category and include projects in:

- Agriculture
- Animal Science
- Biotechnology
- Environmental Science
- Forestry
- Food Science
- Horticulture

- Plant Science
- Veterinary Science

The Young Scientists event, sponsored by the BT Ireland communications corporation, began in 1965. Since its inception, the event has hosted over 50,000 science projects by over 100,000 participants from both the Republic of Ireland and Northern Ireland (UK). Winners of each category can advance to the European Union Science Contest.

Rural Youth Europe

Rural Youth Europe (RY Europe) is an umbrella organization for youth organizations across the continent. The non-governmental organization (NGO) was established in 1957 to promote and activate young people in the rural countryside in Europe. RY Europe provides international training events and works as a conduit between national organizations, youth organizations, and public institutions across Europe (Rural Youth Europe, n.d.).

Macra na Feirme

Within the Republic of Ireland, Macra na Feirme, or Macra for short (www.macra.ie), is an educational and social organization for Irish farmers. It is similar to the Farm Bureau Young Farmers groups or the National Young Farmers Education Association in the United States. Macra was founded in 1944 by a group of agricultural advisors (extension agents), rural science teachers, and farmers. Macra means "stalwarts or the elite" and na Feirme means "of the land." There are Macra groups established in rural areas throughout the country. Figure 8 shows the emblem for Macra na Feirme (Macra na Feirme, n.d.).

Figure 8. Macra na Feirme Emblem

At the national level, Macra has four subcommittees:

- National Executive Committee
- Competitions Committee
- Agricultural Affairs Committee
- Rural Youth (RY) Committee

In the past, Macra used to host agricultural competitions for agricultural science students enrolled in the Teagasc agricultural colleges across Ireland. However, these activities no longer take place. Macra currently does not have any local groups at the Teagasc agricultural colleges. There is discussion that Macra may try to expand and establish local groups at the colleges in the near future.

The agricultural competitions that Macra currently hosts include:

- Public Speaking
- Impromptu Public Speaking
- Debating
- Impromptu Debating
- Know your Agriculture Quiz
- Farm skills
- Welding
- Sheep Stock Judging
- Beef Stock Judging
- Dairy Stock Judging

Young Farmers' Clubs of Ulster (YFCU)

Within Northern Ireland (UK), the Young Farmers' Clubs of Ulster (YFCU) is a vibrant organization that provides educational and social activities for agriculturalists from 12 to 30 years of age. The 90-year-old organization has over 3,000 members in 51 local clubs across the six counties in Northern Ireland. The mission of the organization is to encourage individual development, creativity, initiative, and contribution for the benefit of the member, the Charity, Industry and the Community (Young Farmers' Clubs of Ulster, n.d.).

The Young Farmers' Clubs of Ulster conduct over 30 different competitions for its members in the areas of agriculture, sports, arts, and drama. The agricultural competitions include:

- Machinery Handling
- Sheep Shearing
- Silage Making
- Silage Assessment
- Soil Assessment
- Stock Judging: Beef
- Stock Judging: Dairy
- Stock Judging: Sheep
- Floral Art

- Build It
- Ulster Young Farmer
- Junior Ulster Young Farmer
- Public Speaking

JAPAN

Agricultural Education Programs

For over a century, Japan has had an extensive system of agricultural education in place throughout the country. Instruction in agriculture in Japanese schools begins in the Lower Secondary Schools which includes grades 7 through 9. Agricultural subjects are offered as optional electives in the lower secondary schools for students are interested in a career in agriculture after leaving school (Miyayama, n.d.).

Upper Secondary Schools in Japan include grades 10, 11, and 12 for students up to 18 years of age. These schools may include either full-time or part-time instruction. There are over 800 upper secondary schools in Japan that offer instruction in agricultural subjects. Approximately 500 of these schools offer full-time courses. However, only 5% of students in upper secondary schools enroll in agricultural courses.

Agricultural subjects taught in both lower and upper secondary schools include agriculture, horticulture, livestock breeding, forestry, agricultural manufacturing, and agricultural civil engineering. About half the instruction time in upper secondary schools is devoted to general subjects of Japanese language, social studies, mathematics, and science while the other half of the instruction time is allotted to agricultural subjects.

Agricultural education courses are included in both comprehensive high schools and agricultural high schools throughout Japan. The following is a brief list of some of the agricultural high schools throughout the country:

- Kumamoto Agricultural High School
- Shizuoka Prefectural Iwata Agricultural High School
- Affiliated Taoyuan Agricultural and Industrial High School

Table 3. Tokyo Metropolitan high schools that offer agricultural education

Tokyo Metropolitan High School	Agricultural Subjects Taught
Engei	Agriculture, Horticulture, Food Science, Animal Science
Nogei	Horticulture, Food Science, Green Environment
Nosan	Horticultural Design, Food Science
Nogyo	Urban Horticulture, Food Science, Landscape Planning
Mizuho Nogei	Horticultural Science, Pastoral Science, Food Science
Oshima	Agriculture and Forestry
Miyake	Agriculture
Hachijo	Horticulture

- Saga Prefectural Saga Agricultural Senior High School
- Fukuoka Prefectural Miyako High School
- Goshogawara Agriculture and Forestry High School
- Tajima Agricultural High School

In the Tokyo metropolitan area there are numerous high schools that offer instruction in agricultural subjects. Table 3 includes a list of the Tokyo Metropolitan High Schools that offer agricultural education.

Goshogawara Agriculture and Forestry High School

Goshogawara Agriculture and Forestry High School is in the Aomori Prefecture, a northern apple producing region of Japan. The school was established in 1902 and enrolls about 500 students from the ages of 16-18. The school is divided into four departments: Bioproduction, Forest Science, Environmental Engineering, and Food Science. The students in the Bioproduction department produce a variety of crops on the school farm. The school was the first in Japan to receive Global G.A.P. (Good Agricultural Practice) Certification for apples, rice, melon, potatoes, and onions produced on the farm.

The students at the school are highly active in the Japan School Agricultural Club. This organization is also known as the Future Farmers of Japan (FFJ). Students attend the National Convention of the organization and have competed in livestock judging including dairy and beef cow evaluation, project presentations, and opinion presentations.

Kumamoto Agricultural High School

The school was originally named the Kumamoto Agricultural School when it was opened in 1899. The agricultural departments at the school include Agriculture, Horticulture, Livestock, Living for Agriculture, Agricultural Economics, Food Industry, and Agricultural and Civil Engineering. Within the Livestock department instruction is offered in cattle, pigs, and chickens. The Living for Agriculture area includes the production of flowers and vegetables and instruction in irrigation, fertilization, and crop seeding. It also includes home economics related instruction in cooking, clothing production, childcare, and home nursing.

Students at the Kumamoto Agricultural High School may participate in extracurricular activities including the Flower Arranging Club, Agricultural Machinery Club, Livestock Club, Gardening Club, or the Agricultural Engineering Club.

Teacher Education

With the large number of agricultural high schools and agricultural programs throughout Japan, there is a large need for trained agricultural educators. Japan is blessed with a comprehensive system of higher agricultural education institutions across the country. In 2007 there were 53 universities that offered bachelor's degrees in agricultural subjects in Japan.

The Tokyo University of Agriculture (Nodai) is a private university of agriculture in Japan. The Faculty of Agriculture offers degree programs in agriculture, animal science, bioresource development, and agricultural innovation for sustainable society. Students in the Faculty of Agriculture may complete the Teacher Education Course to be certified to teach science in Junior High Schools (lower secondary) or science/agriculture in senior high schools (upper secondary).

Another Japanese university that offers a teacher education program in agriculture is the Fukui Prefectural University. While there is no Faculty of Agriculture at Fukui, agricultural subjects are taught in the Department of Bioscience and Biotechnology. The department offers a teacher training course to prepare students to obtain a high school teaching license to teach science/agriculture in Japanese secondary schools.

Youth Organizations

The Future Farmers of Japan (FFJ) was established in 1950 in an effort to improve agricultural education programs after World War II. A former American high school agriculture teacher, Ivan Nelson, was attached to General MacArthur's staff in the Army of Occupation. He was instrumental in helping to establish the FFJ. It remains the most successful future farmer organization outside of the United States.

While it is known as the Future Farmers of Japan to agricultural education professionals outside of Japan, it is often referred to as the Japan School Agricultural Club within Japan. In the late 1970s, the Future Farmers of America started an exchange program with the FFJ. Each year the six National FFA Officers travel to Japan to meet with FFJ members and visit agricultural high schools and cultural attractions throughout the country. FFJ members also travel to the United States to participate in the National FFA Convention, tour agricultural education programs and visit with local FFA members. Figure 9 shows the emblem of the Future Farmers of Japan.

Figure 9. Future Farmers of Japan Emblem

SUMMARY

Agricultural education, in one form or another, is taught in schools in virtually every country around the world. Instruction in agricultural subjects can start as early as primary school. Youth are often introduced to agriculture through hands-on gardening programs incorporated into their primary school curriculum.

As students advance, more formalized instruction in agriculture is often included in secondary or higher secondary school programs. Agricultural instruction at the secondary level can be focused either on agricultural science principles or vocational skill development instruction.

Agricultural instruction is offered in rural, suburban, and urban schools. These schools can be either government funded, private, or parochial schools affiliated with a specific religious denomination. Some of the institutions offer residential schools where students live and learn at the campus for weeks or months at a time.

While not a focus of this chapter, technical and scientific agricultural education programs are also offered at colleges, universities, and technical schools of higher education in countries around the world. However, few higher education institutions offer specific teacher education programs for agricultural teachers.

As is the practice in the United States, schools in other countries obtain agricultural teachers a wide variety of ways. These include graduates of university technical agriculture programs, university teacher education programs for general education or science teachers, or individuals with experience in the agricultural industry who have not completed formal university programs in agriculture or education.

Regardless of the country or the type of instruction, there is an important emphasis on teaching practical agricultural skills through hands-on instruction. Many agricultural education programs include instructional laboratory facilities including gardens, land laboratories, livestock facilities, nurseries, or greenhouses. The educational philosophy of "learning by doing," or "learning while doing" is practiced in many institutions that offer agricultural education programs.

The use of youth agricultural organizations differs widely from country-to-country. Some countries have very well-organized youth organizations for students studying agriculture. The Future Farmers of Japan is one organization that has a long and distinguished record of accomplishment within agricultural education programs in Japan. Other countries have organizations such as Young Farmer programs that include programs for both school-aged youth and young adults in higher education or entering the agricultural industry.

Government agriculture and education agencies, public and private schools, and non-government agencies all know the importance of agricultural education in a comprehensive educational program. Agricultural education programs introduce students to hands-on agricultural skills, interpersonal leadership skills, and community development initiatives that help to improve the livelihoods of all citizens in countries around the world. Agricultural education will continue to play a key role as countries continue to improve their educational systems and community development activities for centuries to come.

REVIEW QUESTIONS

1. What is the common educational philosophy followed in secondary agricultural education programs around the world that prepares students to learn hands-on agricultural skills?
2. What soft skills such as leadership, communications, teamwork, problem-solving, etc. are included in secondary agricultural education programs around the world?
3. What type of teacher education programs exist at colleges and universities around the world?
4. How do future agriculture teachers obtain technical agricultural experience and skills?
5. What are some different youth organizations available for students studying agriculture in other countries?
6. In what type of schools are agricultural education programs commonly found in other parts of the world (e.g., public vs. private, rural vs. urban, technical vocational vs. comprehensive schools, etc.)

REFERENCES

Barrick, R. K., Samy, M. M., Gunderson, M. A., & Thoron, A. C. (2011b). A needs assessment to determine knowledge and ability of Egyptian agricultural technical school teachers related to supervised agricultural experience. *Journal of Agricultural Education, 52*(2), 1–11. doi:10.5032/jae.2011.02001

Barrick, R. K., Samy, M. M., Roberts, T. G., Thoron, A. C., & Easterly, R. G. (2011a). Assessment of Egyptian agricultural technical school instructors' ability to implement experiential learning activities. *Journal of Agricultural Education, 52*(3), 6–15. doi:10.5032/jae.2011.03006

BT Young Scientist & Technology. (n.d.). *Our story*. https://btyoungscientist.com/about/

Daly, M. E. (2002). *The first department: A history of the department of agriculture*. Institute of Public Administration.

de Sá Barretto, E. S. (2014). Teacher education policies in Brazil: Contemporary challenges*. *Revista Brasileira de Educação, 20*(62), 679–700.

Education Encyclopedia. (2022). *Egypt – Teaching Profession (2022)*. https://education.stateuniversity.com/pages/417/Egypt-TEACHING-PROFESSION.html

Falvey, L., Barlow, S., Beard, J., Hickey, M., Larkins, F., Dow, K. L., Topp, J., White, R., & Wood, N. (2017). Agricultural education in Victoria & the faculty of agriculture, University of Melbourne. Victoria, Australia: Society of old agriculture fellows.

Faust, A. F. (1959). *Brazil: Education in an expanding economy*. U.S. Department of Health, Education, and Welfare.

Food and agriculture organization of the United Nations. (2014). *Youth and agriculture: Key challenges and concrete solutions*. https://www.fao.org/3/i3947e/I3947E.pdf

Goris, M. B., Lopes, I. S., Verschoor, G., Behagel, J., & Botelho, M. I. V. (2021). Popular education, youth and peasant agroecology in Brazil. *Journal of Rural Studies, 87*, 12–22. https://doi.org/10.1016/j.rurstud.2021.08.003

Irish Agricultural Science Teachers' Association. (n.d.). *About agricultural sciences*. https://www.iasta.ie/about/

LaTowsky, R. J. (1997). Egypt's NGO sector: A briefing paper. University of Reading.

Learning while doing. (n.d.). *About Us*. http://learningwhiledoing.in/InnerPages/About.aspx

Leite, F. C., & Radhakrishna, R. B. (2004). *Profile of agricultural education and extension: Challenges from a changing Brazilian rural milieu*. doi:10.5191/jiaee.204.11302

Leite, F. C. T. (2003). *Factors influencing the adoption of educational reform in Brazil's federally supported agricultural schools* [Unpublished doctoral dissertation]. Pennsylvania State University.

Leite, F. C. T., Baggett, C. D., & Radhakrishna, R. B. (2004). Educational reform as innovation: A qualitative study on the perceptions of Brazilian agricultural school teachers. *Journal of Agricultural and Extension Education, 10*(4), 171–179.

Lend a hand India. (n.d.). *Our Focus Area.* https://lendahandindia.org/

Macra na feirme. (n.d.). *About Us.* https://macra.ie/pages/about

Maguire, C. J. (2021). *Vocational training in the Arab republic of Egypt combines technical and innovation skills for agriculture. In Agricultural innovations systems: An investment sourcebook.* The World Bank.

Mangione, T. (2017). *The famine and the advance of agricultural education in Ireland.* doi:10.13140/RG.2.2.32040.67842

Mendonca, S. R. (2006). The two dichotomies of agricultural education in Brazil (1930-1960). *Estudos Sociedade e Agricultura, 14*(1), 88–113.

Miyayama, H. (n.d.). *Agricultural education in Japan.* Japanese Ministry of Education. https://www.jircas.go.jp/sites/default/files/publication/jarq/04-1-046-049_0.pdf

Mokkapati, S. R. V. (2018, January 12). Time we taught farming in schools [Editorial]. *The Hindi Business Line.* https://www.thehindubusinessline.com/opinion/time-we-taught-farming-in-schools/article9715292.ece

Naik, K. C. (1961). *Agricultural education in India: Institutes and organizations (I.C.A.R. Education Series No. 4).* Indian Council of Agricultural Research.

Narendra, P. M. (2020, August 29). Modi pitches for taking farm education to middle schools level. *The Times of India.* https://timesofindia.indiatimes.com/india/pm-modi-pitches-for-taking-farm-education-to-middle-school-level/articleshow/77823659.cms

Osava, M. (2009). *Brazil: Agricultural school cultivates pride in family farming.* www.ipsnews.net/2009/07/brazil-agricultural-school-cultivates-pride-in-family-farming/

Pollard, L. (2014). Egyptian by association: Charitable states and service societies, circa 1850-1945. *International Journal of Middle East Studies, 46*(2), 239–257. https://jstor.org/stable/43303141

Pratley, J. (2013). *Review into agricultural education and training in New South Wales report.* NSW Government.

Rajghat Besant School. (2017). *School history.* https://www.rajghatbesantschool.org/school/history

Rural Youth Europe. (n.d.). *About Rural Youth Europe.* https://www.ruralyoutheurope.com/about/

Spencer, J. D. (2000). *Principals and principles in agricultural education.* Author.

Vivekandand Uchchatar Uttar Buniyadi Vidyalaya. (2022). *Education.* http://munisevaashram.org/education/vivekanand-uchchatar-uttar-buniyadi-vidyalaya

Young farmers' clubs of Ulster. (n.d.). *About Us.* https://www.yfcu.org/about

ADDITIONAL READING

Barrick, R. K., Samy, M. M., Gunderson, M. A., & Thoron, A. C. (2009). A model for developing a well-prepared agricultural workforce in an international setting. *Journal of International Agricultural and Extension Education, 16*(3), 25–31. doi:10.5191/jiaee.2009.16303

Chesterfield, R., & Schutz, P. (1977). School – Community integration in rural Brazil. *Community Development Journal: An International Forum, 12*(2), 112–115. doi:10.1093/cdj/12.2.112

Flannery, S., Keaveney, K., & Murphy, F. (2019). An exploration of the professional development needs of agricultural educators within the VET sector: A mixed methods study. *International Journal of Agricultural Extension, 7*(3), 247–256. doi:10.33687/ijae.007.03.2988

Tajima, A., & Ito, M. (2009). The historical development and current status of the higher agricultural education system in Japan. *Journal of Developments in Sustainable Agriculture, 4*(1), 29–40.

KEY TERMS AND DEFINITIONS

Agricultural High School: A secondary or higher secondary school that offers educational programming that focuses on agricultural science or vocational agricultural skill development.

Learning to Do – Learning While Doing: An educational philosophy followed by agricultural educators that focuses on teaching students agricultural skills through hands-on, field-based, agricultural projects, community development activities, or social activities.

Soft Skills: Also called employability, people, essential, or interpersonal skills. An example of some of these skills includes leadership, communications – listening - public speaking, teamwork, problem-solving, critical thinking, adaptability, collaboration, integrity/dependability, time management, and work ethic.

Teacher Education Programs: Postsecondary teacher education programs offered at institutions of higher education in other countries around the world. The teacher education programs prepare general education teachers, science teachers, or vocational education teachers.

Youth Organizations in Agriculture: Youth organizations in agriculture can follow the future farmers, 4-H youth development, or young farmers model. The organizations often provide educational, social, cultural, and leadership activities of individuals from youth ages 15 to young adults as old as 30 years of age.

Compilation of References

Agricultural Education About FFA. (n.d.). https://www.ffa.org/agricultural-education/

Alston, A. J., Miller, W. W., & Williams, D. L. (2003). The future role of instructional technology in agricultural education in North Carolina and Virginia. *Journal of Agricultural Education*, *44*(2), 38–49. doi:10.5032/jae.2003.02038

American Association for Agricultural Education (AAAE). (2001). *National Standards for Teacher Education in Agriculture.* Retrieved October 3, 2021, from http://www.aaaeonline.org

American Association for Agricultural Education. (2017). *Standards for School-Based Agriculture Teacher Preparation.* http://aaaeonline.org/Standards-for-Teacher-Preparation

Anderson, T. J., Barrick, R. K., & Hughes, M. (1992). Responsibilities of teacher education for vocational teacher professional development programs. *Journal of Agricultural Education*, *33*(2), 43–50. doi:10.5032/jae.1992.02043

Association for Advancing Quality in Educator Preparation (AAQEP). (2022). https://aaqep.org/

Association for Career and Technical Education (ACTE). (2021). *What is CTE?* Author. https://www.acteonline.org/why-cte/what-is-cte/

Association for Career and Technical Education. (2021). https://www.acteonline.org/about/

Baker, M. A., & Robinson, J. S. (2016). Aligning Kolb's experiential learning theory with a comprehensive agricultural education model. *Journal of Agricultural Education, 52*(4), 1–16. doi: .2012.04001 doi:10:5032/jae

Baker, M. A., & Robinson, J. S. (2016). The effects of Kolb's experiential learning model on successful intelligence in secondary agriculture students. *Journal of Agricultural Education*, *57*(3), 129–144. doi:10.5032/jae.2016.03129

Baker, M. A., Robinson, J. S., & Kolb, D. A. (2012). Aligning Kolb's experiential learning theory with a comprehensive agricultural education model. *Journal of Agricultural Education*, *53*(4), 1–16. doi:10.5032/jae.2012.04001

Ball, A. L., & Knobloch, N. A. (2005). A document analysis of the pedagogical knowledge espoused in agriculture teaching methods courses. *Journal of Agricultural Education*, *46*(2), 47–57. doi:10.5032/jae.2005.02047

Banchi, H., & Bell, R. (2008). The many levels of inquiry. *Science and Children*, *46*(2), 26–29.

Banks, T. (2014). Creating positive learning environments: Antecedent strategies for managing the classroom environment & student behavior. *Creative Education*, *5*(7), 519–524. doi:10.4236/ce.2014.57061

Barrick, R. K., & Thoron, A. C. (2019). *Principles of Teaching and Learning*. Retrieved from University of Florida website: https://edis.ifas.ufl.edu/pdf/WC/WC24300.pdf

Barrick, R. K., & Thoron, A. C. (2019). *Teaching Behavior and Student Achievement*. Retrieved from University of Florida website: https://edis.ifas.ufl.edu/pdf/WC/WC24400.pdf

Barrick, K. R., Ladewig, H. W., & Hedges, L. E. (1983). Development of a systematic approach to identifying technical inservice needs of teachers. *Journal of Agricultural Education, 24*(1), 13–19. doi:10.5032/jaatea.1983.01013

Barrick, R. K., Heinert, S. B., Myers, B. E., Thoron, A. C., & Stofer, K. (2018). Integrating disciplinary core ideas, the agriculture, food and natural resources career pathways and Next Generation Science Standards. *Career and Technical Education Research, 43*(1), 41–56. doi:10.5328/cter43.1.41

Barrick, R. K., Samy, M. M., Gunderson, M. A., & Thoron, A. C. (2011b). A needs assessment to determine knowledge and ability of Egyptian agricultural technical school teachers related to supervised agricultural experience. *Journal of Agricultural Education, 52*(2), 1–11. doi:10.5032/jae.2011.02001

Barrick, R. K., Samy, M. M., Roberts, T. G., Thoron, A. C., & Easterly, R. G. (2011a). Assessment of Egyptian agricultural technical school instructors' ability to implement experiential learning activities. *Journal of Agricultural Education, 52*(3), 6–15. doi:10.5032/jae.2011.03006

Bear, W. F., & Hoerner, T. A. (1986). *Planning, Organizing, and Teaching Agricultural Mechanics*. Hobar Publications.

Birkenholz, R. J., & Deeds, J. (2010). National Standards for teacher education in agriculture. In R. M. Torres, T. Kitchel, & A. L. Ball (Eds.), Preparing and advancing teachers in agricultural education (pp. 221-230). The Ohio State University.

Blackburn, J. J. (2018). Teaching STEM in agricultural mechanics. *The Agricultural Education Magazine, 90*(5), 24–25.

Blackburn, J. J., & Robinson, J. S. (2016). Determining the effects of cognitive style, problem complexity, and hypothesis generation on the problem solving ability of school-based agricultural education students. *Journal of Agricultural Education, 57*(2), 46–59. doi:10.5032/jae.2016.02046

Bloom, B. S. (1956). Taxonomy of educational objectives. Vol. 1: Cognitive domain. *McKay, 20*(24), 1.

Borich, G. D. (n.d.). *Effective Teaching Methods: Research-Based Practice* (10th ed.). Academic Press.

Borich, G. (1980). A needs assessment model for conducting follow-up studies. *Journal of Teacher Education, 31*(3), 39–42. doi:10.1177/002248718003100310

Boyer, E. L. (1990). *Scholarship reconsidered: Priorities of the professoriate*. Princeton University Press.

Bransford, J. D., Brown, A. L., & Cocking, R. R. (2000). *How people learn*. National Academy Press.

Brinkley, H. R. (1977). Teacher education programs in agricultural education should be located in colleges of education. *The Journal of the American Association of Teachers in Agriculture, 18*(3), 2–26. https://www.jae-online.org/attachments/article/1159/Binkley,H_Vol18_3_2,4-6.pdf

Brooks, D. C. (2012). Space and consequences: The impact of different formal learning spaces on instructor and student behavior. *Journal of Learning Spaces, 1*(2).

Brower, R., & Keller, A. (2006). *Empowering Students: Seven Strategies for a Smart Start in School and Life*. Rowman & Littlefield Education.

BT Young Scientist & Technology. (n.d.). *Our story*. https://btyoungscientist.com/about/

Bureau of Labor Statistics. (2018). *Occupational outlook handbook*. Author. https://www.bls.gov/ooh/#tab-6

Burris, S. (2008). PBL: Less content, more process. *The Agricultural Education Magazine, 80*(6), 6–7, 10.

Carl D. Perkins Vocational Education Act of 1984, *20 U.S.C. § 2301* et seq.

Chen, H. T. (2015). *Practical Program Evaluation: Theory-Driven Evaluation and the Integrated Evaluation Perspective* (2nd ed.). Sage Publications, Inc.

Chi, M. T. H., & Glaser, R. (1985). Problem-solving ability. In R. J. Sternberg (Ed.), *Human abilities: An information processing approach* (pp. 227–250). Freeman.

Cohen, L. E., & Waite-Stupiansky, S. (2017). *Early Childhood Education: Developmental, Behaviorist and Critical.* Routledge. doi:10.4324/9781315641560

Coleman, B. M., Bunch, J. C., & Thoron, A. C. (2020). Identifying agriscience teachers' instructional practice professional development needs by certification type. *Journal of Agricultural Education, 61*(3), 86–100. doi:10.5032/jae.2020.03086

Common Core State Standards Initiative. (n.d.). *Development process.* Common core state standards initiative: Preparing America's students for college & career. http://www.corestandards.org/about-the-standards/development-process/

Cooper, J. M., & Alvarado, A. (2006). *Preparation, recruitment and retention of teachers.* International Academy of Education.

Council for the Accreditation of Education Preparation (CAEP). (2022). https://caepnet.org

Council for the Accreditation of Educator Preparation (CAEP). (2021). *Initial level standards.* Retrieved November 6, 2021, from http://caepnet.org/standards/2022-itp/introduction

Council of Chief State School Officers. (2013, April). *Interstate Teacher Assessment and Support Consortium InTASC Model Core Teaching Standards and Learning Progressions for Teachers 1.0: A Resource for Ongoing Teacher Development.* Washington, DC: Author.

Cowan, J., & Goldhaber, D. (2016). National Board Certification and teacher effectiveness: Evidence from Washington state. *Journal of Research on Educational Effectiveness, 9*(3), 233–258. doi:10.1080/19345747.2015.1099768

Croom, B., & Stair, K. (2005). Getting from q to a: Effective questioning for effective learning. *Agricultural Education Magazine, 78*(1), 12.

Croom, D. B. (2008). The development of the integrated three-component model of agricultural education. *Journal of Agricultural Education, 49*(1), 110–120. doi:10.5032/jae.2008.01110

Croom, D. B., & Moore, G. E. (2003). Student misbehavior in agricultural education: A comparative study. *Journal of Agricultural Education, 44*(2), 14–26. doi:10.5032/jae.2003.02014

Cruickshank, D. R. (1996). *Preparing America's Teachers.* Phi Delta Kappa Educational Foundation.

Crunkilton, J. R., & Krebs, A. H. (1982). *Teaching agriculture through problem solving.* Interstate Printers & Publishers, Inc.

D'Angelo, T., Barry, D., Bunch, J. C., & Thoron, A. (2018). *Selecting educational resources.* AEC 640. http://edis.ifas.ufl.edu

Daly, M. E. (2002). *The first department: A history of the department of agriculture.* Institute of Public Administration.

Danielson, C. (1996). *Enhancing professional practice: A framework for teaching.* Association for Supervision and Curriculum Development.

Darling-Hammond, L., Hyler, M. E., & Gardner, M. (2017). *Effective Teacher Professional Development.* Learning Policy Institute. https://learningpolicyinstitute.org/sites/default/files/product-files/Effective_Teacher_Professional_Development_REPORT.pdf

De Lay, A. M., & Washburn, S. (2013). The role of collaboration in secondary agriculture teacher career satisfaction and career retention. *Journal of Agricultural Education, 54*(4), 104–120. doi:10.5032/jae.2013.04104

de Sá Barretto, E. S. (2014). Teacher education policies in Brazil: Contemporary challenges*. *Revista Brasileira de Educação, 20*(62), 679–700.

Desimone, L. M. (2009). Improving impact studies of teachers' professional development: Toward better conceptualizations and measures. *Educational Researcher, 38*(3), 181–199. doi:10.3102/0013189X08331140

Dewey, J. (1904). The relation of theory to practice in education. National Society for the Study of Education. Public School Publishing Co.

Dewey, J. (1902). *The child and the curriculum.* University of Chicago Press.

Dewey, J. (1938). *Experience and education.* Macmillan.

Dewey, J. (1997). *How we think.* Dover Publications. (Original work published 1910)

DiBenedetto, C. A., Willis, V. C., & Barrick, R. K. (2018). Needs assessments for school-based agricultural education teachers: A review of literature. *Journal of Agricultural Education, 59*(4), 52–71. doi:10.5032/jae.2018.04052

Dillman, D. A., Smyth, J. D., & Christian, L. M. (2014). *Internet, Phone, Mail, and Mixed-Mode Surveys: The Tailored Design Method* (4th ed.). John Wiley & Sons, Inc.

Doss, W., Rayfield, J., Burris, S., & Lawver, D. (2021). A quantitative content analysis of survey research methods over a 40-year time period in the *Journal of Agricultural Education. Journal of Agricultural Education, 62*(3), 310–328. doi:10.5032/jae.2021.03310

Dougherty, S. M., Gottfried, M. A., & Sublett, C. (2019). Does increasing career and technical education coursework in high school boost educational attainment and labor market outcomes? *Journal of Education Finance, 44*(4), 423–447. https://muse.jhu.edu/article/738163/pdf

Education Encyclopedia. (2022). *Egypt – Teaching Profession (2022).* https://education.stateuniversity.com/pages/417/Egypt-TEACHING-PROFESSION.html

Egbert, J. (2017). *Methods of education technology: Principles, practice, and tools.* Pressbooks. https://opentext.wsu.edu/tchlrn445/front-matter/introduction/

Epps, R.B. (2021). *Evaluating curriculum material.* AED 583 Course Material.

Falvey, L., Barlow, S., Beard, J., Hickey, M., Larkins, F., Dow, K. L., Topp, J., White, R., & Wood, N. (2017). Agricultural education in Victoria & the faculty of agriculture, University of Melbourne. Victoria, Australia: Society of old agriculture fellows.

Fanghanel, J., Prittchard, J., Potter, J., & Wisker, G. (2016). *Defining and supporting the Scholarship of Teaching and Learning (SoTL): A sector-wide study.* High Education Academy.

Faust, A. F. (1959). *Brazil: Education in an expanding economy.* U.S. Department of Health, Education, and Welfare.

Fellenz, M. R. (2004). Using assessment to support higher level learning: The multiple-choice item development assignment. *Assessment & Evaluation in Higher Education, 29*(6), 703–719. doi:10.1080/0260293042000227245

Ferand, N. K., DiBenedetto, C. A., Thoron, A. C., & Myers, B. M. (2020). Agriscience teacher professional development focused on teaching STEM principles in the floriculture curriculum. *Journal of Agricultural Education, 61*(4), 189–202. doi:10.5032/jae.2020.04189

Fincher, R. M. E., & Work, J. A. (2006). Perspectives on the scholarship of teaching. *Medical Education-Oxford*, *40*(4), 293–295. doi:10.1111/j.1365-2929.2006.02404.x PMID:16573663

Fisher, D. (2006). Checking for understanding: Strategies for ensuring learning. *Thinking Classroom*, *7*(4), 48.

Food and agriculture organization of the United Nations. (2014). *Youth and agriculture: Key challenges and concrete solutions*. https://www.fao.org/3/i3947e/I3947E.pdf

Foster, D. D., Lawver, R. G., & Smith, A. R. (2020). *National Agricultural Education Supply and Demand Study, 2019 Executive Summary*. http://aaaeonline.org/Resources/Documents/NSD2019Summary.pdf

Foster, D. D., Lawver, R. G., & Smith, A. R. (2020). *National Agricultural Education Supply and Demand Study, 2019 Executive Summary*. Retrieved from: http://aaaeonline.org/Resources/Documents/NSD2019Summary.pdf

Fournier, D. M. (2005). Evaluation. In S. Mathison (Ed.), *Encyclopedia of Evaluation*. Sage Publications, Inc. doi:10.4135/9781412950558.n179

Franklin, E. A., & Molina, Q. F. (2012). Teacher induction programs in Agricultural Education: Description of the role of AAAE Higher Education Teacher Preparation Programs. *Journal of Agricultural Education*, *53*(1), 123–135. doi:10.5032/jae.2012.01123

Gage, N. L., & Berliner, D. C. (1998). *Educational psychology* (6th ed.). Houghton Mifflin.

Gagne, R. W. (1967). Curriculum research and the pro-motion of learning. In R. W. Tyler, R. M. Gagne, & M. Scriven (Eds.), *Perspectives of curricular evaluation*. Rand McNally.

Gardner, H. (1991). *The unschooled mind: How children think and how schools should teach*. Basic Books.

Georgia Senate Bill 330, Quality Basic Education Act. (n.d.). https://tinyurl.com/2dntpjpj

Goe, L., Bell, C., & Little, O. (2008). *Approaches to evaluating teacher effectiveness: A research synthesis*. National Comprehensive Center Quality. Retrieved November 6, 2021, from https://gtlcenter.org/sites/default/files/docs/EvaluatingTeachEffectiveness.pdf

Goe, L., Bell, C., & Little, O. (2008). *Approaches to evaluating teacher effectiveness: A research synthesis*. National Comprehensive Center for Teacher Quality.

Goodlett, T. L. (2019). Assignment for AED 583: Curriculum Writing in Career and Technical Education. Community and Leadership Development Department, University of Kentucky.

Gordon, H. R., & Schultz, D. (2020). *The history and growth of career and technical education in America*. Waveland Press.

Goris, M. B., Lopes, I. S., Verschoor, G., Behagel, J., & Botelho, M. I. V. (2021). Popular education, youth and peasant agroecology in Brazil. *Journal of Rural Studies*, *87*, 12–22. https://doi.org/10.1016/j.rurstud.2021.08.003

Graham, C. M., & Edwards, M. C. (2018). Supervision of school-based, agricultural education: A historical review. *Journal of Research in Technical Careers*, *2*(1), 42–51. doi:10.9741/2578-2118.1030

Gray, L., & Lewis, L. (2018). *Career and technical education programs in public school districts: 2016-17. First look*. National Center for Education Statistics. https://files.eric.ed.gov/fulltext/ED582132.pdf

Guyton, E., & Byrd, D. (Eds.). (2000). *Standards for field experience in teacher education*. Association of Teacher Educators.

Hamlin, H. M. (1956). Fifty years of progress in agricultural education. *American Vocational Journal*, *31*(9), 39–46.

Harden, R. M., & Crosby, J. (2000). AMEE guide no 20: The good teacher is more than a lecturer-the twelve roles of the teacher. *Medical Teacher*, *22*(4), 334–347. doi:10.1080/014215900409429

Herren, R. V., & Hillison, J. (1996). Agricultural education and the 1862 land-grant institutions: The rest of the story. *Journal of Agricultural Education*, *37*(3), 26–32. doi:10.5032/jae.1996.03026

Herring, D. R. (1999). Is state supervision of agricultural education needed in the 21st century? A teacher educator's perspective. *The Agricultural Education Magazine*, *71*(6), 12–21.

Hillison, J. (1989). Congressional district schools: Forerunner of federally supported vocational agriculture. *Journal of Agricultural Education*, *30*(4), 7–13. doi:10.5032/jae.1989.04007

Hillison, J. (1998). The role of the agricultural education teacher educator yesterday, today, and tomorrow. *Journal of Agricultural Education*, *39*(1), 1–7. doi:10.5032/jae.1998.01001

Hillison, J. (1999). Whatever happened to the supervisor of the Smith-Hughes man? *Journal of Agricultural Education*, *40*(2), 55–63. doi:10.5032/jae.1999.02055

Hunter, M. (1993). *Mastery teaching*. TIP Publications.

Hutchings, P., Huber, M. T., & Ciccone, A. (2011). *The scholarship of teaching and learning reconsidered: Institutional integration and impact* (Vol. 21). John Wiley & Sons.

Illinois State Board of Education. (n.d.). *Example assessment: Shielded metal arc welding*. Illinois State Board of Education Local Assessment Support Project. https://www.isbe.net/Documents/shielded-metal-arc-welding.pdf

Improving Teacher Preparation: Building on Innovation. (n.d.). Retrieved November 6, 2021, from https://www.ed.gov/teacherprep

International Test Commission. (2006). International guidelines on computer-based and internet-delivered testing. *International Journal of Testing*, *6*(2), 143–171. doi:10.120715327574ijt0602_4

Irish Agricultural Science Teachers' Association. (n.d.). *About agricultural sciences*. https://www.iasta.ie/about/

Jaworski, J. M. (2019). Assignment for AED 583: Curriculum Writing in Career and Technical Education. Community and Leadership Development Department, University of Kentucky.

Johnson, D. W., Johnson, R. T., & Stanne, M. E. (2000). *Cooperative Learning Methods: A meta-analysis. University of Minnesota*. Cooperative Learning Center. http://www.cooperation.org/pages/cl-methods.html

Joint Committee on Standards for Educational Evaluation. (1994). *The Program Evaluation Standards*. Sage Publications, Inc.

Kagan, S. (1994). *Cooperative Learning*. Kagan Publishing.

Kagan, S. (1996). Avoiding the Group-Grades Trap. *Learning*, *24*(4), 56–58.

Kerr, H. T. (1987). With a view to the future. *SAGE Journals*, *44*(2), 164–164. 10.1177%2F004057368704400202

Kids Health. (n.d.). *504 Education Plans: Tips for Teachers*. Jacksonville, FL: The Nemours Foundation. www.kidshealth.org

Kids Health. (n.d.). *Individualized Education Programs: Tips for Teachers*. Jacksonville, FL: The Nemours Foundation. www.kidshealth.org

Killian, J. E., & McIntyre, D. J. (1983). Quality in the early field experiences: A product of grade level and cooperating teachers' training. *Teaching and Teacher Education*, *2*(4), 367–376. doi:10.1016/0742-051X(86)90029-6

Knowles, M. S. (1980). The modern practice of adult education: From pedagogy to andragogy (revised and updated). Englewood Cliffs, NJ: Cambridge Adult Education.

Kolb, A., & Kolb, D. A. (2005). Learning styles and learning spaces: Enhancing experiential learning in higher education. *Academy of Management Learning & Education*, *4*(2), 193–212. doi:10.5465/amle.2005.17268566

Kolb, D. A. (1984). *Experiential learning: Experience as the source of learning and development.* Prentice Hall.

Kolb, D. A. (1984). *Experiential Learning: Experience as the source of learning and development.* Prentice-Hall.

Krathwohl, D. R. (2002). A revision of Bloom's taxonomy: An overview. *Theory into Practice*, *41*(4), 212–218. doi:10.120715430421tip4104_2

Krisco, K. H. (1997). *Leadership and the art of conversation.* Prima Publishing.

Larmer, J., & Mergendoller, J. R. (2010). Seven essentials for project-based learning. *Educational Leadership*, *68*(1), 34–37.

LaTowsky, R. J. (1997). Egypt's NGO sector: A briefing paper. University of Reading.

Lawver, R. G., Foster, D. D., & Smith, A. R. (2018). *Status of the U.S. Supply and Demand for Teachers of Agricultural Education, 2014 – 2016.* Retrieved November 6, 2021, from http://aaaeonline.org/Teacher-Supply-and-Demand

Learning while doing. (n.d.). *About Us.* http://learningwhiledoing.in/InnerPages/About.aspx

Leite, F. C. T. (2003). *Factors influencing the adoption of educational reform in Brazil's federally supported agricultural schools* [Unpublished doctoral dissertation]. Pennsylvania State University.

Leite, F. C., & Radhakrishna, R. B. (2004). *Profile of agricultural education and extension: Challenges from a changing Brazilian rural milieu.* doi:10.5191/jiaee.204.11302

Leite, F. C. T., Baggett, C. D., & Radhakrishna, R. B. (2004). Educational reform as innovation: A qualitative study on the perceptions of Brazilian agricultural school teachers. *Journal of Agricultural and Extension Education*, *10*(4), 171–179.

LeJeune, C. A., & Roberts, R. (2020). Reflections on reform: A historical narrative of the relocation of the Louisiana FFA Office. *Journal of Agricultural Education*, *61*(3), 276–293. doi:10.5032/jae.2020.03276

Lend a hand India. (n.d.). *Our Focus Area.* https://lendahandindia.org/

Livingston, K., Hayward, L., Higgins, S., & Wyse, D. (2015). Multiple influences on curriculum decisions in a super-complex world. *Curriculum Journal*, *26*(4), 515–517. doi:10.1080/09585176.2015.1095015

Loews, L. H. (2015, April 2). When did science education become STEM? *Education Week.* http:// blogs.edweek.org/edweek/curricu-lum/2015/04/when_did_science_ed- ucation_become_STEM.html

Macra na feirme. (n.d.). *About Us.* https://macra.ie/pages/about

Mager, R. F. (1984). *Preparing instructional objectives* (2nd ed.). David S. Lake.

Maguire, C. J. (2021). *Vocational training in the Arab republic of Egypt combines technical and innovation skills for agriculture. In Agricultural innovations systems: An investment sourcebook.* The World Bank.

Maheshwari, V. K. (2017). The problem-solving method in education. *Philosophical commentary on issues of today.* http://www.vkmaheshwari.com/WP/?p=2375

Manathunga, C., & Lant, P. (2006). How do we ensure good PhD student outcomes? *Education for Chemical Engineers*, *1*(1), 72–81. doi:10.1205/ece.05003

Mangione, T. (2017). *The famine and the advance of agricultural education in Ireland*. doi:10.13140/RG.2.2.32040.67842

Marzano, R. J. (2018). *Marzano focused evaluation models*. https://www.marzanocenter.com

McBrien, J. L. (2019). *The Nature of Curriculum*. SAGE Publications, Inc. www.sagepub.com. https://pdf4pro.com/view/chapter-1-the-nature-of-curriculum-sage-b784c.html

McBrien, J. L., & Brandt, R. (Eds.). (1997). *The language of learning: A guide to educational terms*. ASCD.

McIntyre, D. J., Byrd, D. M., & Foxx, S. M. (1996). Field laboratory experiences. In J. P. Sikula, T. J. Buttery, & E. Guyton (Eds.), *Handbook of research on teacher education* (2nd ed., pp. 171–193). Macmillan.

McKendree, R. B., & McKim, A. J. (2021). Teacher changing the discipline: A case study of participatory professional development. *Journal of Agricultural Education*, *62*(3), 72–84. doi:10.5032/jae.2021.03072

McKim, A. J., Raven, M. R., Palmer, A., & McFarland, A. (2019). Community as context and content: A land-based learning primer for agriculture, food, and natural resources education. *Journal of Agricultural Education*, *60*(1), 172–185. doi:10.5032/jae.2019.01172

McTighe, J., & Wiggins, G. (2013). *Essential questions: Opening doors to student understanding*. ASCD.

Mendonca, S. R. (2006). The two dichotomies of agricultural education in Brazil (1930-1960). *Estudos Sociedade e Agricultura*, *14*(1), 88–113.

Michaelsen, L. K., Knight, A. B., & Fink, L. D. (Eds.). (2004). *Team-based learning: A transformative use of small groups. Sterling*. Stylus Publishing, LLC.

Miller, E. C., & Krajcik, J. S. (2019). Promoting deep learning through project-based learning: A design problem. *Disciplinary and Interdisciplinary Science Education Research*, *1*(1), 1–10. doi:10.118643031-019-0009-6

Miller, L. E., & Warmbrod, J. R. (1982). Research in teacher education in agriculture. In A. L. Berkey (Ed.), *Teacher education in agriculture* (pp. 247–262). The Interstate Printers & Publishers.

Miller-Young, J. E., Anderson, C., Kiceniuk, D., Mooney, J., Riddell, J., Schmidt-Hanbidge, A., Ward, V., Wideman, M. A., & Chick, N. (2017). Leading up in the scholarship of teaching and learning. *The Canadian Journal for the Scholarship of Teaching and Learning*, *8*(2), 1–14. doi:10.5206/cjsotl-rcacea.2017.2.4

Miyayama, H. (n.d.). *Agricultural education in Japan*. Japanese Ministry of Education. https://www.jircas.go.jp/sites/default/files/publication/jarq/04-1-046-049_0.pdf

Mokkapati, S. R. V. (2018, January 12). Time we taught farming in schools [Editorial]. *The Hindi Business Line*. https://www.thehindubusinessline.com/opinion/time-we-taught-farming-in-schools/article9715292.ece

Myers, B. E., & Dyer, J. E. (2004). Agriculture teacher education programs: A synthesis of the literature. *Journal of Agricultural Education*, *45*(3), 44–52. doi:10.5032/jae.2004.03044

Myung, J., Martinez, K., & Nordstrum, L. (2013). *A human capital framework for a stronger teacher workforce*. White paper. Carnegie Foundation for the Advancement of Teaching.

Naik, K. C. (1961). *Agricultural education in India: Institutes and organizations (I.C.A.R. Education Series No. 4)*. Indian Council of Agricultural Research.

Narendra, P. M. (2020, August 29). Modi pitches for taking farm education to middle schools level. *The Times of India*. https://timesofindia.indiatimes.com/india/pm-modi-pitches-for-taking-farm-education-to-middle-school-level/articleshow/77823659.cms

National Academy Press. (1995). *Colleges of agriculture at the land grant universities: A profile.*, doi:10.17226/4980

National Association for Agricultural Education. (n.d.). *NAAE membership benefits, service, and programs.* Author. https://www.naae.org/resources/index.cfm

National Association of Agricultural Educators (NAAE). (2017). *Agricultural education in the United States.* Author. Retrieved from https://www.naae.org/

National Association of Agricultural Educators (NAAE). (2021). *Who we are.* Author. Retrieved from https://www.naae.org/whoweare/index.cfm

National Board for Professional Teaching Standard. (2021, November 6). *NBPTS.* https://www.nbpts.org/

National Council for Agricultural Education. (2015). *Agriculture, food, and natural resources (AFNR) career cluster content standards.* https://www.ffa.org/SiteCollectionDocuments/council_afnr_career_cluster_content_standards.pdf

National Council for Agricultural Education. (2015). *National AFNR content standards.* Author. https://thecouncil.ffa.org/afnr/

National Council on Teacher Quality. (2021a, November 5). *Teacher preparation policy.* https://www.nctq.org/publications/State-of-the-States-2021:-Teacher-Preparation-Policy

National Council on Teacher Quality. (2021b, November 5). *What makes teacher preparation traditional.* https://www.nctq.org/dmsView/NCTQ_-_What_Makes_Teacher_Prep_Traditional_or_Non_Traditional

National F. F. A. Organization. (2021). *National FFA career and leadership development events handbook.* Agricultural Sales Career Development Events. https://www.ffa.org/participate/cdes/agricultural-sales/

National F. F. A. Organization. (2021a). *Constitution, bylaws, and federal charter.* Author. https://www.ffa.org/constitution-bylaws/

National F. F. A. Organization. (2021b). *National FFA leadership.* Author. https://www.ffa.org/our-leadership/

National F. F. A. Organization. (2021c). *Official FFA manual.* Author. www.ffa.org/official-manual/

National F. F. A. Organization. (2021d). *State FFA associations.* Author. https://www.ffa.org/state-ffa-associations/

National F. F. A. Organization. (2022). *The three-component model.* Agricultural Education About FFA. https://www.ffa.org/agricultural-education/

National Institute of Food and Agriculture. (n.d.). *College Partners Directory.* National Institute of Food and Agriculture. https://nifa.usda.gov/land-grant-colleges-and-universities-partner-website-directory?field_u_s_states_and_territories_target_id=All&field_map_filter_value=All

National Research Council. (1988). *Understanding agriculture: New directions for education.* Author.

National Research Council. (2000). *How People Learn: Brain, Mind, Experience, and School: Expanded Edition.* The National Academies Press., doi:10.17226/9853

National Research Council. (2012). *A framework for K-12 science education: Practices, crosscutting concepts, and core ideas.* National Academies Press., doi:10.17226/13165

National Young Farmer Educational Association. (2013). http://www.nyfea.org/educational-overview.html

NationalF. F. A. (2021). *About FFA.* Author. Retrieved from https://www.ffa.org/about/

New Horizons, F. F. A. (2018). *Frequently asked questions: National FFA charter revision*. National FFA Organization. https://www.ffa.org/the-feed/frequently-asked-questions-national-ffa-charter-revision/

Newcomb, L. H., McCracken, J. D., Warmbrod, J. R., & Whittington, M. S. (2003). *Methods of Teaching Agriculture* (3rd ed.). Pearson Prentice Hall.

Newcomb, L. H., McCracken, J. D., Warmbrod, J. R., & Whittington, M. S. (2004). *Methods of teaching agriculture* (3rd ed.). Pearson.

Osava, M. (2009). *Brazil: Agricultural school cultivates pride in family farming*. www.ipsnews.net/2009/07/brazil-agricultural-school-cultivates-pride-in-family-farming/

Park, T. D., & Osborne, E. (2006). Content area reading strategies and textbooks in agricultural education. *Journal of Agricultural Education*, *47*(4), 1–14. doi:10.5032/jae.2006.04001

Parr, B., & Edwards, M. C. (2004). Inquiry-based instruction in secondary agricultural education: Problem-solving – An old friend revisited. *Journal of Agricultural Education*, *45*(4), 106–117. doi:10.5032/jae.2004.04106

Patton, M. Q. (2008). *Utilization-Focused Evaluation* (4th ed.). Sage Publications, Inc.

Pauk, W., & Owens, R. J. Q. (2010). How to Study in College (10th ed.). Boston, MA: Wadsworth.

Pedaste, M., Maeots, M., Siiman, L. A., deJong, T., van Riesen, S. A. N., Kamp, E. T., Manoli, C. C., Zacharia, Z. C., & Tsourlidaki, E. (2015). Phases of inquiry-based learning: Definitions and the inquiry cycle. *Educational Research Review*, *14*, 47–61. doi:10.1016/j.edurev.2015.02.003

Peters, J. L. (1980). *The effects of laboratory teaching experience (microteaching and reflective teaching) in an introductory teacher education course on students' views of themselves as teachers and their perceptions of teaching* [Unpublished dissertation]. The Ohio State University, Columbus, OH, United States.

Pew Research Center. (2021). *Who owns cell phones and smartphones?* https://www.pewresearch.org/internet/fact-sheet/mobile/

Phipps, L. J., Osborne, E. W., Dyer, J. E., & Ball, A. (2008). *Handbook on agricultural education in public schools* (6th ed.). Thomson Delmar Learning.

Pollard, L. (2014). Egyptian by association: Charitable states and service societies, circa 1850-1945. *International Journal of Middle East Studies*, *46*(2), 239–257. https://jstor.org/stable/43303141

Popham, W. J. (2020). *Classroom assessment: What teachers need to know* (9th ed.). Pearson.

Pratley, J. (2013). *Review into agricultural education and training in New South Wales report*. NSW Government.

Quigley, A., & Coleman, R. (2019). *Improving Literacy in Secondary Schools. Guidance Report*. Education Endowment Foundation.

Rajghat Besant School. (2017). *School history*. https://www.rajghatbesantschool.org/school/history

Rateau, J. R., Kaufman, E. K., & Cletzer, D. A. (2015). Innovative Classroom Strategies that Prepare College Graduates for Workplace Success. *Journal of Agricultural Education*, *56*(3), 52–68. doi:10.5032/jae.2015.03052

Retallick, M. S. (2005). Early field experiences in agricultural teacher education. *Retrospective Theses and Dissertations*. 17206. https://lib.dr.iastate.edu/rtd/17206

Retallick, M. S., & Miller, G. (2007). Early field experience documents in agricultural education. *Journal of Agricultural Education*, *48*(4), 20–31. doi:10.5032/jae.2007.04020

Rice, A. H., & Kitchel, T. (2016). Influence of knowledge of content and students on beginning agriculture teachers' approaches to teaching content. *Journal of Agricultural Education*, *57*(4), 86–100. doi:10.5032/jae.2016.04086

Riley, R. (1997). The role of the federal government in education- Supporting a national desire for support for state and local education. *Saint Louis University Public Law Review*, *17*(1), 29–54. https://scholarship.law.slu.edu/cgi/viewcontent.cgi?article=1479&context=plr

Riser-Kositsky, M. (2019, January). *Education statistics: Facts about American schools*. https://www.edweek.org/leadership/education-statistics-facts-about-american-schools/2019/01

Roberts, R., Wittie, B. M., Stair, K. S., Blackburn, J. J., & Smith, H. E. (2020). The dimensions of professional development needs for secondary agricultural education teachers across career stages: A multiple case study comparison. *Journal of Agricultural Education*, *61*(3), 128–143. doi:10.5032/jae.2020.03128

Roberts, T. G., & Ball, A. L. (2009). Secondary agricultural science a content and context for teaching. *Journal of Agricultural Education*, *50*(1), 81–91. doi:10.5032/jae.2009.01081

Roberts, T. G., & Dyer, J. E. (2004). Inservice needs of traditionally and alternatively certified agriculture teachers. *Journal of Agricultural Education*, *45*(4), 57–70. doi:10.5032/jae.2004.04057

Roberts, T. G., Harder, A., & Brashears, M. T. (Eds.). (2016). *American Association for Agricultural Education national research agenda: 2016-2020*. University of Florida Department of Agricultural Education and Communication.

Roddy, C., Amiet, D.L., Chung, J., Holt, C., Shaw, L., McKenzie, S., Garivaldis, F., Lodge, J.M., & Mundy, M.E. (2017). Applying Best Practice Online Learning, Teaching, and Support to Intensive Online Environments: An Integrative Review. *Frontiers in Education*. doi:10.3389/feduc.2017.00059

Rugg, H. O. (Ed.). (1927). The foundations of curriculum-making (Twenty-Sixth Yearbook of the National Society for the Study of Education, Part II). Bloomington, IL: Public School Publishing.

Rural Youth Europe. (n.d.). *About Rural Youth Europe*. https://www.ruralyoutheurope.com/about/

Sagen, H. B., Dallam, J. W., & Laverty, J. R. (2000). Effects of Career Preparation Experiences on the Initial Employment Success of College Graduates. *Research in Higher Education*, *41*(6), 753–767. doi:10.1023/A:1007072705601

Saricoban, A. (2002). Reading strategies of successful readers through the three phase approach. *The Reading Matrix*, *2*(3).

Scott, D. (2016). *Ph.D. vs. Ed.D.: Which terminal degree is right for you?* Retrieved from: https://www.insidehighered.com/blogs/gradhacker/phd-vs-edd

Shoulders, C. W., & Meyers, B. E. (2012). Teachers' use of agricultural laboratories in agricultural education. *Journal of Agricultural Education*, *53*(2), 124–138. doi:10.5032/jae.2012.02124

Shulman, L. (2001). From Minsk to Pinsk: Why a scholarship of teaching and learning? *The Journal of Scholarship of Teaching and Learning*, 48–53.

Shulman, L. S. (1986). Those Who Understand: Knowledge Growth in Teaching. *Educational Researcher*, *15*(2), 4–14. doi:10.3102/0013189X015002004

Smalley, S. W., Retallick, M. S., & Paulsen, T. H. (2015). Relevance of student teaching skills and activities from the perspective of the student teacher. *Journal of Agricultural Education*, *56*(1), 73–91. doi:10.5032/jae.2015.01073

Smith-Hughes Act of 1917, 65 U.S.C. § 2301 *et seq.*

Sousa, D. (2017). *How the Brain Learns* (5th ed.). Corwin Press.

Spencer, J. D. (2000). *Principals and principles in agricultural education.* Author.

Steffe, L. P., & Gale, J. (2012). *Constructivism in Education.* Routledge. doi:10.4324/9780203052600

Stimson, R. W. (1919). *Vocational agricultural education by home projects.* The Macmillan Company.

Stimson, R. W., & Lathrop, F. W. (1954). *History of agricultural education of less than college grade in the United States.* Unites States Government Printing Office.

Stone, J. R. III, Alfeld, C., & Pearson, D. (2008). Rigor and relevance: Enhancing high school students' math skills through career and technical education. *American Educational Research Journal, 45*(3), 767–795. doi:10.3102/0002831208317460

Talbert, B. A., Vaughn, R. C., Croom, D. B., & Lee, J. S. (2014). *Foundations of agricultural education.* Pearson Education.

Taz Networks. (2016). *If your computer was human, how old would it be?* https://www.taznetworks.com/computer-human-years/

Teacher Certification Degrees. (2021, August 24). Retrieved November 6, 2021, from https://www.teachercertification-degrees.com/#guide

The National Council for Agricultural Education. (2015). *Agriculture, food, and natural resources (AFNR) career cluster content standards.* https://thecouncil.ffa.org/afnr/

The PPAT® assessment from ETS. (n.d.) retrieved November 6, 2021, from https://www.ets.org/ppa/test-takers/teachers/about

Torres, R. M., & Garton, B. L. (1991). *A program for the preparation of preservice teachers of agricultural education.* https://files.eric.ed.gov/fulltext/ED347382.pdf

True, A. C. (1929). *A history of agri-cultural education in the United States: 1785-1925.* Washington, DC: United States Government Printing Office.

Tyler, R. W. (1957). The curriculum then and now. In *Proceedings of the 1956 Invitational Conference on Testing Problems.* Princeton, NJ: Educational Testing Service.

Tyler, R. W., Gagné, R. M., & Scriven, M. (1967). *Perspectives of Curriculum Evaluation.* Rand McNally & Company.

U. S. Department of Education. (n.d.). *Individuals with Disabilities Education Act.* Washington, DC: Author. www.sites.ed.gov/idea

U.S. Equal Employment Opportunity Commission. (n.d.). *The Rehabilitation Act of 1973 Sections 501 and 505.* Washington, DC: Author. www.eeoc.gov/statutes/rehabilitation-act-1973

United States Department of Agriculture. (2019). *1994 land-grant colleges and universities map.* National Institute of Food and Agriculture. https://nifa.usda.gov/resource/1994-land-grant-colleges-and-universities-map

United States Department of Education. (2008, February). *Organization of U.S. education.* Https://Www2.Ed.Gov/About/Offices/List/Ous/International/Usnei/Us/Edlite-Org-Us.Html

United States Department of Education. (2010, February). *What we do.* https://www2.ed.gov/about/what-we-do.html

United States Department of Education. (2014, September). *10 facts about k-12 education funding.* https://www2.ed.gov/about/overview/fed/10facts/index.html

United States Department of Education. (2021, June). *The federal role in education.* https://www2.ed.gov/about/overview/fed/role.html

University of New South Wales. (2012, August 10). *Assessing laboratory learning.* Assessment Toolkit. https://teaching.unsw.edu.au/assessing-laboratory-learning

Vatterott, C. (2015). *Rethinking grading: Meaningful assessment for standards-based learning.* ASCD.

Vescio, V., Ross, D., & Adams, A. (2008). A review of research on the impact of professional learning communities on teaching practice and student learning. *Teaching and Teacher Education, 24*(1), 80–91. doi:10.1016/j.tate.2007.01.004

Vithal, R. (2018). Growing a scholarship of teaching and learning institutionally. *Studies in Higher Education, 43*(3), 468–483. doi:10.1080/03075079.2016.1180350

Vivekandand Uchchatar Uttar Buniyadi Vidyalaya. (2022). *Education.* http://munisevaashram.org/education/vivekanand-uchchatar-uttar-buniyadi-vidyalaya

Wallace, T., Stariha, W. E., & Walberg, H. J. (2004). *Teaching speaking, listening and writing.* International Academy of Education.

Warner, D. B., & Koeppel, K. (2009). General education requirements. A comparative analysis. *The Journal of General Education, 58*(4), 241–258. doi:10.2307/jgeneeduc.58.4.0241

Weller, C., Rewind, C., & Huber, B. (2013). *A Guide for Adult Agricultural Education and Young Farmer Programs in Pennsylvania.* A project of Vision for Pennsylvania Agricultural Education in Cooperation with Pennsylvania Department of Education Pennsylvania State University Department of Agricultural and Extension Education.

Wells, T., & Hainline, M. (2021). Examining teachers' agricultural mechanic's professional development needs: A national study. *Journal of Agricultural Education, 62*(2), 217–238. doi:10.5032/jae.2021.02217

Wiggins, G. P., & McTighe, J. (2005). *Understanding by design.* ASCD.

Williams, M. R., Warner, W. J., Flowers, J. L., & Croom, D. B. (2014). Teaching with technology: North Carolina agriculture teachers' knowledge acquisition, attitudes, and identified barriers. *Journal of Agricultural Education, 55*(5), 1–15. doi:10.5032/jae.2014.05001

Young farmers' clubs of Ulster. (n.d.). *About Us.* https://www.yfcu.org/about

About the Contributors

Andrew Thoron is the Department Head for the Agricultural Education and Communication Department at Abraham Baldwin Agricultural College in Tifton, Georgia. Andrew began teaching school-based agricultural courses at Mt. Pulaski High School in Illinois. He then took on a state leadership role with the Illinois State Board of Education for agricultural education where his primary responsibility was to aid agriscience teachers, administrators, and community members in developing and maintaining agricultural education programs. Dr. Thoron served on the faculty at the University of Illinois at Urbana-Champaign and University of Florida where he served as the teacher education coordinator and provided state-wide leadership for teacher professional development. His research expertise is in the area of teaching and learning that focuses on effective teaching and argumentation skill development. He has authored/co-authored over 50 referred journal articles, over 125 paper/poster presentations, and has been awarded over $3 million in grants. Dr. Thoron has been recognized as the Journal of Agricultural Education author of the year four times and been recognized nationally and internationally for his teaching, advising, and research. Dr. Thoron holds degrees from Lincoln Land Community College (AS), Illinois State University (BS) – Agricultural Education, and the University of Florida (MS & Ph.D.) in Agricultural Education & Communication – Teacher Preparation.

R. Kirby Barrick began his career as a high school teacher of vocational agriculture. After serving as an assistant state supervisor with the Ohio Department of Education, he earned the Ph.D. in Agricultural Education from The Ohio State University and joined the faculty where he taught courses in teaching and learning. He was awarded the OSU Distinguished Teaching Award, the Gamma Sigma Delta Teaching Award, and the NACTA Distinguished Educator Award. His research interests include teaching and learning theory and practice in domestic and international settings. He has authored/presented more than 275 articles, papers, posters and workshops, and served as principal investigator for more than $8 million in grants. He is a Fellow in the AAAE and is a Fulbright Senior Specialist. Barrick was a department chair Ohio State, Associate Dean for Academic Programs at the University of Illinois, and Dean of the College of Agricultural and Life Sciences at the University of Florida. He has taught graduate-level courses in teacher education, instructional design, and university governance, and conducted professional development activities for practicing educators. He has worked extensively internationally, conducting teaching and learning and curriculum development workshops for university and technical school faculty.

* * *

Joey Blackburn is an Associate Professor of Agricultural Education at Louisiana State University where he focuses on preservice agriculture teacher preparation and agricultural systems and technology. Dr. Blackburn teaches undergraduate and graduate courses in the areas of instructional design, technological change, agricultural systems and technology, and research methodology. His research focuses on student learning outcomes, specifically critical thinking and problem solving in authentic learning environments. Dr. Blackburn has been recognized for his teaching and research efforts at Louisiana State University, as well as through the American Association for Agricultural Education.

Amanda Bowling is an assistant professor of Agriscience Education in the Department of Agricultural Communication, Education, and Leadership at The Ohio State University. Dr. Bowling also serves as Honors and Undergraduate Research Coordinator for the Department of Agricultural Communication, Education, and Leadership. She received her bachelors, master's, and Ph.D. in Agricultural Education from the University of Missouri. Prior to earning her Ph.D., she taught middle and high school agriculture at South Callaway Public Schools in Mokane, Missouri. Dr. Bowling's research focuses on fostering youth motivation within formal and non-formal educational settings through effective teaching practices. AT OSU she teaches courses which center around teaching methods, research methods and design, quantitative instrumentation, and statistical analyses. In 2021, Dr. Bowling, along with her co-authors, was awarded First Runner Up for Outstanding Journal of Agricultural Education Manuscript and she has received five other research conference presentation awards.

James Connors is a Professor and Head of the Department of Agricultural Education, Leadership and Communications in the College of Agricultural & Life Sciences at the University of Idaho. He has been a teacher educator in agriculture for over 35 years and has worked at four land-grant universities: Michigan State University (1989-92), the University of Arizona (1992-1995), the University of Idaho (1996-2000 & 2008 to present), and The Ohio State University (2000-2008). He is also an Affiliate Faculty member in the Department of Curriculum and Instruction in the College of Education, Health, and Human Sciences at the University of Idaho. Dr. Connors began his career in agricultural education as a vocational agriculture teacher at Brown City High School in Brown City, Michigan from 1984-1989. He has three degrees from Michigan State University including: BS in Agricultural & Natural Resources Education (1984), M.S. (1990) and Ph.D. (1992) in Agricultural & Extension Education. In 2017, he was named a Senior Fellow in the American Association for Agricultural Education for his service to agricultural teacher education on the local, state, national, and international levels.

Kevin Curry Jr. earned undergraduate degrees in animal science and agricultural education, and a master's and Doctoral degree in agricultural education from North Carolina State University in Raleigh, N.C. He taught high school agriculture in Randolph County, N.C., during which time he facilitated the chartering of a new FFA chapter and oversaw the development of barn, greenhouse, and laboratory facilities for courses in animal science, horticulture, and biotechnology. Dr. Curry is currently an Assistant Professor of Agricultural Education at Penn State. His primary teaching responsibilities lie in the preparation of preservice school-based agriculture teachers. Dr. Curry's research agenda centers on student motivation and agricultural education's role in developing the science literacy of students. He serves on the Pennsylvania FFA board of directors and the Pennsylvania Commission for Agriculture Excellence.

Catherine A. DiBenedetto, Ph.D., is a teacher educator in the Agricultural Education teacher preparation program at Clemson University. She teaches courses focused in the areas of instructional methods, curriculum design, and teaching in agriscience laboratories. She is passionate about advising, mentoring, and working with pre-service teachers to prepare them to be effective instructors in successful school-based agricultural education programs. Her research agenda is focused on the study of student readiness in the 21st century to effectively prepare students with the 21st century employability skills needed to become career ready. Her research is centralized around inquiry-based instructional methods, self-efficacy, and interdisciplinary STEM education. She is the Executive Director of STEM it Up, a professional development program that provides strategies to agriscience teachers to enhance effective teaching practices, and design innovative instructional materials and curriculum specifically for plant science career pathways. She disseminates her research to practice by supporting pre-service and in-service agriscience teachers with resources to increase their instructional self-efficacy and highlight inquiry-based instruction and STEM education in their school-based agricultural education classrooms and laboratories.

Will Doss is currently an Assistant Professor of Agricultural Education and Mechanics at Texas A&M University – Commerce. He received his Ph.D. from Texas Tech University in Agricultural Communications and Education with prior bachelor's and master's degrees from Texas A&M University focused on agricultural education. He currently teaches courses in research methodology to master's and doctoral students, undergraduate and graduate agricultural education, and undergraduate agricultural mechanics. He has presented several studies at American Association for Agricultural Education research conferences and has published studies related to survey research methodology and school-based agricultural education in the Journal of Agricultural Education and Career and Technical Education Research. Professional research interests include effective teaching strategies to increase learning and retention, effective preservice teacher preparation, making supervised agricultural experiences relevant, and research methodology within agricultural education. Prior to working at Texas A&M University – Commerce, Dr. Doss was a secondary school agricultural education teacher in Texas for five years.

Christopher J. Eck is an assistant professor of Agricultural Education in the department of Agricultural Sciences at Clemson University. His research focuses on identifying and development the necessary human capital for school-based agricultural education teachers to be effective across all three components of an agricultural education program (i.e., classroom/laboratory instruction, FFA, and SAE).

Rebekah Epps's primary research focus is in the area of STEM education. She has recently completed a research project involving CASE Institute participants. Her secondary research focus is the development and retention of secondary agricultural educators and the assessment of student learning. Dr. Epps understands and values the importance of having a high quality educator in the front of every classroom. She is further studying what barriers exist inside the classroom, school, or community that hinders individuals from entering and staying in the teaching profession. Dr. Epps is also developing a leadership curriculum in conjunction with the UK Army ROTC program. As a teacher educator, Dr. Epps is responsible for the Early Field Experience Program in Career and Technical Education. Dr. Epps also teaches the Curriculum Planning and Assessment and Methods of Teaching Course. She strives to ensure her students are prepared to create valuable and engaging lesson plans that will truly challenge and assess secondary students. She also teaches a CLD course entitled Schools, Community, and Society, Experiential Education, Advanced Teaching Methods, and a graduate statistics course.

Jeremy Falk is an Associate Professor of Agricultural Education at University of Idaho. He teaches courses, conducts research, and provides outreach in agricultural education around the region. Jeremy grew up near Oxford, Ohio and was involved in 4-H and FFA. He attended The Ohio State University and received his bachelor's degree in Agricultural Education. Jeremy then gained experience as a high school agricultural education at Federal Hocking High School in Ohio. He learned about student and community engagement and how the three-circle model of agricultural education can develop students and agriculture. Jeremy returned to Ohio State and completed a master's degree and PhD in Agricultural Education. His current research involves community supporters of agricultural education programs, as well as coaching behaviors of teachers. He is active in the agriculture teacher organizations in the region and also remains involved with raising sheep and serving as a 4-H leader.

Rebecca (Becki) Lawver is currently a Professor in Agricultural Education and Department Head for the Applied Science, Technology & Education Department in the College of Agriculture and Applied Sciences at Utah State University. Becki taught high school agricultural education for nine years in Nebraska while earning an M.S. degree in Leadership Education from the University of Nebraska-Lincoln. She earned a Ph.D. in Agricultural Education from the University of Missouri - Columbia, specializing in Teacher Education and Leadership. Becki provides leadership to a department of 28 faculty and 560 students in eight undergraduate majors, three M.Sc. degrees in Agricultural Extension and Education, Family and Consumer Sciences Extension and Education, and Technology and Engineering Education, and M.Ed. and Ph.D. in Career and Technical Education. Becki advises graduate and undergraduate students, has taught undergraduate courses in agricultural education and leadership, and currently teaches graduate courses in the history of career and technical education, leadership and administration, and the professoriate. Her concentrated area of research focuses on the professional development of formal and non-formal educators in agriculture.

Sallie McHugh serves as an Assistant Professor in the Department of Agricultural Education and Communication at Abraham Baldwin Agricultural College in Tifton, GA. Prior to teaching at the collegiate level, she served ten years as a school-based agricultural education teacher at the middle, high, and adult education levels in Georgia. Dr. McHugh has been recognized state, nationally, and internationally for her high-quality classroom instruction and student engagement. Dr. McHugh holds degrees from Abraham Baldwin Agricultural College, the University of Georgia, and her Ph.D from Auburn University.

Robert Bud (Buddy) McKendree is an assistant professor in Agricultural Education within the Department of Community Sustainability at Michigan State University. Previously, Buddy taught middle and high school agriculture for three years in Florida and Indiana. He then earned an M.S. and Ph.D. from Kansas State University in Agricultural Education (M.S.) and Curriculum and Instruction (Ph.D.). In his role at MSU Buddy works with preservice and in-service Agricultural Educators, teaching undergraduate and graduate courses including teaching methods, reflection and practice, and Agricultural Education induction. Other components of his appointment include supervising student teachers and providing outreach activities to Michigan Agricultural Educators. His research area of interest is teacher understanding of student thinking.

Aaron J. McKim is an Assistant Professor in the Department of Community Sustainability at Michigan State University. Before his current role, Aaron was a graduate student at Oregon State University and high school Agricultural Educator at Switzerland County High School in Vevay, Indiana. Currently, Aaron's research focuses on the facilitation of interdisciplinary learning spaces within the context of agriculture, food, and natural resources (AFNR) education. In this work, Aaron explores the role educators play in making connections between AFNR and science, mathematics, leadership, and engineering. This work is compelled by the critical need to prepare individuals to think outside traditional disciplinary boundaries to solve complex problems at the nexus of social and ecological interactions. In addition to his work in AFNR education, Aaron has a passion for research and teaching within leadership.

Greg Miller is a professor in the Department of Agricultural Education and Studies and in the School of Education at Iowa State University. Professor Miller earned Bachelor of Science in Education and Master of Education degrees from Auburn University. He earned the Doctor of Philosophy degree in Agricultural Education from The Ohio State University. Miller has held numerous leadership roles including National President of the American Association for Agricultural Education, Editor of the Journal of Agricultural Education, and Chair of the National Agricultural Education Research Meeting. Professor Miller is a Senior Fellow in the American Association for Agricultural Education. He is also a North American Colleges and Teachers of Agriculture Teacher Fellow. Miller has received many other awards and recognitions including the American Association for Agricultural Education Distinguished Research Award, the Iowa Educational Research and Evaluation Association Distinguished Research Award, the Journal of Agricultural Education Author of the Year Award, and the Outstanding Agricultural Educator Award for the North Central region of the American Association for the Agricultural Education.

Michael L. Pate is an Associate Professor of Agricultural Systems Technology in the Department of Applied Sciences, Technology, and Education at Utah State University. Michael taught high school agricultural education for two years and one year as a science instructor in Arkansas while earning a M.S. degree in Agricultural Education from the University of Arkansas. He earned a Ph.D. in Agricultural Education from Iowa State University, focusing on Teacher Education. Michael serves as the program coordinator for the agricultural systems technology major. Michael mentors undergraduate researchers and teaching fellows. He serves on numerous graduate student committees and teaches courses in research methods and data analysis. His concentrated area of research focuses on agricultural safety education and teacher professional development.

Jon Ramsey's experience in the classroom guides his work with agricultural educators. As a professor, Dr. Ramsey's support of agricultural educators begins with high school students interested in teaching careers and continues decades into their careers. Dr. Ramsey anchors his teaching philosophy on student success. He believes students are a direct reflection of the program, institution, and faculty. In addition, he focuses his teaching on preparing students who have identified school-based agricultural education as their college major and career choice. Specifically, equipping them to effectively deliver instruction and leadership to agricultural programs in Oklahoma and other states. During his twenty years at Oklahoma State University, Dr. Ramsey has invested in over five hundred pre-service teachers, conservatively 75% have entered the teaching profession as school-based agricultural education instructors. Dr. Ramsey has been recognized for his teaching and academic advising at the departmental, university and national level. Most recently he received the Oklahoma State University Award for Academic Advisement and the American Association for Agricultural Education Distinguished Agricultural Educator Award. Dr. Ramsey's contribution to the scholarship of teaching and learning includes 20 peer reviewed journal articles and abstracts, 70 peer reviewed conference presentations and poster presentations, and several popular press articles focused on teaching and learning in the context of school-based agricultural education.

John Rayfield is a professor and serves as the program coordinator for Agricultural Education in the Department of Agricultural Education and Communications at Texas Tech University. Dr. Rayfield earned his bachelor's degree in agricultural education from Auburn University, his master's degree in agricultural extension from the University of Georgia, and his doctorate of education in agricultural education from Texas Tech University. In 2001, his school's FFA chapter was recognized as the Outstanding Middle School Program in the nation by National FFA. Dr. Rayfield has been recognized at many levels for his teaching, research and service to agricultural education. He is a Texas Tech Integrated Scholar, Auburn University College of Education Outstanding Alumnus, and Journal of Agricultural Education Author of the Year. Dr. Rayfield is actively involved in livestock showing and judging. He serves as superintendent for the National 4-H Livestock Judging Contest and judges junior and open livestock shows across the United States.

Richie Roberts is an Assistant Professor of Agricultural Education in the Department of Agricultural and Extension Education and Evaluation at Louisiana State University. His research has focused on human capital development in agricultural education. Dr. Roberts has a history of working with non-traditional transfer students in North Carolina. Further, Dr. Roberts has also been engaged in curriculum development and evaluation. In total, he has authored 42 peer-refereed journal publications, 107 peer-refereed publications, and accrued nearly $1 million in grants and extramural funding. Dr. Roberts has also authored the Article of the Year in the Journal of Agricultural Education, Journal of International Agricultural and Extension Education, and the Journal of Southern Agricultural Education Research.

J. Shane Robinson is a professor of agricultural education and the associate director of the Institute for Teaching and Learning Excellence (ITLE) at Oklahoma State University. As a faculty member, Dr. Robinson teaches undergraduate courses related to the teaching major for pre-service teachers as well as graduate courses for masters and doctoral students who aspire to enter the professoriate. His research focuses on assessing and developing the human capital necessary for graduates' (both secondary and college) employability in various sectors of the agricultural industry and their level of success in those positions once they are employed. His role with ITLE includes directing the Teaching and Learning Support staff in identifying, planning, and delivering high-quality, timely, and relevant professional development on effective teaching for all faculty instructors at Oklahoma State University.

Eric D. Rubenstein earned an undergraduate degree in Agricultural and Extension Education from The Pennsylvania State University, and a master's and doctoral degree in Agricultural Education and Communication from the University of Florida. Prior to beginning his graduate degrees, Dr. Rubenstein taught middle and high school agricultural education in Pennsylvania. During his teaching experience, he was awarded over $10,000 in grant funding to enhance the technology accessible to student in the agriculture department and taught all areas of agricultural education. Currently, Dr. Rubenstein is an Associate Professor and Program Coordinator in Agricultural Education at the University of Georgia and serves as the AAAE Representative to the National FFA Board of DIrectors. Within his current position,

Dr. Rubenstein is responsible for teaching courses in teacher education at both the undergraduate and graduate levels. Dr. Rubenstein's research agenda focuses on the development and integration of supervised agricultural experienced programs in school-based agricultural education programs to ensure that all students across the country have the opportunity to engage and learn through this unique experience.

Scott Smalley is currently an Associate Professor of Agricultural Education at Iowa State University. He received his Ph.D. and master's degree from Iowa State University in Agricultural Education and bachelor's degrees from Michigan State University focused on agricultural education. He currently teaches courses in instructional methods for adult and higher education in agriculture and natural resources to masters and doctoral students, and foundational courses to undergraduate education students. He has presented several studies at American Association for Agricultural Education research conferences and has published studies related to agricultural education focusing on pre-service teachers, professional development, CASE, beginning and midcareer teachers. Prior to working at Iowa State University and South Dakota State University in agricultural education, Dr. Smalley was a secondary school agricultural education teacher in Iowa for six years.

Jay Solomonson grew up on a 1000-acre family farm in rural, Coal Valley, Illinois. After high school, he attended Western Illinois University where he received his B.S. degree in Agriculture- Teacher Certification. He began his career at Orion High School where he taught agricultural education classes and was the FFA advisor for 17 years. During that time, he earned his M.S. degree in Agricultural Education from the University of Illinois and his Ph.D. in Agricultural Education from Iowa State University. In 2019, he accepted a position as the District 1 Program Advisor for the Facilitating Coordination in Agricultural Education (FCAE)- a project of the Illinois State Board of Education. He was in that position for a year prior to joining the faculty at Illinois State University as an Assistant Professor of Agricultural Education in 2020. Dr. Solomonson teaches courses in agricultural education, agricultural leadership & communications, and agricultural mechanization. His research interests include agricultural education teacher recruitment and retention and the scholarship of teaching and learning.

Tyson J. Sorensen is currently an Associate Professor of Agricultural Education at Utah State University. His primary role is teaching. He directs the pre-service agricultural education program, teaches a variety of undergraduate and graduate courses, and mentors student researchers. He earned Bachelor's and Master's degrees at Utah State University in Agricultural Education and a Doctoral degree in Science Education from Oregon State University. Dr. Sorensen taught high school agriculture at Red Mountain High School in Mesa, Arizona for seven years. He taught a variety of courses in agriculture, science, and cooperative education, and he co-advised the FFA chapter. Over the course of his career, Dr. Sorensen has formally taught courses at various levels and in a variety of settings and formats. He has developed several programs, including international programs, focused on agriculture teacher skill development and educational improvement. Dr. Sorensen has a passion for teaching and helping students and educators be successful through contextualized experiential learning. As such, his approach to teaching often involves challenging students through engagement with real-world issues, hands-on experiential learning and practice, and critical reflection. His primary goal as an educator is to provide high impact experiences for students and teachers and to help students and teachers experience success. Dr. Sorensen's current scholarly efforts and teaching programs are intertwined, focusing on improving agricultural education at all levels and ensuring, through education, that well-trained agriculturists and scientists can address important global and socio-scientific challenges of our day. He has received numerous regional and national awards throughout his career for excellence in teaching, mentorship, and research.

Kristin Stair is a faculty member in the Department of Agricultural and Extension Education and Evaluation at Louisiana State University. Her primary role within the department is focused on teacher education coursework and coordinating the student teaching program. Prior to coming to LSU, she served as a faculty member at New Mexico State University and as a lecturer at North Carolina State University. She completed her doctorate at North Carolina State University in 2009 with a dissertation focused on identifying the confidence levels and educational strategies of agricultural education teachers when working with students with special needs. Her research focuses on the exploration of various teacher concerns within agricultural education and several of her research studies have served as the basis for professional development workshops for teachers. Prior to completing her doctorate, she was a high school agriculture teacher in North Carolina.

Jonathan Ulmer specializes in agricultural education. He teaches a variety of classes preparing students for a career as a High School Agriculture Teacher. Prior to coming to KSU, Dr. Ulmer served on the faculty at Texas Tech University and the University of Missouri. Growing up in Nebraska he earned his Bachelor's degree in Agricultural Education at the University of Nebraska and taught Agri-Science at Medicine Valley High School in Curtis, Nebraska. He earned his Master's in Agricultural Education at Oklahoma State University and his Ph.D. in Agricultural Education at the University of Missouri-Columbia.

Shannon Washburn is currently Professor of Agriscience Education and Chair of the Department of Agricultural Communication, Education, and Leadership (ACEL) in the College of Food, Agricultural, and Environmental Sciences at The Ohio State University. Shannon taught high school agricultural education in Kansas for three years before serving as an instructor at Kansas State University for two years. He earned a Doctor of Philosophy in Agricultural Education from the University of Missouri – Columbia, a Master of Science in Curriculum and Instruction and a Bachelor of Science in Agricultural Education both from Kansas State University. The ACEL Department serves nearly 200 undergraduate students and over 60 graduate students. Graduate programs include a Ph.D., M.S., and M.Ed. and the M.S. degree is delivered both online and in residency. Washburn previously served as Assistant Dean in Academic Programs and Professor of Agricultural Education at Kansas State University and as Assistant Professor of Agricultural Education at the University of Florida.

Index

D

E

F

G

H

I

L

M

N

O

P

Q

R

S

T

U

V

W

Y

Recommended Reference Books

IGI Global's reference books are available in three unique pricing formats: Print Only, E-Book Only, or Print + E-Book.

Shipping fees may apply.

www.igi-global.com

ISBN: 9781522589648
EISBN: 9781522589655
© 2021; 156 pp.
List Price: US$ **155**

ISBN: 9781799840756
EISBN: 9781799840763
© 2021; 282 pp.
List Price: US$ **185**

ISBN: 9781799855989
EISBN: 9781799856009
© 2021; 382 pp.
List Price: US$ **195**

ISBN: 9781799864745
EISBN: 9781799864752
© 2021; 271 pp.
List Price: US$ **195**

ISBN: 9781799875710
EISBN: 9781799875734
© 2021; 252 pp.
List Price: US$ **195**

ISBN: 9781799857709
EISBN: 9781799857716
© 2021; 378 pp.
List Price: US$ **195**

Do you want to stay current on the latest research trends, product announcements, news, and special offers?

Join IGI Global's mailing list to receive customized recommendations, exclusive discounts, and more.

Sign up at: **www.igi-global.com/newsletters.**

Publisher of Timely, Peer-Reviewed Inclusive Research Since 1988

IGI Global
PUBLISHER of TIMELY KNOWLEDGE

www.igi-global.com Sign up at www.igi-global.com/newsletters facebook.com/igiglobal twitter.com/igiglobal linkedin.com/igiglobal

www.ingramcontent.com/pod-product-compliance
Lightning Source LLC
LaVergne TN
LVHW062118300926
849161LV00010B/2487
* 9 7 8 1 6 6 8 4 3 4 2 0 8 *